Supported Catalysts and Their Applications

Supported Catalysts and Their Applications

Edited by

D.C. Sherrington
University of Strathclyde, Glasgow, UK

A.P. Kybett
Royal Society of Chemistry, Cambridge, UK

ROYAL SOCIETY OF CHEMISTRY

The proceedings of the 4th International Symposium on Supported Reagents and Catalysts in Chemistry held on 2–6 July 2000 at the University of St Andrews, UK

The front cover illustration is taken from p. 204 in the paper on Organic Modification of Hexagonal Mesoporous Silica by D.B. Jackson, D.J. Macquarrie and J.H. Clark

Special Publication No. 266

ISBN 0-85404-880-4

A catalogue record for this book is available from the British Library

Published by The Royal Society of Chemistry,
Thomas Graham House, Science Park, Milton Road,
Cambridge CB4 0WF, UK

Registered Charity No. 207890

For further information see our web site at www.rsc.org

Printed by Bookcraft Ltd, UK

Preface

The drive to develop increasingly active and selective heterogeneous catalysts continues with considerable vigour. In the case of large and medium scale production processes the stimulation remains the need to increase profitability and to improve process environmental acceptability. In the speciality, fine chemicals and pharmaceuticals businesses the drivers are the same, but include also the need to develop more efficient and faster methods for synthesising potential new products. This is particularly the case in the pharmaceuticals and agrochemicals areas where the high throughput synthesis and screening of potentially active compounds has become an economic imperative.

Traditionally heterogeneous catalysts have been based primarily on inorganic oxide materials, and attempts to construct molecularly well-defined metal complex centres have been fewer in number. In contrast the much less used polymer-based heterogeneous catalysts have focussed more on immobilising well-defined catalytic entities. Interestingly these two areas are now moving closer towards each other, such that a healthy overlap has started to develop. This trend seems set to continue and can only benefit the whole heterogeneous catalysis field.

This development was certainly apparent at the 4th International Symposium on Supported Reagents and Catalysts in Chemistry held at the University of St Andrews, 2–6 July 2000. Six Keynote and nine Invited Lectures were presented, along with 22 Oral Contributions. In addition 40 posters were presented. Keeping up the tradition of this meeting, therefore, a large proportion of the participants were active in presenting their work in one format or another.

The present text contains the written form of 31 of the presentations, and is representative of the coverage of the meeting. The collection of papers is also a good indication of the state-of-the-art of this rapidly moving field.

David C Sherrington
Glasgow, Scotland
November 2000

Contents

SELECTIVITY IN OXIDATION CATALYSIS

B. K. Hodnett

Department of Chemical and Environmental Sciences
and
The Materials and Surface Science Institute
University of Limerick, Limerick, Ireland

ABSTRACT

Selectivity in oxidation catalysis has been reviewed for conventional catalysts used for the production of bulk chemicals and epoxidations. The point of activation of the substrate is identified as a key factor identifying three mechanistic features. These are (i) activation of the weakest C-H bond in a substrate, (ii) activation of the strongest C-H bond and (iii) electrophilic attack in olefins. Key features of each type of reaction are identified and new catalyst types needed to break through existing selectivity barriers are discussed.

1 INTRODUCTION

It has been established for some time that the chemical structure of substrates (reactants) is important in determining reactivity over heterogeneous catalysts. Yao[1] established the following order of reactivity for alkane total oxidation over supported platinum catalysts $n - C_4H_{10} > C_3H_8 > C_2H_6 > CH_4$ (> more reactive) and there is a body of work which indicates that C-H bond strength is an important factor in determining reactivity; molecules with weak C-H bonds tend to be more reactive.[2,3]

The situation with respect to selectivity is less clear. Some examples of selective oxidation catalysts used in commercial practice are listed in Table 1. A consistent feature is that many oxidation catalysts are not highly dispersed when viewed on the atomic scale. Hence particle sizes tend to be large, even for supported precious metal catalysts.[1,4-7] This feature, in turn, has led to descriptions of active site structures on these catalysts that are extensions of bulk structures.

Table 1
Structural Features Oxidation Catalysts

Reaction	Catalyst	Phase Composition	Surface Area / $m^2 g^{-1}$	Texture	Ref
$H_2C=CH_2$ + 0.5 O_2 → (ethylene oxide, H_2C–CH_2 with O bridge)	Ag/α-Al$_2$O$_3$	Silver Metal and α-Alumina	<1	Large silver particles > 100nm covering the alumina support	4
(butene) + 3.5 O_2 → (maleic anhydride) + 4 H_2O	Vanadium Phosphorus Oxide	$(VO)_2P_2O_7$	<30	Crystalline platelet morphology Dimensions (70 x 70 x 10 nm)	5
(propene) + O_2 → (acrolein, $C=O$ / H) + H_2O	Bismuth molybdates promoted with cobalt and iron	Mixture of cobalt and iron molybdates covered by a bismuth molybdate layer	<15	Surface layer of bismuth molybdate 0.2-1 nm in depth. Particle diameter ca. 300nm	6
(isobutane) + 5 O_2 → 3CO_2 + 4H_2O	Pt/ Al$_2$O$_3$	Pt particles on carrier		1-30 nm	1
(o-xylene, CH_3 / CH_3) + 3 O_2 → (phthalic anhydride) + 3 H_2O	V$_2$O$_5$/ TiO$_2$	Vanadium oxide layer spread on a titanium dioxide	<30	Surface layer thickness 1-5 nm. Support Diameter ca 100 nm	7
(phenol, OH) + H_2O_2 → (hydroquinone, OH / OH) or (catechol, OH / OH)	Titanium silicalite	solid solution of titanium in silicalite	micro porous	Isolated titanium sites in TS-1, to a maximum of 2.5 mole %	8

2 SUBSTRATE ACTIVATION BY C-H BOND RUPTURE

The term activation is often used in relation to hydrocarbon reactivity over heterogeneous catalysts. Here, it is defined as identifying the primary point of attack on a reacting molecule. Literature evidence relating to kinetic isotope effect (KIE) studies of selective oxidation and ammoxidation reactions are listed in Table 2.[9-11] In each case, a KIE is observed only in relation to a specific C-H bond in the substrate. For example, in n-butane oxidation to maleic anhydride, a KIE is observed only when the methylene (-CH$_2$-) hydrogens are replaced by deuterium, consistent with these C-H bonds being the point of activation in n-butane and their rupture being the slow step in the overall reaction. Further analysis of these results indicate that the point of activation is the weakest C-H bond available in substrate. Individual C-H bond strengths are annotated in Column 2 of Table 2.

This activation feature identifies one class of selective oxidation catalyst namely, those that activate the substrate through rupture of the weakest C-H bond. The performance of these selective oxidation catalysts is best presented in terms of selectivity-conversion plots. Using this approach, multiple selectivity-conversion plots can be generated, such as that shown in Figure 1 for isobutene and isobutane oxidation to methacrolein.[12] These plots are intended to illustrate that there exists in relation to each selective oxidation reaction an upper performance limit beyond which experimental studies have not yet progressed.

Table 2
Kinetic Isotope Effects in Selective Oxidation and Ammoxidation Reactions

Reaction, Catalyst and Temperature	C-H Bond Energies / kJ mol^{-1}	Isotopic form of the Substrate	Exptl k_H/k_D	Ref
n-C$_4$H$_{10}$ +3.5 O$_2$ → C$_4$H$_2$O$_3$ + 4 H$_2$O		CH$_3$CD$_2$CD$_2$CH$_3$	2.18	9
Catalyst (VO)$_2$P$_2$O$_7$ Temp 673K		CD$_3$CH$_2$CH$_2$CD$_3$	1.05	
C$_3$H$_6$ + O$_2$ → C$_3$H$_4$O + H$_2$O		CH$_2$=CH-CH$_2$D	2.04	10
		CHD=CH-CH$_3$	1.02	
Catalyst BiMoCoFeOx, Temp 723K		CH$_2$=CD-CD$_3$	1.78	
CH$_3$CH$_2$CH$_3$ + 2 O$_2$ + NH$_3$ → CH$_2$=CH-CN + 4 H$_2$O		CH$_3$CD$_2$CH$_3$	1.7	11
		CD$_3$CH$_2$CD$_3$	1.1	
Catalyst 50%VSb$_{3.5}$P$_{0.5}$WO$_x$ -50%Al$_2$O$_3$ Temp 743K		CD$_3$CD$_2$CD$_3$	2.0	

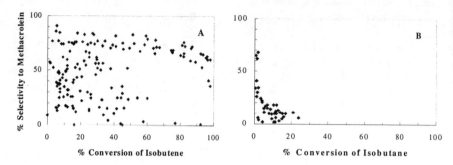

Figure 1 *Multiple selectivity-conversion plots for (A) isobutene and (B) isobutane oxidation to methacrolein[12]*

Consideration of this mechanism leads logically to the conclusion that if activation in the direction of selective oxidation results from rupture of the weakest C-H bond in the substrate, then the selective oxidation product so formed must be subject to attack (destruction) via the same process. In general, rupture of any bond in the selective oxidation product would lead to its destruction.[13] Hence, the function D^OH_{C-H} REACTANT $- D^OH_{C-H \text{ or } C-C \text{ PRODUCT}}$, namely the difference in bond strengths between the weakest C-H bond in the reactant and the weakest bond in the product has been evaluated for 24 oxidation reactions. Figure 2 presents a plot of the selectivity at 30% conversion for a wide range of oxidation reactions against the function $D^OH_{C-H \text{ REACTANT}}$ $- D^OH_{C-H \text{ or } C-C \text{ PRODUCT}}$. The point zero on this scale represents the situation where the weakest C-H bond in a given substrate has the same bond strength as the weakest bond in the selective oxidation product. The data in Figure 2 clearly shows that active sites in conventional oxidation catalysts are capable of selectively activating a C-H bond in a substrate in the presence of similar bonds in the selective oxidation product provided that there is no bond in the product with a bond strength less than 30-40 kJ mol[-1] of the value for the weakest C-H bond in the substrate.[14]

In recent years, a new class of commercial oxidation catalyst has emerged, namely the Fe-ZSM5 catalysts used for phenol production from benzene using nitrous oxide as oxidising agent.[15] This system is said to generate the so-called α-oxygen species. Since this is a zeolite based catalyst in which diffusion limitations can normally be expected, kinetic isotope effect studies are not useful. However, the α-oxygen does appear to have a different reactivity pattern to conventional oxidation catalysts. In a study of the reactivity of α-oxygen towards isopropylbenzene the product distribution shown in Figure 3 was observed, namely that the preferred point of activation of the hydrocarbon is the strongest available C-H bond. This feature identifies a second class of selective oxidation reaction, much less common, namely where the preferred point of activation of the hydrocarbon is the strongest C-H bond in the structure.

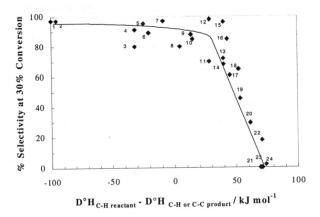

Figure 2 *Selectivity at 30% conversion for the reactions indicated as a function of $D^O H$ $_{C\text{-}H(reactant)}$ – $D^O H_{C\text{-}H \text{ or } C\text{-}C \text{ (product)}}$. 1. ethylbenzene to styrene; 2. 1-butene to 1, 3-butadiene; 3. toluene to benzoic acid; 4. acrolein to acrylic acid; 5. ethane to enthylene; 6. n-butane to maleic anhydride; 7. benzene to phenol; 8. toluene to benzaldehyde; 9. propene to acrolein; 10. 1-butene to 2-butanone; 11. isobutene to isobutene; 12. methanol to formaldehyde; 13. methacrolein to methacyclin acid; 14. propane to propene; 15. ethanol to acetaldehyde; 16. isobutene to methacrolein; 17. n-butane to butene; 18. benzene to maleic anhydride; 19. propane to acrolein; 20. methane to ethane; 21. ethane to acetaldehyde, 22. isobutane to methacrylic acid; 23. methane to formaldehyde; 24. isobutane to methacrolein.*

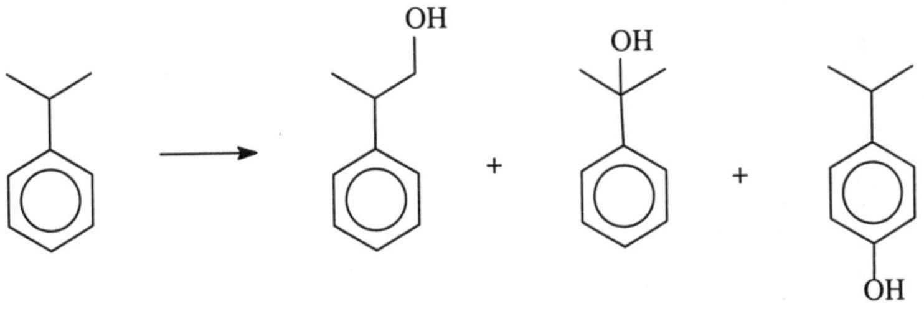

Product Distribution   2   1   12

Figure 3 *Reactivity of α-oxygen towards isopropylbenzene[15]*

3 OLEFIN EPOXIDATION

Olefin epoxidation may be viewed as a third class of selective oxidation reaction in that it does not involve C-H bond rupture in the substrate. A feature of commercial operation of this type of chemistry has been the use of silver catalysts for ethylene epoxidation by oxygen. Another consistent feature is the inability of this same catalyst system to epoxidize propene. Indeed further analysis of a range of substrates over the silver-oxygen system, some of which are presented in Table 3[16-19], indicates that substrate structure is important in determining selectivity in epoxidation. Good selectivities in the silver-oxygen system are possible only for substrates without allylic C-H bonds. Hence, in Table 3, 1-3 butadiene and styrene can be selectively epoxidized in the silver-oxygen system but propene and 1-butene cannot. This data is further analysed in Figure 4 which plots selectivity to epoxide for a range of olefin substrates against the bond dissociation enthalpy of the weakest C-H bond in the olefin.[20] For the silver-oxygen system, the presence of a C-H bond in the olefin with a bond energy below 400 kJ mol^{-1} leads to a very low selectivity, presumably because of activation of a weak C-H bond rather than by electrophilic attack at the double bond.

The situation when a TS-1 peroxide catalyst system is used is entirely different. The temperatures involved are lower and the oxidizing species involved here appears to be much more electrophilic and capable of epoxidizing those substrates in Table 3 (propene and 1-butene) where the silver-oxygen system failed. When the selectivity to epoxide is plotted against the weakest C-H bond in the olefin (Figure 4), there is a clear increase in the range of application of this system over the silver-oxygen system with the oxidizing species on TS-1 being capable of electrophilic attack even when very weak bonds (340 kJmol^{-1}) are present in the olefin.[16-19]

Table 3
KinetComparison of Olefin Epoxidation with Oxygen and Peroxides as Oxidant [16-19]

Substrate	Weakest C-H Bond in Olefin kJ mol^{-1}	Oxidant	Temp / K	%Sel to Epoxide	% Conv of Olefin
Propene	361	O_2	523	15	20
		H_2O_2	323	97	97
1-3 Butadiene	409	O_2	523	96	21
		H_2O_2	323	100	92
1-Butene	345	O_2	523	0	8
		H_2O_2	273	97	90
Styrene	410	O_2	523	95	19
		TBHP	298	94	96

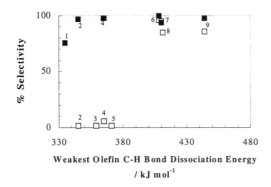

Figure 4 *Selectivity in epoxidation for a range of substrates plotted against the dissociation enthalpy of the weakest C-H bond in the olefin (■) TS-1 peroxide system (□) silver-oxygen system. 1. 1-octene, 2. 1-butane, 3. 2-butane, 4. propene, 5. 4-unyltoluene, 6. 1-3 butadiene, 7. styrene, 8. 4-vinylpyridine, 9. ethylene.[20]*

Clearly, the level of sophistication involved in the TS-1 catalyst is greater than that involved with the other catalyst systems listed in Table 1. The generation of 2.5 mol% titanium in solid solution in silicalite makes for a very dilute system with a limited number of active sites per unit volume.[8] However, this approach seems to be necessary to expand the range and applicability of selective oxidation catalysis.

4 CONCLUSIONS

Selective oxidation has been reviewed and points to a mature technology associated with conventional selective oxidation catalysts where substrate activation occurs via the weakest C-H bond. Discriminating capacity and selectivity of active sites on these catalysts is limited to being able to activate a C-H bond in a substrate that is 30-40 kJmol^{-1} weaker than a similar bond in the selective oxidation product. There are a number of emerging iron-based systems where the strongest C-H bond in a given substrate is activated. Selectivity in olefin epoxidation is related to competition between electrophilic attack and C-H bond rupture. The more electrophilic nature of the oxidizing species in the TS-1 peroxide system gives it a much greater range of applicability by comparison with the silver-oxygen system.

References

1.   Y-F. Y. Yao, *Ind. Eng. Chem. Prod. Res. Dev.,* 1980, **19**, 293.
2.   V. D. Sokolovskii, *Catal. Rev.-Sci. Eng.,* 1990, **32**, 1.
3.   A. O'Malley and B. K. Hodnett, *Catal. Today,* 2000, **54**, 31.
4.   K. P. de Jong, *CATTECH,* 1998, **2**, 87.
5.   H. S. Horowitz, C. M. Blackstone, A. W. Sleight and G. Teufer, *Appl. Catal.,* 1988, **38**, 193.
6.   D-H. H. He, W. Ueda and Y. Moro-oka, *Catal. Lett.,* 1992, **12**, 35.
7.   M. Sanati, L. R. Wallenberg, A. Anderson, S. Jansen and Y. Tu, *J. Catal.,* 1991, **132**, 128.
8.   R. Millini, E. Previde Massara, G. Perego and G. Bellussi, *J. Catal.,* 1992, **137**, 497.
9.   M. A. Pepera, J. L. Callaghan, M. J. Desmond, E. C. Milberger, P. R. Blum and N. J. Bremer, *J. Am. Chem. Soc.,* 1985, **107**, 4883.
10. G. W. Keulks and L. D. Krenzke, Proceedings of the Sixth International Congress on Catalysis, London, 1976, 806.
11. L. A. Bradzil, A. M. Ebner and J. F. Bradzil, *J. Catal.,* 1996, **163**, 117.
12. F. E. Cassidy and B. K. Hodnett, *Erdol Erdgas Kohle,* 1998, **114**, 256.
13. C. Batiot and B. K. Hodnett, *Appl. Catal. A: General,* 1996, **137**, 179.
14. F. E. Cassidy and B. K. Hodnett, *CATTECH,* 1998, **2**, 173.
15. G. I. Panov, A. K. Uriarte, M. I. Rodkin and V. I. Sobolev, *Catal. Today.,* 1998, 41, 365.
16. J. R. Monnier, Proceedings 3rd World Congress on Oxidation Catalysis, Grasseli R. K., Oyama, S. T., Gaffney, A. M. and Lyons J. E. (Eds), Elsevier Science (1997), 135.
17. M. G. Clerici and P. Ingallina, *J. Catal,* 1993, **140**, 71.
18. M. G. Clerici, G. Bellussi and U. Romano, *J. Catal.,* 1991, **129**, 159.
19. A. Corma, M. T. Navarro and J. P. Pariente, *J. Chem. Soc. Chem. Commun.,* 1994, 147.
20. B. K. Hodnett, *Heterogeneous Catalytic Oxidation: Fundamental and Technological Aspects of the Selective and Total Oxidation of Organic Compounds,* J. Wiley and Sons, 2000, 317.

THE DEVELOPMENT AND APPLICATION OF SUPPORTED REAGENTS FOR MULTI-STEP ORGANIC SYNTHESIS

Steven V. Ley* and Ian R. Baxendale

Department of Chemistry
University of Cambridge
Lensfield Road
Cambridge, CB2 1EW, UK

1. Introduction
Synthetic organic chemistry is a continuously evolving subject with new techniques, reactions and methods being developed at an ever increasing rate. In an era when the world is becoming increasingly aware of the limits of its natural resources and the environmental impact of disposing of waste materials, the chemical industries are under considerable pressure to discover, develop and utilise more efficient manufacturing protocols. The areas which have seen the most change in recent years have been the pharmaceutical and agrochemical sectors. These communities are constantly seeking new ways to meet the demand for new, diverse and structurally interesting molecules for biological evaluation. Their traditional approaches to lead compound discovery and optimisation have been both expensive and time consuming. The challenge is therefore to find more efficient and cost-effective methods to produce an ever-increasing number of chemical entities as quickly and as cleanly as possible. This has led to the emergence of combinatorial chemistry and related automation technologies as essential components of the discovery process.[1] Owing to the development of high throughput screening techniques, the speed of biological evaluation of potential drug candidates has increased dramatically. In order to match these advances it is necessary to develop suitable protocols for the fast and efficient generation of chemical libraries. These libraries of small molecules have normally been prepared either in solution or assembled on solid support. The greater flexibility offered by solution phase chemistry is outweighed by the need for time consuming work-up and purification of the individual library components. As a consequence, solid supported reagents[2] have been developed and are becoming increasingly popular since they combine the advantages of polymer-supported chemistry with the versatility of solution phase reactions, allowing clean reactions and removal of contaminating by-products by simple filtration.

2. Polymer-supported reagents
The concept of immobilising reagents on a support material is not new; catalytic hydrogenation and numerous other processes that occur on a solid surface can be classified as examples of supported-reagent systems. It is conceivable that with the appropriate choice of support a diverse variety of reagents could be tethered. Indeed, not only have supported variants of many commonly used reagents been prepared, but also a growing number of scavenging agents capable of sequestering unwanted by-products and excess reactants from solution have also been described.[3, 4] A typical example of how these concepts work in practice to give clean products is shown in Scheme 1. Although the idea of using solid-supported reagents has been known for a long time their specific application in the generation of large chemical arrays via organised multi-step syntheses has to date been little explored. Studies such as these are required to demonstrate the full range of advantages that these reagents offer such as ease of handling, low toxicity and simple reaction monitoring. Furthermore, the increased speed of purification gained by

their use means that this application of chemistry is of special significance in multi-parallel syntheses. We describe below some of our efforts in this area and illustrate how these methods have a broad ranging potential for organic synthesis in the future.

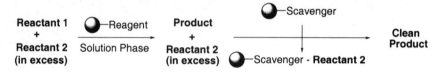

Scheme 1, *Polymer-supported reagents in clean synthesis*

2.1. The evolution of supported reagents – The perruthenates

The reagent tetra-*n*-propylammonium perruthenate (TPAP) is a mild, catalytic, room temperature oxidant and has become one of the principal reagents for the conversion of primary and secondary alcohols to their corresponding aldehydes and ketones.[5] This reagent therefore posed an ideal candidate for immobilisation onto solid support to enable facile work-up and purification of reactions. The polymer-supported perruthenate (PSP) was easily generated by an ion exchange reaction of a commercially available Amberlyst resin, functionalised as the quaternary ammonium chloride, with an aqueous solution of potassium perruthenate.[6] The PSP material was shown to be effective for the stoichiometric oxidation of alcohols to their corresponding carbonyl compounds at room temperature and in high yields (Scheme 2, Conditions 1).[2i, 7h]

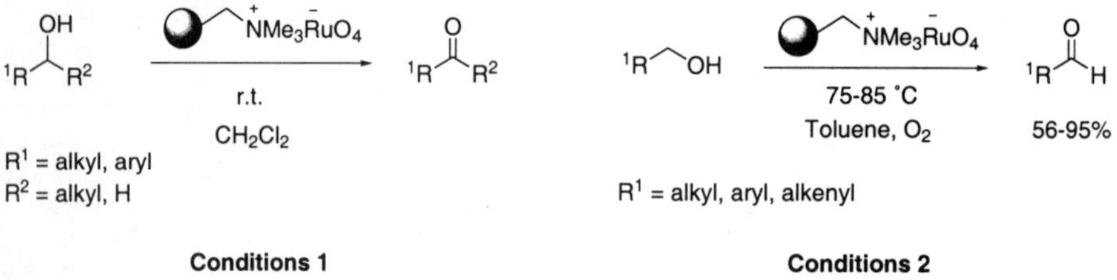

Conditions 1

R^1 = alkyl, aryl
R^2 = alkyl, H

r.t.
CH_2Cl_2

Conditions 2

R^1 = alkyl, aryl, alkenyl

75-85 °C
Toluene, O_2 56-95%

Scheme 2, Oxidation of alcohols using the PSP reagent

A further important development of this process enabled the reaction to proceed catalytically, using atmospheric air or molecular oxygen as the co-oxidant, in toluene at ~80 °C (Scheme 2, Conditions 2).[2e, 7i-7j] This had the additional benefit of greatly simplifying the work-up which was especially useful in the generation of monomer building blocks that are useful in a vast range of combinatorial chemistry programmes. There was, however, a problem associated with the multiple recycling of the PSP reagent due to a small amount of decomposition of the polymer (the tetra-alkylammonium polymer beads are prone to Hoffman elimination) this prompted an investigation into other supporting materials. It was discovered that the perruthenate could alternatively be tethered within the cavity of the mesoporous solid MCM-41 (Figure 1).[2a] This produced a remarkably clean and efficient catalyst with none of the previous stability problems. Even after repeated recycling (up to 15 times) the catalyst showed no loss of activity. We, however, speculate that the actual nature of the active catalyst under consecutive recycling would not survive; rather this species is likely to be some form of cyclic ruthenate silicone oxide intermediate. Work is currently underway to fully characterise this important catalyst.

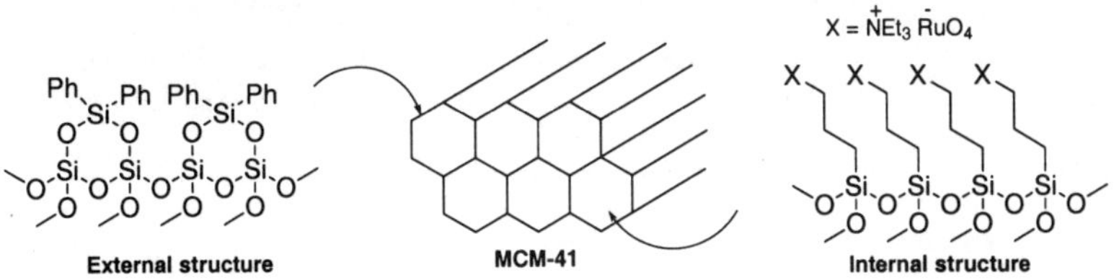

Figure 1, *Structure of the mesoporous silicate MCM-41*

The efficiency of the mesoporous catalyst was demonstrated by the oxidation of 1 g of the alcohol **1** to its corresponding aldehyde using only 25 mg of the supported catalyst, giving the oxidation product cleanly in quantitative yield (Scheme 3).[2a] Again, this was achieved using oxygen as the cooxidant.

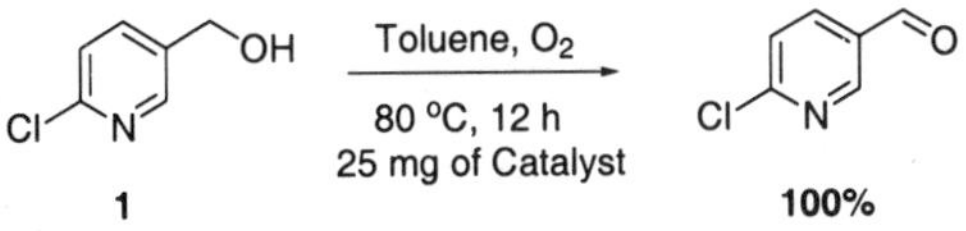

Scheme 3, *The clean, catalytic oxidation of alcohol **1***

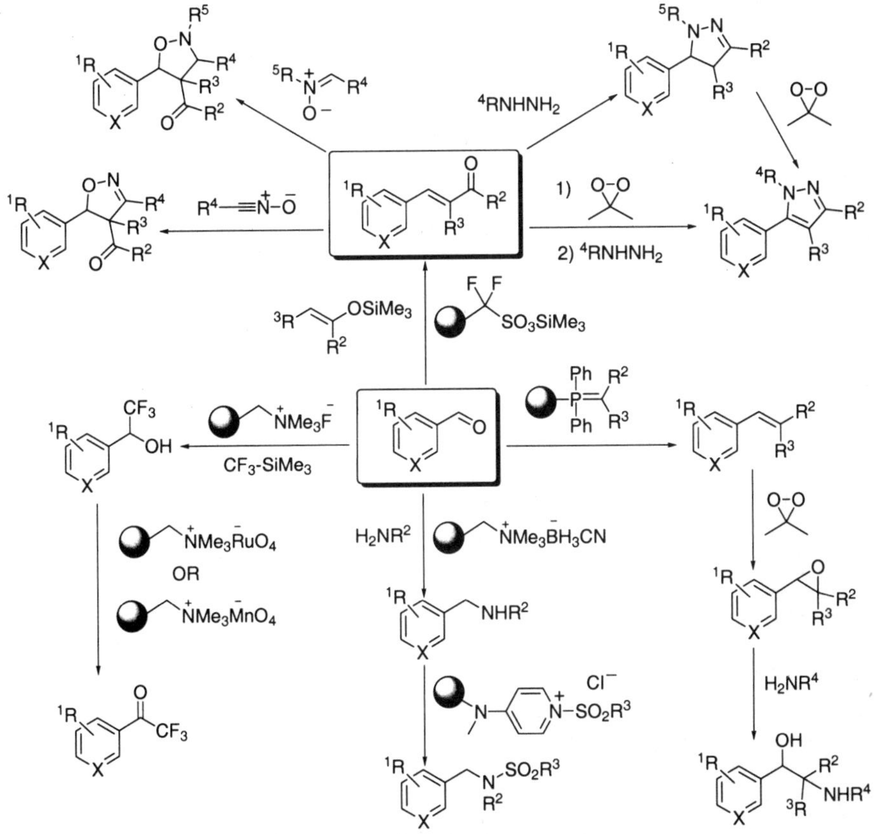

Scheme 4, *Construction of molecular diversity from simple monomers*

3. Synthesis of small molecules – Construction of novel building blocks

Many commercially valuable molecules such as painkillers, antidepressants, cold/flu prescriptions, pesticides, herbicides and fungicides are relatively small in size yet have wide ranging properties.[8] For their synthesis it is desirable to have highly convergent routes which are amenable to combinatorial change so as to produce analogues to elucidate the structure activity relationships in a chemically diverse fashion. We have shown that using polymer-supported reagents it is possible, through only simple chemical manipulations, to construct a number of novel chemical arrays from readily available starting materials (Scheme 4).[7c, 7h-7k] The products, in turn, may be incorporated into more elaborate synthetic constructs. Furthermore these ideas may be extended to the preparation of a range of functionalised heterocycles.[7] One such example from our laboratory was the synthesis of a small library of pyrrole derivatives using polymer-supported reagents (Scheme 5).[7d] This route exemplifies how relatively diverse compounds can be generated from simple building materials in a fast and efficient manner. All of the intermediates in the synthesis can additionally be split and diverted to other synthetic programmes.

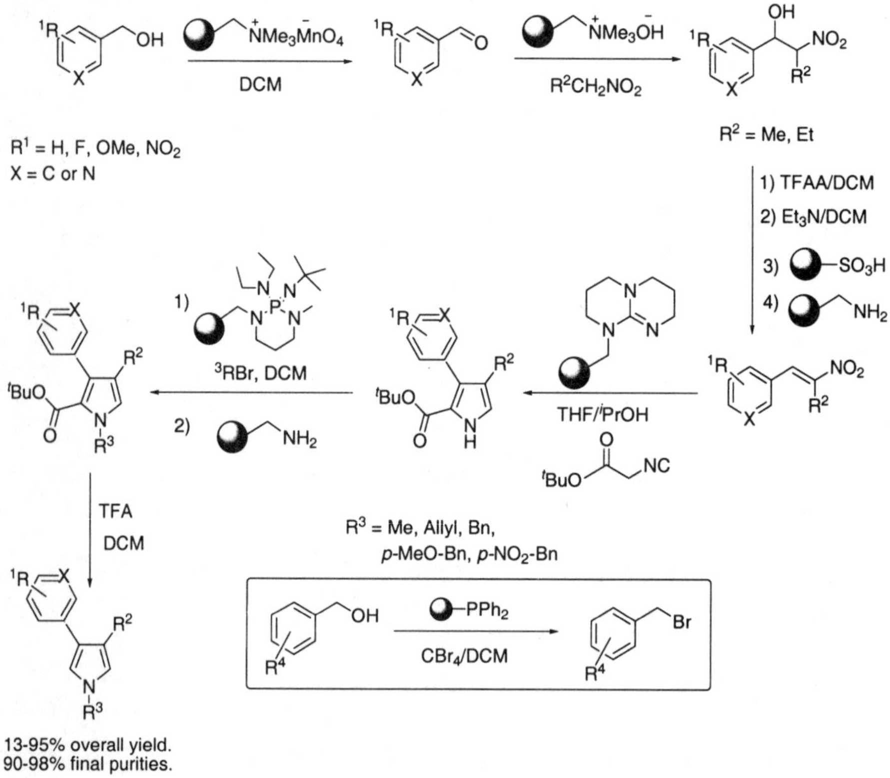

13-95% overall yield.
90-98% final purities.

Scheme 5, *Preparation of an array of tri-substituted pyrrole derivatives*

4. An example of increased efficiency - Library Synthesis

In another study we have constructed a bicyclo[2.2.2]octane library using a tandem Michael addition of enolates of 2-cyclohexenones with various substituted acrylates.[7a, 9] In this way it was possible to prepare a rigid scaffold, from readily available substrates, which could be further elaborated by transformation of the functional groups to give a large array of compounds (Scheme 6). This synthesis required minimal optimisation and was a considerable improvement over a previous route which had been developed with the substrate supported on a Wang resin.[7a, 9a]

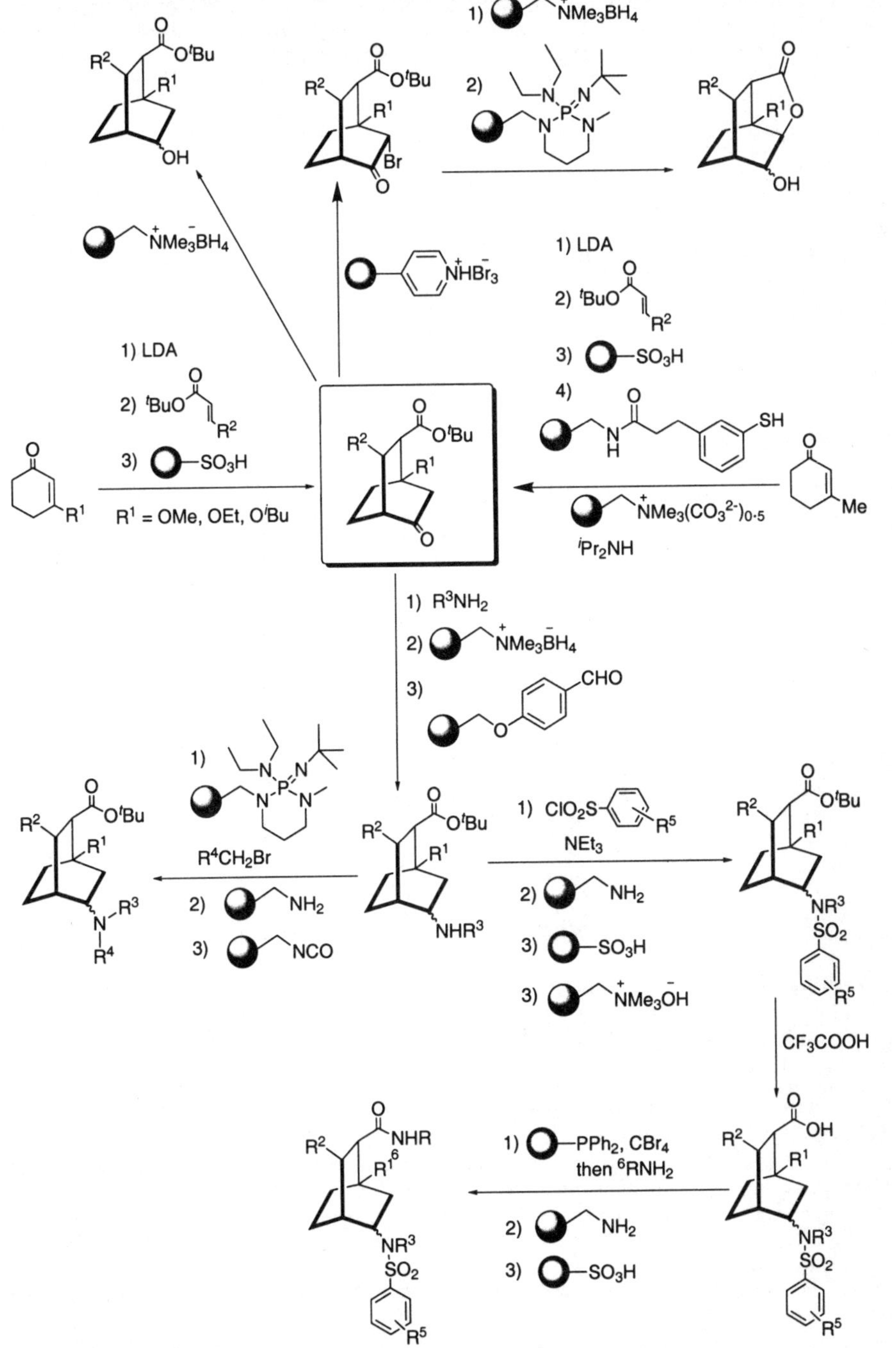

Scheme 6, *Rapid library generation using polymer-supported reagents*

5. Synthesis of alkaloid natural products – epimaritidine and epibatadine

The absence of conventional work-up and purification requirements combined with the ease of optimisation suggests that using polymer-supported reagents would be useful in the assembly of more complex structures such as natural products. We have therefore investigated the synthesis of the alkaloid (±)-epimaritidine.[10a]

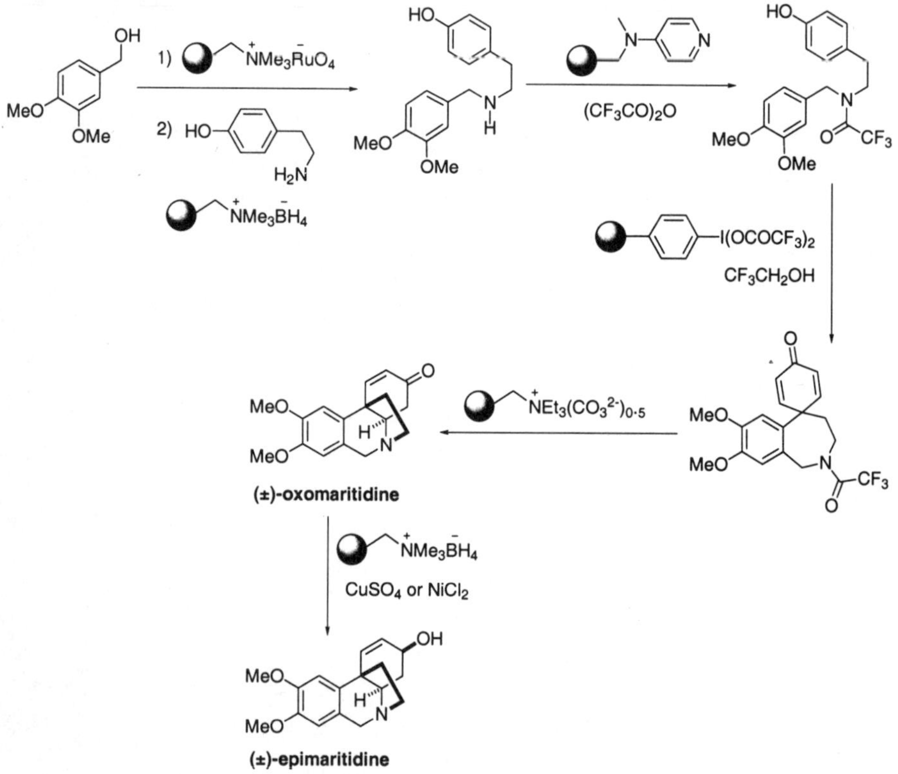

Scheme 7, *Synthesis of the alkaloid natural product (±)-epimaritidine*

Epimaritidine was obtained through a linear six step reaction sequence involving only filtration of the spent reagents at each step in an overall yield of 50%. This short synthetic route allows direct access to (±)-epimaritidine (or its precursor (±)-oxomaritadine) in multigramme quantities, which can be further decorated, in a combinatorial fashion to provide large numbers of analogous compounds for biological evaluation.

The power of these multi-step processes, using supported reagents, has again been demonstrated by the synthesis of the potent analgesic (±)-epibatidine (Scheme 8).[10b] This compound was obtained in an overall yield of 32% and in >90% purity. The combination of polymer-supported reagents and scavengers in this linear ten step sequence highlights the tremendous opportunities for complex molecule synthesis. As most drug substances on the market require at least ten steps for their preparation these new methods become especially attractive.

Furthermore, in this synthesis the polymer-supported reagents were encapsulated in sealed pouches to aid work-up. The reaction sequence starting from the acid chloride **2** through to the intermediate nitroalkene **3** could thus be performed in a one-pot procedure. The reaction progress as effected by the pouched reagent could be easily monitored by

TLC. When the reaction was judged to have reached completion the pouch was removed, with washing, and the next set of pouched reagents were added, thus eliminating the need for individual filtrations between steps. Clearly these processes lend themselves to automation techniques.

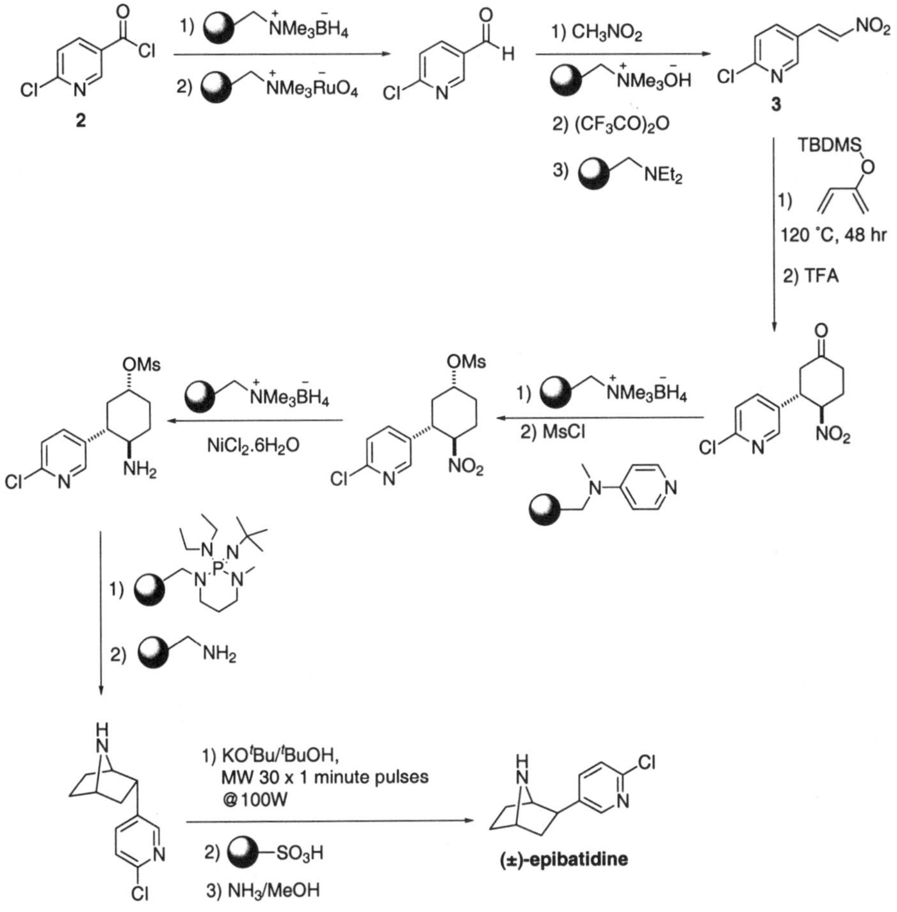

Scheme 8, *Synthesis of the potent analgesic (±)-epibatidine*

6. Supported reagents for the development of drug targets – Sildenafil

Pfizer's Sildenafil (Viagra™) has attracted world-wide attention as a drug for the treatment of male erectile dysfunction.[11, 12] Our synthesis of this important molecule (Scheme 9) demonstrates the principles of using supported reagents in both a sequential and convergent fashion.[13] We believe that these concepts could be easily extended to encompass the synthesis of many other chemical substances in target directed synthesis or in a multi-parallel mode.

Scheme 9, *Preparation of Sildenafil using polymer-supported reagents*

7. Comments on the future

So far only we have scratched the surface of what might be possible with polymer-bound reagents in multi-step organic synthesis. We believe it is possible to use these reagents in many elaborate one-pot multi-reagent combinations, even with the aim of discovering new chemical reactions.

What is required to drive this science forward is the development of many more reagents for the synthesis tool-kit. We require these to be catalytic or readily recyclable and available at a low cost. This will invariably mean that new polymers and support materials will have to be developed. There is also the need for greatly improved and more efficient scavenging and quenching agents to be developed in order to allow a wider range of chemistry to be carried out in a more efficient fashion.

As the true scope and versatility of the supported-reagents are realised the development of automated reactors is becoming increasingly important. We believe that the compatibility of these reagents in automated reaction formats will allow integration with existing reactors for flow processing. Other novel reactor packs, reagent chips or plug-in reagent cartridges will also become prominent in organic synthesis programmes of the future. Inevitably, owing to advances in analytical and separation techniques, intelligent synthesis feedback loops will aid in process optimisation.

We believe that it is only by embracing all of these new techniques and technologies that real advances in the multi-parallel assembly of molecules will be made.

Acknowledgements

We would like to acknowledge the contributions and commitment of all the members of the Polymer-Supported Reagents Group at the University of Cambridge and to thank Pfizer (Sandwich) Postdoctoral Fellowship (to IRB), the BP endowment, Cambridge Discovery Chemistry and the Novartis Research Fellowship (to SVL) for their financial support.

References

1. (a) Hill, D. C. *Curr. Opin. Drug Discovery Dev.* **1998**, *1*, 92-97 (b) Joyce, G. F.; Still, W. C.; Chapman, K. T. *Curr. Opin. Chem. Biol.* **1997**, *1*, 3-4.
2. (a) Bleloch, A.; Johnson, B. F. G.; Ley, S. V.; Price, A. J.; Shephard, D. S.; Thomas, A. W. *Chem. Commun.* **1999**, 1907-1908 (b) Ley, S. V.; Thomas, A. W.; Finch, H. *J. Chem. Soc., Perkin Trans. 1* **1999**, 669-671 (c) Parlow, J. J.; Devraj, R. V.; South, M. S. *Tetrahedron* **1999**, *55*, 6785-6796 (d) Drewry, D. H.; Coe, D. M.; Poon, S. *Med. Res. Rev.* **1999**, *19*, 97-148 (e) Hinzen, B.; Lenz, R.; Ley, S. V. *Synthesis* **1998**, 977-979 (f) Flynn, D. L.; Devraj, R. V.; Naing, W.; Parlow, J. J.; Weidner, J. J. Yang, S. L. *Med. Chem. Res.* **1998**, *8*, 219-243 (g) Flynn, D. L.; Devraj, R. V.; Parlow, J. J. *Curr. Opin. Drug Discovery Dev.* **1998**, 41-50 (h) Shuttleworth, S. J.; Allin, S. M.; Sharma, P. K. *Synthesis* **1997**, 1217-1239 (i). Hinez, B.; Ley, S. V. *J. Chem. Soc., Perkin Trans. 1* **1997**, 1907-1908 (j) Kaldor, S. W.; Siegel, M. G.; Fritz, J. E.; Dressmann, B. A.; Hahn, P.J. *Tetrahedron Lett.* **1996**, *37*, 7193-7196 (k) Akelah, A.; Sherrington, D. C. *Synthesis.* **1981**, *6*, 413-438 (*l*) Akelah, A.; Sherrington, D. C. *Chem. Rev.* **1981**, *81*, 555-600.
3. (a) Ley, S. V.; Baxendale, I. R.; Bream, R. N.; Jackson, P. S.; Leach, A. G.; Longbottom, D. A.; Nesi, M.; Scott, J. S.; Storer, I.; Taylor, S. J.; *J. Chem. Soc., Perkin Trans. 1* **2000**, under preparation. (b) Thompson, L. A. *Curr. Opio. Chem.* **2000**, *4*, 324-3337 (c) Kobayahi, S. *Curr. Opin. Chem.* **2000**, *4*, 338-345.
4. (a) Booth, R. J.; Hodges, J. C. *J. Am. Chem. Soc.* **1997**, *19*, 4882 (b) Kaldor, S. W.; Siegel, M. G.; Dressman, B. A.; Hahn, P. J. *Tetrahedron Lett.* **1996**, *37*, 7193-7196 .
5. (a) Ley, S. V.; Norman, J.; Griffith, W. P.; Marsden, S. P. *Synthesis* **1994**, 639-666 (b) Griffith, W. P.; Ley, S. V.; Whitcomb, G. P.; White, A. D.; *J. Chem. Soc., Chem. Commmun.* **1987**, 1625-1634.
6. For examples of ion exchange resins in organic synthesis see: (a) Parlow, J. J.; *Tetrahedron Lett.* **1996**, *37*, 5257-5260 (b) Bandgar, B. P.; Ghorpade, P. K.; Shrotri, N. S.; Patil, S. V. *Indian J. Chem.* **1995**, *34B*, 153-155 (c) Cainelli, G.; Contento, M.; Manescalchi, F.; Regnoti, R. *J. Chem. Soc., Perkin Trans. 1*, **1980**, *11*, 2516-2519.

7. (a) Ley, S. V.; Massi, A. *J. Comb. Chem.* **2000**, *2*, 104-107 (b) Caldarelli, M.; Baxendale, I. R.; Ley, S. V. *J. Green Chem.* **2000**, 43-45. (c) Ley, S. V.; Lumeras, L.; Nesi, M.; Baxendale, I. R. *Comb. Chem. High Throughput Screening* **2000**, under preparation (*d*) Caldarelli, M.; Habermann, J.; Ley, S. V. *J. Chem. Soc., Perkin Trans. 1* **1999**, 107-110 (*e*) Caldarelli, M.; Habermann, J.; Ley, S. V. *Biorg. Med. Chem. Lett.* **1999**, *9*, 2049-2052 (f) Habermann, J.; Ley, S. V.; Smits, R. *J. Chem. Soc., Perkin Trans. 1* **1999**, 2421-2423. (g) Habermann, J.; Ley, S, V.; Scicinski, J. J.; Scott, J. S.; Smits, R.; Thomas, A. W. *J. Chem. Soc., Perkin Trans. 1* **1999**, 2425-2427 (*h*) Hinzen, B.; Ley, S. V. *J. Chem. Soc., Perkin Trans. 1* **1998**, 1-2 (*i*) Haunert, F.; Bolli, M. H.; Hinzen, B.; Ley, S. V. *J. Chem. Soc., Perkin Trans. 1* **1998**, 2235-2237 (*j*) Ley, S. V.; Bolli, M. H.; Hinzen, B.; Gervois, A. -G.; Hall, B. J. *J. Chem. Soc., Perkin Trans. 1* **1998**, 2239-2241. (*k*) Bolli, M. H.; Ley, S. V. *J. Chem. Soc., Perkin Trans. 1* **1998**, 2243-2246. (*l*) Habermann, J.; Ley, S. V.; Scott, J. S. *J. Chem. Soc., Perkin Trans. 1* **1998**, 3127-3130

8. (a) Uttley, N. *Agro Food Ind. Hi-Tech,* **2000**, 11, 42 (b) Snyder, S. H. *Philosophical Trans. R. C. S. London Series B,* **1999**, *354*, 1985-1994 (c) Collins, A. N.; Sheldrake, G. N.; Crosby, J. *Chirality in Industry II,* **1997**, John Wiley & sons Ltd., *Chp. 3*, 19-38 (d) Ware, M. R. *J. Clin. Psychrarit.* **1997**, *58*, 15-23 (c) Hawkins, J. R.. *J. Clin. Psychrarit.* **1997**, *58*, 324-324.

9. (a) Ley, S. V.; Mynett, D. M.; Koot, W.-J. *Synlett* **1995**, 1017 (b) Bateson, J. H.; Smith, C. F.; Wilkinson, J. B. *J. Chem. Soc. Perkin Trans. 1* **1991**, 651-653 (b) Spitzner, D.; Wagner, P.; Simon, A.; Peters, K. *Tetrahedron Lett.* **1989**, *30*, 547-550 (c) White, B. K.; Reusch, W. *Tetrahedron* **1978**, *34*, 2439-2443 (d) Lee, R. *Tetrahedron Lett.* **1973**, *35*, 3333-3336.

10. Ley, S. V.; Schucht, O.; Thomas, A. W.; Murray, P. J. *J. Chem. Soc., Perkin Trans. 1* **1999**, 1251-1252 (b) Habermann, J.; Ley, S. V.; Scott, J. S. *J. Chem. Soc., Perkin Trans. 1* **1999**, 1253-1255.

11. (a) Dale, D. J.; Dunn, P. J.; Golightly, C.; Huges, M. L.; Pearce, A. K.; Searle, P. M.; Ward, G.; Wood, A. S. *Org. Process Res. Dev.* **2000**, *4*, 17-22 (b) Terrett, N. K.; Bell, A. S.; Brown, D.; Ellis, P. *Bioorg. Med. Chem. Lett.* **1996**, *6*, 1819-1824 (c) Bell, A. S.; Brown, D. European Patent 0 463 756 A1, **1992**.

12. (a) Stief, C. G.; Ückert, S.; Becker, A. J.; Harringer, W.; Truss, M. C.; Forssmann, W. –G.; Jonas, U. *Urology,* **2000**, *55*, 146-150. (b) Schulthesis, D.; Schlote, N.; Steif, C. G.; Jonas, U. *Eur. Urology,* **2000**, *37*, A1-A11 (c) Manecke, R. G.; Mulhall, J. P. *Ann. Med.* **1999**, *31*, 388-398

13. Baxendale, I. R.; Ley, S. V. *Biorg. Med. Chem. Lett.,* **2000**, Submitted.

MESOPOROUS MOLECULAR SIEVE CATALYSTS: RELATIONSHIPS BETWEEN REACTIVITY AND LONG RANGE STRUCTURAL ORDER/DISORDER

Thomas J. Pinnavaia, Thomas R. Pauly, and Seong Su Kim

Department of Chemistry
Michigan State University
East Lansing, MI 48824

1 INTRODUCTION

The electrostatic assembly of mesoporous molecular sieve catalysts typically leads to ordered framework structures with hexagonal, cubic or lamellar symmetry.[1,2] Long range order is facilitated by coulombic interactions occurring among the ionic reagents being assembled at the surface of the electrically charged micelle. Another characteristic property of electrostatically assembled mesostructures is the monolithic nature of the resulting particles. The M41S[1] and SBA[2] families of mesostructures, for example, typically possess particle sizes in the 1-10 μm range and beyond. This feature can impose severe diffusion restrictions and greatly limit access to catalytic sites on the framework walls.

We have developed alternative pathways for the assembly of mesoporous molecular sieves based on hydrolysis of an electrically neutral inorganic precursor (I°) in the presence of a neutral amine (S°) surfactant as the predominate structure directing agent. This S°I° pathway was first used to prepare a mesoporous molecular sieve silica and a Ti-substituted analog. A small amount of protonated amine was used as a co-surfactant in the original synthesis[3], but subsequent studies[4] showed that the protonated co-surfactant component was not needed to achieve framework assembly. Electrostatic forces do not play an important role in S°I° assembly. Instead, the assembly forces at the surfactant-inorganic precursor interfaces are based on hydrogen bonding. An equivalent H-bonding pathway, denoted N°I°, has also been demonstrated for nonionic polyethylene oxide surfactants and I° precursors[5].

Mesostructures prepared through S°I° or N°I° pathways have either wormhole[6,7] or lamellar framework structures[8,9]. The wormhole structures possess a three-dimensional channel structure. Moreover, the framework domain size can be made very small (e.g., 20-200 nm), which introduces an intraparticle textural porosity that is complementary to the framework porosity. The combination of wormhole framework pores and textural pores can greatly facilitate access to catalytic centers in the framework walls. Lamellar frameworks, on the other hand, can be folded into vesicular particles with very thin mesostructured shells and hollow cores. These hierarchical structures also can facilitate access to reactive catalytic centers in the framework walls by minimizing the diffusion path length.

The present paper provides an overview of the physical properties of silica mesostructures with representative wormhole and lamellar framework structures assembled through S°I° pathways. The wormhole framework silica, denoted HMS silica, was assembled using an alkylamine surfactant as the structure director and a silicon alkoxide as the inorganic precursor. The lamellar framework silica with a vesicular hierarchical structure, denoted MSU-G silica, was obtained from tetraethylorthosilicate (TEOS) as the silica precursor and a bi-functional gemini amine surfactant of the type $RNH(CH_2)_2NH_2$. We then provide examples of the catalytic activity of these disordered mesostructures in comparison to more ordered framework mesostructures such as hexagonal MCM-41.

2 PHYSICAL PROPERTIES OF HMS AND MSU-G SILICAS

2.1 HMS Silicas with Wormhole Framework Structures

The first example of an HMS silica was prepared at ambient temperature in the presence of a 13.5:1 molar mixture of dodecylamine (DDA) and dodecylammonium ion as the structure directing co-surfactants[3]. The product formed under these reaction conditions exhibited only one resolved XRD reflection, which precluded the assignment of a long range ordered structure. Selected area electron diffraction studies provided evidence for the occasional occurrence of very small domains of hexagonal symmetry, but the vast majority of the sample was highly disordered and lacking in a long range regular structure.

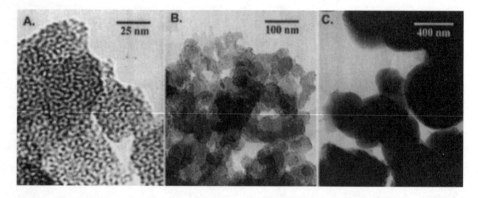

Figure 1: *TEM images of HMS silicas showing (A) the wormhole framework, (B) the intraparticle texture of HMS-HTx (C) the monolithic particles of HMS-LTx*

Subsequent studies revealed that equivalent HMS silicas could be prepared by omitting the onium ion form of the reaction mixture and using only the neutral amine as the structure director[4]. This S°I° pathway afforded silicas with N_2 adsorption properties, pore sizes, and XRD patterns virtually identical to the original HMS products formed using a mixture of S° and S⁺ surfactants. Also, the sparsely occurring small domains of hexagonal order were absent. In fact, hexagonal regions are very rarely formed even when protonated surfactant is present. Instead, the wormhole channel motif shown in Figure 1A is formed almost exclusively[10] even when up to 15% of the amine is protonated. The onium ion can be introduced by adding a protonic acid. Alternatively, the introduction of certain Lewis acid centers, as in the replacement of some Si^{4+} sites by Al^{3+}, Fe^{3+} or B^{3+}, will result in the formation of some protonated amine surfactant during the assembly process in order to balance the resulting framework. However, this small electrostatic participation of the surfactant is structurally inconsequential, and does not alter the wormhole channel motif.

The particle texture of HMS silica depends critically on the polarity of the solvent used to assemble the mesostructure. Water-rich solvent mixtures afford highly textured particles (denoted HMS-HTx) formed by the intergrowth of nanoscale framework domains (cf., Figure 1B), whereas less polar alcohol-rich solvents afford more monolithic particles (denoted HMS-LTx) with little or no textural mesoporosity (cf., Figure 1C). The difference in particle texture is readily distinguished in the N_2 adsorption-desorption isotherms shown in Figure 2 for HMS silicas with high and low textural mesoporosities (denoted HMS-HTx and HMS-LTx, respectively). Both materials exhibit a BET surface area of ~ 1100 m^2/g and a Horvath-Kowazoe framework pore size of ~ 2.8 nm, as judged by the position of the adsorption step at a P/P_o value of ~ 0.3. However, the HMS-HTx sample exhibits an additional uptake of N_2 above P/P_o ~ 0.9, indicative of the presence of intraparticle mesopores in the 15-50 nm range. For the HMS-HTx sample, the textural pore volume (0.94 cc/g) is larger than the framework pore volume (0.73 cc/g), but the HMS-LTx has almost no textural porosity (0.05 cc/g). The difference in textural porosity has no affect on the powder x-ray diffraction patterns. As shown in Figure 3, the XRD patterns are equivalent, exhibiting a single, broad diffraction line indicative of the pore-pore correlation distance and a shoulder at higher 2θ value.

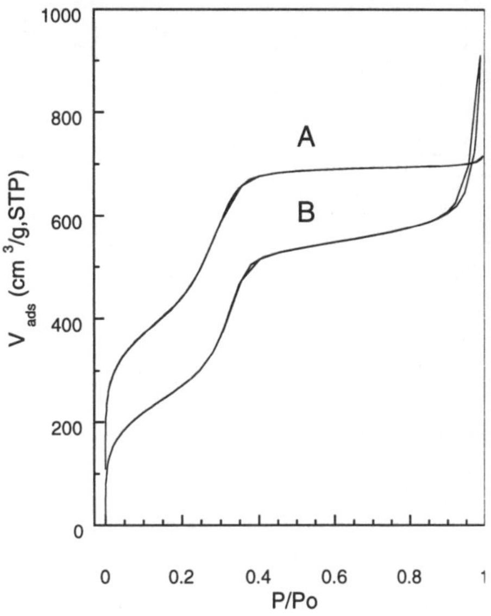

Figure 2: *N_2 adsorption – desorption isotherms for (A) HMS-LTx and (B) HMS-HTx*

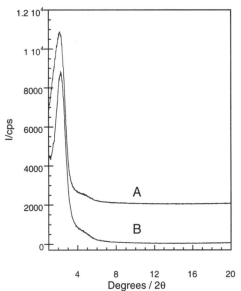

Figure 3: *XRD patterns for (A) HMS-LTx and (B) HMS-HTx*

2.2 Lamellar MSU-G Silicas with Vesicular Hierarchical Structures

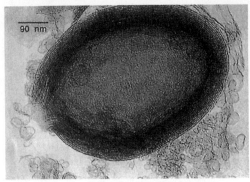

MSU-G mesostructures are obtained through the S°I° assembly of silica in the presence of a $RNH(CH_2)_2NH_2$ gemini amine surfactant in which the R group contains between 8 and 22 carbon atoms[9]. One of the most remarkable properties of these lamellar mesostructures, in addition to the vesicular hierarchical structure, is the high degree of framework crosslinking. Typically, more than

Figure 4: *TEM showing the hierarchical vesicular structure of a MSU-G silica*

85% of the SiO_4 units of an as-made MSU-G silica are Q^4 centers that are fully crosslinked to four adjacent SiO_4 units. In comparison, in as-made HMS and MCM-41 silicas, <65% of the SiO_4 centers are fully crosslinked. This added crosslinking imparts exceptional thermal and hydrothermal stability to the framework, properties that can be especially important in catalytic applications.

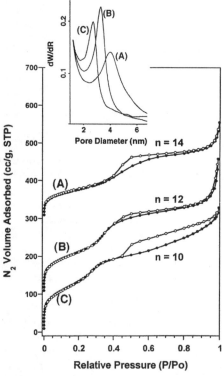

Figure 5: *N_2 adsorption-desorption isotherm and (inset) pore size distributions of MSU-G silicas assembled from Gemini surfactants containing (A) 14, (B) 12, (C) 10 carbon atoms*

The vesicular particle morphology of a typical MSU-G silica is shown in the TEM micrograph of Figure 4. The size of the vesicles varies over a wide range (20-1500 nm). The vesicle shells may consist of a single silica nanolayer, or they may be composed of several nanolayers. Each silica layer is ~3 nm in thickness. Mesopores oriented both parallel and orthogonal to the lamellae are apparent in the image.

Figure 5 presents the N_2 adsorption/desorption isotherms for calcined (650 °C) MSU-G silicas assembled from $C_nH_{2n+1}NH(CH_2)_2NH_2$ surfactants with n = 10, 12, and 14. The inset to the figure provides the framework pore distributions. The maxima in the Horvath-Kawazoe pore size distributions increase in the order 2.7, 3.2, 4.0 nm as the surfactant chain length increases. The textural porosity evident from the hysteresis loop at P/P_o

> 0.9 arises from the filling of the central voids of the smaller vesicles.

3 CATALYTIC PROPERTIES

3.1 Functionalized HMS Silicas

Pure mesoporous silicas have little or no intrinsic catalytic activity, but catalytic centers can be introduced by grafting organic ligands onto the framework walls[11] or by incorporating metal ions into the framework[12,13]. Ti-functionalized derivatives are especially effective in demonstrating the importance of framework access in determining the catalytic activity of a mesostructure.

We consider first the textural properties of a typical wormhole Ti-HMS in comparison to well ordered hexagonal Ti-MCM-41 and Ti-SBA-3 analogs prepared by S^+T and $S^+X^-T^+$ electrostatic assembly pathways. Table 1 provides the surface areas and pore volumes that characterize the framework mesoporosity (V_{fr}) and textural porosity (V_{tx}). The total mesoporosity (V_{total}) is the sum of these two values. Each mesostructure contains ~2 mole % Ti and exhibits a HK pore size near 2.8 nm. The values in parenthesis in the table are for the corresponding pure silicas. Note the very high ratio of textural to framework mesoporosity for the HMS molecular sieves ($V_{tx}/V_{fr} \equiv 1.06$) compared to the hexagonal molecular sieves ($V_{tx}/V_{fr} \equiv 0.03$). As will be shown below, the textural porosity of HMS catalysts can improve catalytic activity by facilitating substrate transport to the active sites in the mesostructure framework.

The catalytic properties of mesoporous Ti-HMS and of hexagonal Ti-MCM-41 and Ti-SBA-3 mesostructures for the liquid phase oxidations of methylmethacrylate, styrene and 2,6-di-*tert*-butylphenol are described in Table 2. Included in the table for comparison are the conversions and selectivities obtained with microporous TS-1 as the catalyst. As expected based on pore size considerations, the conversions observed for all three substrates are substantially larger than for TS-1. But the most efficient mesoporous catalyst is Ti-HMS.

The differences in catalytic reactivity between Ti-HMS, Ti-MCM-41, and Ti-SBA-3 cannot be attributed to differences in Ti siting. XANES and EXAFS studies showed that the titanium center adopts primarily tetrahedral coordination in all three catalysts[12]. Also, the coordination environment is very similar for the three catalysts, as judged from the similarities in the EXAFS features. UV-VIS adsorption spectra showed no phase segregation of titania, the spectral features being consistent with site-isolated titanium centers. Because the framework walls of HMS tend to be thicker than MCM-41, the superior reactivity of Ti-HMS cannot be due to an enhancement in the fraction of Ti available for reaction on the pore walls. Thicker walls should bury more titanium at inaccessible sites within the walls. The most distinguishing feature is the greater textural mesoporosity for Ti-HMS. This complementary textural mesoporosity facilitates substrate transport and access to the active sites in the framework walls.

Others have verified that functionalized wormhole mesostructures are typically more reactive catalysts than well-ordered framework structures[11,13,14,15,16,17,18] In some cases, wormhole frameworks have been found to be less active than hexagonal MCM-41[19,20,21,22]. But in these latter cases the wormhole framework lacked the textural mesoporosity that is characteristic of a HMS mesostructure assembled from a water-rich solvent. If HMS particles are monolithic and comparable in size to a well-ordered hexagonal mesostructure, then framework access is determined primarily by the length of the mesopores and no catalytic advantage is realized.

Table 1 *Textural Properties of Mesoporous Ti-Substituted Silica*

	Ti-MCM-41	Ti-SBA-3	Ti-HMS
Surfactant	$C_{16}H_{33}N(CH_3)_3^+$	$C_{16}H_{33}N(CH_3)_3^+$	$C_{12}H_{25}NH_2$
Ti (mole%)			
gel	2.0	10	2.2
calcined	2.2	2.5	2.4
Parameter[a]			
d(Å)	38.1 (36.0)	36.5 (33.0)	40.2 (36.0)
$S_{BET}(m^2/g)$	859 (923)	1354 (1345)	1075 (1108)
$V_{total}(cm^3/g)$	0.70 (0.72)	0.92 (0.95)	1.40 (1.42)
$V_{fr}(cm^3/g)$	0.68 (0.70)	0.90 (0.92)	0.68 (0.70)
$V_{tx}(cm^3/g)$	0.02 (0.02)	0.02 (0.03)	0.72 (0.72)
V_{tx}/V_{fr}	0.03 (0.03)	0.02 (0.03)	1.06 (1.03)

[a]*The total liquid pore volume, V_{total}, was estimated at a relative pressure of 0.95 assuming full surface saturation. The volume of framework-confined mesopores, V_{fr}, was determined from the upper inflection point of the corresponding adsorption step. The volume of textural mesopores, V_{tx}, was obtained from the equation $V_{tx} = V_{total} - V_{fr}$. The data in parentheses are for the pure silica analogs.*

Table 2 *Catalytic Activity of Ti-substituted (2 mole%) Mesoporous Silicas*

	Catalyst	TS-1	Ti-MCM-41 (S^+I)	Ti-SBA-3 (S^+XI^+)	Ti-HMS (S^0I^0)
MMA	conv. (mol%)	2.5	4.0	6.2	6.8
oxidation	MPV[a]	78	93	93	93
	select. (mol%)				
	conv. (mol%)	8.4	10	23	28
Styrene	PhCHO				
oxidation	select. (mol%)	71	82	78	77
	Epoxide				
	select. (mol%)	14	6.2	4.1	4.7
	Diol				
	select. (mol%)	4.5	3.8	8.2	9.6
2,6-DTBP	conv. (mol%)	5.0	3.9	22	55
oxidation	Quinone				
	select.[b] (mol%)		91	90	91

[a]*MPV is methyl pyruvate;* [b] *Quinone selectivity is expressed as the cumulative selectivity of monomer and dimer quinone.*

3.2 Functionalized MSU-G Silicas

We have recently begun to investigate the catalytic activities of functionalized MSU-G silicas and the preliminary results are especially encouraging[23]. In order to assess the catalytic reactivity of MSU-G in comparison to MCM-41, we have prepared aluminated forms of both mesostructures for the acid-catalyzed conversion of 2,4-di-*tert*-butylphenol to a flavan using cinnamyl alcohol as an alkylating agent. Al-MCM-41 has been shown to be an especially active heterogeneous catalyst for this conversion[24]:

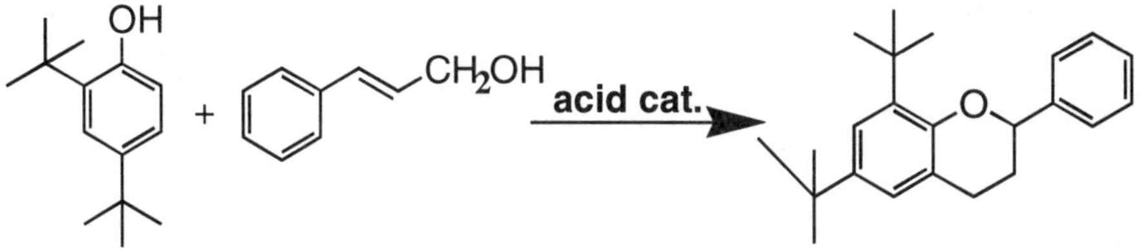

The reactivity of 2%Al-MSU-G as an acid catalyst for flavan synthesis is provided in Table 3. Included for comparison purposes are the yields of flavan obtained using 2%Al-MCM-41 and sulfuric acid. Although 2%Al-MCM-41 has the same framework pore size and twice the surface area as 2%Al-MSU-G, the yield provided by 2%Al-MSU-G (48.8%) is substantially higher obtained with 2%Al-MCM-41 (31.2%) or sulfuric acid (9.4%).

We attribute the enhanced reactivity to more facile access of the reagents to the framework walls of the vesicular MSU-G particles. The framework pores of MSU-G mesostructures are interconnected with pores running orthogonal to the lamellae as well as parallel to the lamellae. Thus, the diffusion path length can be as short as the thickness of the vesicle shells. In contrast, the lengths of the one-dimensional pores of MCM-41 are determined by the size of the monolithic particles, which typically are on a micrometer length scale. In comparison to electrostatically assembled mesostructures with monolithic particle morphologies, the well - expressed, vesicle - like hierarchical structure of MSU-G greatly facilitates access to the framework walls.

Table 3. *The catalytic activity of Al-substituted (2 mol %) MSU-G and MCM-41 materials for the alkylation of 2,4-di-tert-butylphenol (DTBP) with cinnamyl alcohol.[a]*

Catalyst	Aluminum source	Conversion of DTBP (%)	Selectivity of Flavan (%)	Yield of Flavan (%)
Al-MSU-G	Al(NO$_3$)$_3$	76.2	64.1	48.8
Al-MCM-41	Al(NO$_3$)$_3$	50.4	61.9	31.2
H$_2$SO$_4$		25.5	36.9	9.4

[a] *In a typical experiment 250 mg of catalyst or 30 mg of H$_2$SO$_4$ was added to a solution of 2,4-di-tert-butylphenol (1.0 mmol) and cinnamyl alcohol (1.0 mmol) in 50 mL of isooctane as solvent at 90 $^{\circ}$C for 24 h.*

4 ACKNOWLEDGEMENT

The support of this research by the National Science Foundation through CRG grant CHE-9903706 is gratefully acknowledged.

5 REFERENCES

[1] J.S. Beck, J.C. Vartuli, W.J. Roth, M.E. Leonowicz, C.T. Kresge, K.D. Schmit, C.T.-W. Chu, D.H. Olson, E.W. Sheppard, S.B. McCullen, J.B. Higgins and J.L. Schlenker, *J. Am. Chem. Soc.*, 1992, **114**, 10834.

[2] Q. Huo, D.I. Margolese, U. Ciesla, P. Feng, T.E. Gier, P. Sieger, R. Leon, P.M. Petroff, F. Schűth, and G.D. Stucky, *Nature*, 1994, **368**, 317.

[3] P.T. Tanev, M. Chibwe and T.J. Pinnavia, *Nature*, 1994, **368**, 321.

[4] P.T. Tanev and T.J. Pinnavaia, *Science*, 1995, **267**, 865.

[5] S.A. Bagshaw, E. Prouzet and T.J. Pinnavaia, *Science*, 1995, **269**, 1242.

[6] T.R. Pauly, Y. Liu, T.J. Pinnavaia S.J.L. Billinge and T.P. Rieker, *J. Amer. Chem. Soc.*, 1999, **121**, 8835.

[7] S.A. Bagshaw and T.J. Pinnavaia, *Angew. Chem. Intern. Ed. Engl.*, 1996, **35**, 1102.

[8] P.T. Tanev, Y. Liang and T.J. Pinnavaia, *J. Amer. Chem. Soc.*, 1997, **119**, 8616.

[9] S.S. Kim, W. Zhang, and T.J. Pinnavaia, *Science*, 1998, **282**, 1302.

[10] W. Zhang, T.R. Pauly and T.J. Pinnavaia, *Chem. Mater.*, 1997, **9**, 2491.

[11] D. J. McQuarrie, D. B. Jackson, S. Tailland, K. Wilson, and J. H. Clark, *Stud. Surf. Sci. Catal.*, 2000, **129**, 275.

[12] W. Zhang, M. Fröba, J. Wang, P.T. Tanev, J. Wong and T.J. Pinnavaia, *J. Amer. Chem. Soc.*, 1996, **118**, 9164.

[13] A. Sayari, *Chem. Mater.*, 1996, **8**, 1840.

[14] K.M. Reddy, I. Moudrakovski, and A. Sayari, *J. Chem. Soc. Chem. Commun.*, 1994, 1059.

[15] J.S. Reddy and A.J. Sayari, *J. Chem. Soc. Chem. Commun.*, 1995, 2231.

[16] R. Mokaya and W. Jones, *Chem. Commun.*, 1996, 981.

[17] R. Mokaya and W. Jones, *Chem. Commun.*, 1996, 983.

[18] R. Mokaya and W. Jones, *J. Catal.*, 1997, **172**, 211.

[19] T.D. On, M.P. Kapoor, P.N. Joshi, L. Bonneviot, and S. Kaliaguine, *Catal. Lett.*, 1997, **44**, 171.

[20] S. Gontier, A. Tuel, *Zeolites*, 1995, **15,** 601.

[21] S. Gontier, A. Tuel, *J. Catal.*, 1995, **157**, 124.

[22] A. Tuel and S. Gontier, *Chem. Mater.*, 1996, **8**, 114.

[23] S. S. Kim, Y. Liu, and T. J. Pinnavaia, *Micropor. Mesopor. Mater.*, in press.

[24] E. Armengol, M.L. Cano, A. Corma, H. Garcia, M.T. Navarro, *J. Chem. Soc., Chem. Commun.*, 1995, 519.

ZEOLITE BETA AND ITS USES IN ORGANIC REACTIONS

J.C. van der Waal and H. van Bekkum

Department of Applied Organic Chemistry and Catalysis
Delft University of Technology
Julianalaan 136, 2628 BL, Delft, the Netherlands
E-mail: J.C.vanderWaal@tnw.tudelft.nl; H.vanBekkum@tnw.tudelft.nl

1 INTRODUCTION

The first of the high silica zeolites to be prepared by mankind, was Mobil's zeolite beta synthesized by Wadlinger, Kerr and Rosinski[1,2] in 1967. Since then, this zeolite has to some extent been overshadowed by subsequently discovered materials, in particular the medium pore MFI type zeolites. Zeolites have since long attracted the interests of many scientists as selective catalysts for organic reactions.[3] This is due to the wide variety of zeolite pore sizes and geometries, chemical compositions and incorporated catalytically active metals that are available to the chemist.[4] In the last decades it has been recognized that zeolite beta is one of the few large-pore high-silica zeolites with a three-dimensional pore structure containing 12-membered ring apertures[5-7], which makes it a very suitable and regenerable catalyst in organic reactions.

Nowadays many large pore zeolites are known (Table 1). However, only zeolite Beta seems to have the right overall characteristic for organic reactions. Beta is commercially available in various Si:Al ratios. The commercially available Faujasite, Mordenite and Linde type L all have low Si:Al ratios, while the high-silica zeolite ZSM-12 has a parallel channel system giving rise to diffusional problems. The recently discovered zeolites DAF-1, CIT-1 and ITQ-7 require expensive templates and the synthesis is often quite delicate.

Table 1 *Pore geometry and sizes of zeolites with 12-membered ring apertures.*

Zeolite	Pore geometry	Pore dimensions [Å]
Beta	3-D intersecting	7.6 x 6.4 [a] + 5.5
Faujasite	3-D connecting cages	7.4 (windows)
L	Parallel channels	7.1
DAF-1	3-D intersecting	7.3 + 6.4 x 5.4 [b] + 5.6 x 3.4 [c]
CIT-1	3-D intersecting	7.0 x 6.4 + 6.8 + 5.1 [b]
Mordenite	Parallel channels	7.0 x 6.5 + 5.7 x 2.6 [c]
ITQ-7	3-D intersecting	6.2 x 6.1 [a] + 6.3 x 6.1
ZSM-12	Parallel channels	5.9 x 5.5

[a]Twice. [b] 10-Ring channel. [c] 8-Ring channel

2 SYNTHESIS OF ZEOLITE BETA

Zeolite Beta is usually obtained with Si:Al ratios between 5 and infinity using tetraethylammonium (TEA[+]) as the template.[1,2] Important parameters in the TEA-Beta synthesis are the alkali cation concentration and the type of cation used,[8-10] the hydroxide concentration,[8,10,11] the nature and the amount of the organic template,[1,2,8-10,12-17] the temperature[10,18] and the type of silica source used.[19,20]

The synthesis of Si:Al ratios higher than 80 is in general rather difficult and only recently Camblor *et al.*[21] reported the synthesis of the all-silica analogue using TEA[+] as the template. The required synthesis conditions are similar to the all-silica Beta synthesis we reported earlier using dibenzyldimethylammonium (DBDMA[+]),[12] a template which was introduced by Rubin.[22] A defect-free all-silica material was synthesized by Corma *et al.* using TEA[+] in a fluoride medium.[23] At the low Si:Al side the minimum Si:Al ratio attained so far is 5 using TEA[+] as the template. Guisnet *et al.*[24] have reported that a substantial amount of non-framework Aluminum in the low Si:Al materials may be present. However, the natural Beta analogue Tschernichite[25] possesses a Si:Al ratio of 3. The mineral's composition is $Ca_{0.97}Na_{0.05}Mg_{0.08}Al_{2.00}Fe_{0.02}Si_{5.95}O_{16.00}$. This suggests that nature used divalent cations (Ca^{2+}) as the template, offering potential new routes to zeolite Beta.

The aluminium atom is not the only non-silicious metal that can be incorporated in the Beta framework. So far the boron,[26] iron,[27] gallium[28] and titanium[21,29] containing materials have been reported. Especially the Ti-containing analogue has received a lot of attention due to its potential in oxidation chemistry using aqueous hydroperoxide as the oxidant (*qui vivre*). The synthesis of Ti-beta is quite difficult compared to the aluminum analogue

Finally we mention that – under suitable hydrothermal conditions – Beta can be grown onto supports such as pre-shaped alumina[30] or stainless steel monoliths[31]

3 MATERIAL CHARACTERISTICS

The high Si:Al ratio of the material, typically around 10, makes the material inherently hydrophobic. Changing the Si:Al ratio of the material has, however, a marked influence on the hydrophobicity. The increased hydrophobicity of siliceous Beta has been demonstrated by competitive adsorption of toluene and water. The so-called hydrophobicity index[32], the amount of toluene adsorbed divided by the amount of water adsorbed at 25 °C amounts to 1.4 to 2.2 for a Si:Al = 10 sample and increases to 10.8 to 66 for the all-silica material (see Table 2). The large difference observed for the all-silica zeolites is most likely due to differences in the amount of defects in the material. These defects are essentially silanol pairs required for template charge compensation during synthesis as shown by van der Waal *et al.*[12]

Zeolite Beta particles are often conglomerates of very small crystallites (typically around 50 nm). Consequently the outer surface is large; values up to 80 m^2/g, against 680 m^2/g for the internal surface area, have been reported. When applying special synthesis recipes[33,34] or when applying fluoride media[35] large bipyramidal crystals with sizes up to 5 μm can be obtained. The two tops of these 'pyramid' crystals are typically not fully out-grown, resulting in a truncated shape.

Table 2 *Hydrophobicity Index for zeolite Beta with various Si:Al ratios.*

Si: Al ratio	Hydrophobicity Index
5	0.8 [a]
10	2.3 [a] 1.4 [b]
45	4.8 [a]
100	5.4 [a]
Infinity	10.8 [a] 66 [b]

[a] After Verhoef *et al.*[36] [b] After Stelzer *et al.*[37]

3.1 The Dynamic Nature of the Aluminium Atom in the Beta Framework

As demonstrated first by Jia *et al.*[38] using ^{29}Si MAS NMR and Fajula[39] using ^{27}Al MAS NMR, the Al in Beta has a dynamic character and is able to cycle between the tetrahedral full-lattice configuration and octahedral and tetrahedral lattice-grafted forms. As a consequence, the aluminum atom in Beta can exert classical Brønsted acidity as well as Lewis acidity as was encountered by us in the Meerwein-Ponndorf-Verley (MPV) reduction of ketones (see below). For methods of quantitative measurements of Brønsted and Lewis acidity see Guisnet *et al.*.[24]

Mild steaming enhanced the activity in the MPV reaction and was accompanied by an IR band at 3782 cm^{-1} assigned to an OH group attached to an Al.[40] A recent infrared study gives convincing evidence that this 3782 cm^{-1} band is related to a 885 cm^{-1} band and belongs to a defect Lewis-acidic aluminum site grafted to the Beta framework.[41]

The reversible character of the Al transitions was recently confirmed by ^{27}Al-MAS-NMR and Al K-edge XANES studies by van Bokhoven *et al.*.[42] Following heating and steaming up to 450°C the Al reverts completely to its tetrahedral lattice position upon ammonia treatment. At 550°C irreversible transitions of aluminum come to the fore.

3.2 Outer Surface Passivator of Zeolite Beta

Using large probe molecules[43] the activity of the outer surface compared to the microporous activity of Beta materials can be determined. It was shown that for typical TEA-synthesized materials a large fraction of the activity observed could be attributed to the non-shape-selective outer surface acid groups. In order to minimize this non-shape-selective activity various techniques have been employed to passivate the outer surface.

In a recent study the H$^+$-catalyzed rearrangement of allyl (3,5-di-*tert*-butylphenyl) ether was used as selective probe for the outer surface activity.[44] The best passivation results were obtained by treatment of the Beta crystals with tetraethyl orthosilicate (TEOS). Slow hydrolysis of TEOS by traces of water present in the pores of the zeolite provides a thin, porous layer of amorphous silica and leads to complete outer surface passivation on 1 μm crystals.

4 BRØNSTED ACID-CATALYSED REACTIONS

Many examples of H-Beta acting as a Brønsted acid catalyst in organic conversions are known or can easily be envisaged. We mention here etherification, esterification,[45] acetalisation, hydration[46] and hydroalkoxylations of olefins,[47] dehydration of alcohols, aliphatic hydrocarbon alkylation,[48] alkylation of carbohydrates,[49] Fischer Indole synthesis[50] and the C-C coupling of α-pinene and acetone.[51]

In this ariticle we confine ourselves to some classes of electrophilic aromatic substitutions and Lewis acid catalysis.

4.1 Friedel-Crafts Related Reactions.

4.1.1. Alkylation.   In the field of alkylation of benzene with ethene zeolite-based catalysts are used for the past 20 years, replacing the conventional $AlCl_3$- and BF_3-on-alumina based processes. Here the question in case of a new plant is not whether a zeolite-based process will be selected but rather which one to choose. The Mobil-Badger process uses ZSM-5 as the catalyst and is the most widely applied though recently other zeolites (Y, Beta and MCM-22) have come to the fore.

Some excellent reviews exist on benzene alkylation.[52] Comparison of Beta with other zeolites (USY [53] or MCM-22 and ZSM-5) [54] shows that Beta seems to be the most active catalyst whereas MCM-22 shows the best overall properties combining a good activity with an excellent stability. Similar results are found for the zeolite-based cumene processes where zeolite Beta is used in the process developed by Enichem.[52]

Linear alkylbenzenes are made from linear terminal olefins and benzene and are important precursors of biodegradable anionic surfactants (LAS, linear alkylbenzenesulfonates). The conventional catalyst is HF, first to be replaced by a fluorinated silica-alumina in the DETAL process. The DETAL process is safer than the HF process and also more cost-effective because no special metallurgy is required and no calcium fluoride waste stream exists.[52] Zeolites such as Beta may come to the fore here because they display a higher selectivity to the desired 2-phenyl isomers.[55]

Moreau *et al.*[56] obtained unexpected results in the alkylation of naphtalene with 2-propanol over H-Beta in the liquid phase at 200°C. Here a cyclic compound **1** was formed with a selectivity around 40% at 28.5% conversion. When applying HY as the catalyst alkylation to di- and trialkylnaphthalenes was faster but the cyclic compound was not observed. These results illustrate the more confined space within the zeolites Beta channels. The cyclic compound is assumed to be formed through *iso*-propylation of naphthalene followed by a hydride abstraction giving a carbenium ion, reaction with a propylene and finally ring-closure.

1

The Fajula group also recently studied the alkylation of naphthalene with *tert*-butanol.[57] Over H-beta 2-*tert*-butylnaphthalene was obtained as the main product together with relatively small amounts of the di-*tert*-butylnaphthalenes. Over HY 2,6- and 2,7-di-*tert*-butylnaphthalene the major products, demonstrating again the more spacious Y-pore system.

4.1.2 Acylation.   Aromatic acyclation processes are widely applied in the fine chemical industry as reaction steps in the synthesis of pharmaceuticals and fragrances. Present day industrial synthesis generally involves acid chlorides as the acylating agent. Due to the strong coordination of the product ketone to the Lewis-acid "catalyst" ($AlCl_3$,

FeCl$_3$ and TiCl$_4$) more than stoichiometric amounts of the metal chloride are required. Other acylating agents, like anhydrides, exhibit similar disadvantages. Moreover the hydrochloric acid formed in the reaction work-up generates highly corrosive media. A direct and truly catalytic route involving the free carboxylic acid or even the anhydride as the acylating agent would be most attractive. Zeolite catalysts such as H-Beta show great promise in this respect.

Recently Davis *et al.*[58] studied the industrially relevant acetylation of *iso-*butylbenzene with acetic anhydride over H-Beta samples at 100°C in 1,2-dichloroethane as the solvent. Low catalyst acivity together with high (> 99%) *para*-selectivity were observed. The acetic acid formed may act as an inhibitor in this case by selective adsorption in the zeolite pores. Similarly, a patent of Uetikon[59] claims a yield of 80% with 96% *para*-selectivity when operating H-Beta at 140°C.

In zeolite-catalysed acylation of aromatic ethers, zeolites have two advantages with respect to benzene and alkylbenzene acylation:
- the ethers are activated for electrophilic substitution.
- the ethers are in better competition with the acylating agents for adsorption into the zeolite pores.

Following the work of Corma *et al.*[60] on the acylation of anisole, Rhodia (at that time Rhône-Poulenc) workers started a systematic investigation into the acylation of anisole and veratrole with acetic anhydride (Scheme 1).[61] A solvent-free process was developed with both reactants passing a fixed bed of H-Beta. This new process for *para*-acetylanisole is clean and also brings along a substantial process simplification with respect to the conventional AlCl$_3$-catalysed process.[62] This multistage process used 1,2-dichloroethane as the solvent and involved a hydrolysis step, several washings and separations and a recycle of water and solvents.

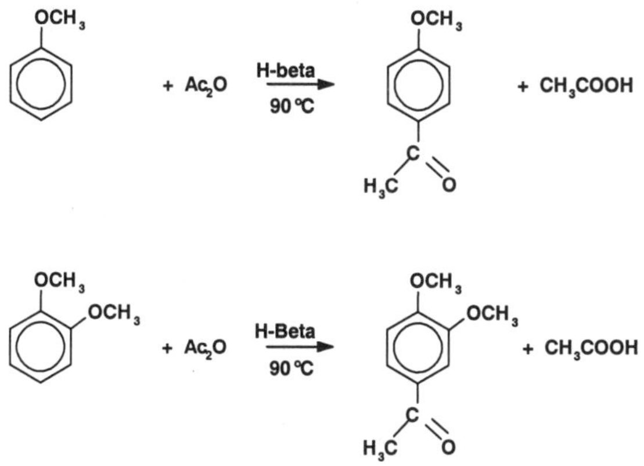

Scheme 1

Compared to the above success story, the selective 6-acylation of 2-methoxynaphthalene – a precursor for the drug Naproxene – is in a much earlier state of development (Scheme 2). Here the kinetically favored substitution is at the 1-position. Two approaches can be applied to enhance the yield of the 6-isomer:
- formation of the 1-isomer followed b isomerisation to the thermodynamically more stable 6-isomer

- application of a zeolite with a pore system too narrow for the 1-isomer to form. Recent evidence suggests that only 6-substitution occurs inside zeolite Beta.[62] Taking into account the large external surface (~60 m^2/g) normally found for standard synthesized zeolite Beta, large crystals combined with outer-surface passivation will be required for this reaction.[44] When applying such catalyst a strongly enhanced selectivity towards 6-acetyl-2-methoxynaphthalene (up to 92%) was observed.

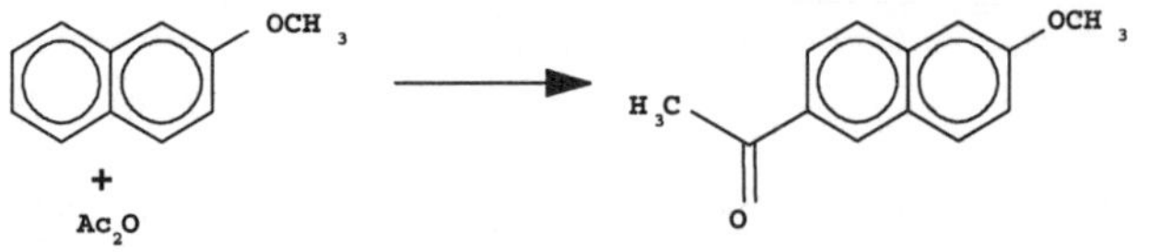

Scheme 2

4.2 The Fries Rearrangement

The Fries rearrangement is the isomerisation of phenyl esters towards *ortho-* and *para-* acylphenols and is usually performed using a relative large amount of AlCl$_3$ as the catalyst. Similar to the acylation reaction the product ketone forms a 1:1 adduct with the AlCl$_3$, requiring more than stoichiometric amounts of catalyst. Heterogeneous catalysts, zeolite in particular, have been suggested as possible alternative catalysts for this reaction. Most of the studies so far concentrate on the synthesis of hydroxyacetophenones either by Fries rearrangement of the phenyl ester or by acetylation of phenol with the acetic acid or the anhydride.[63] H-Beta gives particular good results in this reaction with *o/p* hydroxy-acetophenone ratios of up to 4.7 .[64]

When using resorcinol as the starting material the regioselectivity simplifies and excellent results have been obtained in the benzoylation towards 2,4-dihydroxybenzo-phenone,[65] a precursor of a well-known sunscreen agent (Scheme 3). With H-beta as the catalyst of choice and benzoic acid the acylating agent, yields of almost 90% have been achieved working in the liquid phase. The present industrial process involves the reaction of benzotrichloride (PhCCl$_3$) with resorcinol in the presence of FeCl$_3$.

The limited space inside the H-Beta pores becomes apparent when 2-methyl and 2,6-dimethylbenzoic acid are applied as the reactants with resorcinol and compared to benzoic acid. With 2-methylbenzoic acid the conversion to the benzophenone is accelerated due to electronic effects, with 2,6-dimethylbenzoic acid the reaction slows down because the intermediate ester is too bulky to be formed inside in the pores. [65,66]

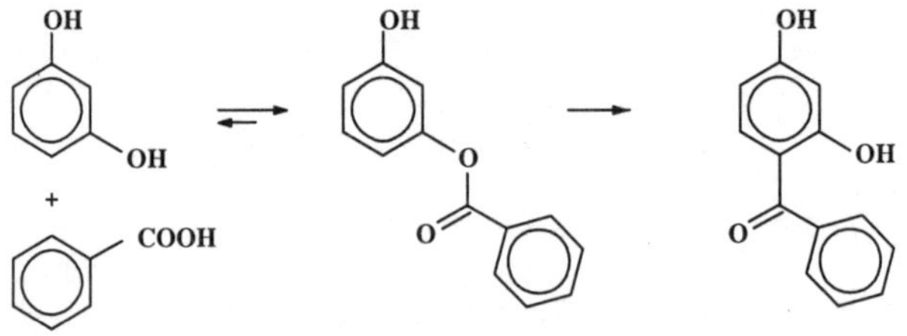

Scheme 3

4.3 Nitration of Aromatics

Nitration of aromatics, toluene in particular, has been extensively studied using zeolites in order to direct the reaction from the normally favoured ortho-product to the desired *para*-product. Application of the industrially preferred nitric acid as the alkylating agent normally deactivates zeolites because of poisoning of the acid site by water.[67] Zeolites can be used if the water present is removed by raising the reaction temperature[67] or by forming *in situ* an acetyl nitrate with acetic anhydride.[68] In the last method safety considerations become prominent. In a recent study it was shown that zeolite Beta initially provides a higher *para/ortho* ratio (1.1) than zeolite ZSM-5 (0.9), Mordenite (0.75) and Deloxan (0.83), a polysiloxane functionalised with alkylsulfonic groups. However, the high *para/ortho* ratio for zeolite beta dropped to 0.8 after 30h reaction time.[69] The initial high selectivity was tentatively ascribed to an adsorption-induced steric blockage of the *ortho* position of the substituted aromatics rather than a classical transition state selectivity.

5 LEWIS ACID CATALYSED REACTIONS

Zeolites are not typically used in Lewis acid type catalysis due to the 'absence' of Lewis acid centers in zeolites. This is due to the coordination of the Al-site to four lattice-oxygens in a perfect zeolite framework. It has, however, been shown for zeolite Beta that the aluminum atom can reversibly move between a framework Brønsted acid site and a framework-grafted Lewis-acid site.[70] Accordingly, Creyghton *et al.* showed that zeolite Beta is active in the Meerwein-Ponndorf-Verley reduction (MPV) of ketones (scheme 4).[71] In this reaction a hydrogen hydride transfer reaction between an alcohol and a ketone takes place.

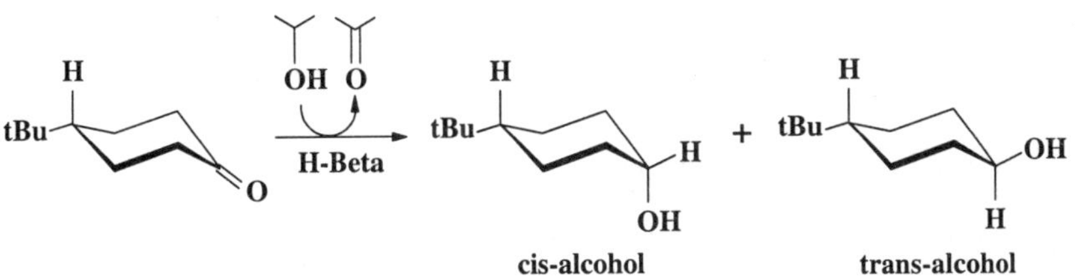

cis-alcohol   trans-alcohol

95 : 5

Scheme 4

The activity of zeolite Beta was found to increase with increasing activation temperature.[71] In an FT-IR study of the applied Al-Betas these authors also acquired convincing data that the observed catalytic activity is related to partially bonded framework Al-atoms.[72] Furthermore, deep-bed calcination conditions, *i.e.* a higher degree of auto-steaming, gave higher catalytic activity than a shallow-bed procedure, indicating a relation between activity and the extent of framework de-alumination. This was further supported by a detailed ^{29}Si and ^{27}Al MAS-NMR and FT-IR study.[73] In the proposed mechanism, the first step consists of the chemisorption of the donating alcohol to a coordinatively unsaturated framework-grafted Aluminum atom, resulting in a surface bonded alkoxide species that is considered to be the hydrogen donor. Attachment of the ketone to the same aluminum atom enables the formation of a six-membered transition state allowing the hydride transfer.

A very high stereoselectivity was observed in the reduction of 4-*tert*-butylcyclohexanone to the *cis*-alcohol (> 95%), which is the industrially relevant product. The observed high selectivity to the thermodynamically unfavorable *cis*-alcohol was explained by a restricted transition-state for the formation of the *trans*-alcohol within the pores of the zeolites (Scheme 5). This reaction was found not only to be catalysed by Al-Beta, van der Waal *et al.* reported the catalytic activity of aluminum-free zeolite titanium beta (Ti-Beta) in the same reaction.[74] Again, a very high selectivity to the *cis*-alcohol was observed indicating similar steric restrictions on the mechanism. Kinetically restricted product distributions were also reported for the 2-,3- and 4-methylcyclohexanone; the *cis, trans-* and *cis*-isomers being the major products, respectively. In this case the tetrahedrally coordinated Ti-atom was assumed to behave as the Lewis acid metal center. Recent quantum-chemical calculations on zeolite TS-1 and Ti-Beta confirm the higher Lewis acidic nature of the latter one.[75]

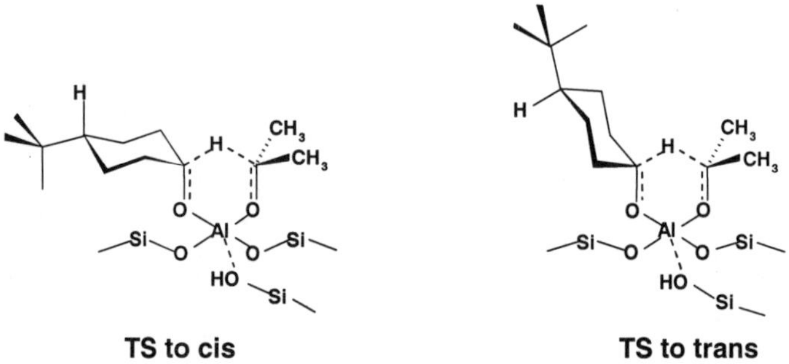

TS to cis   TS to trans

Scheme 5

Further evidence of the mechanism in which both the alcohol and ketone are simultaneously coordinated to a Lewis acid metal center was obtained from the reduction of the pro-chiral phenylacetone with (S)-2-butanol.[76] A nearly identical enantiomeric excess (e.e.) of 34% was observed over both the Al- and Ti-Beta.

Although the activity of Ti-Beta is much lower than that of Al-Beta, it was found to have a high tolerance for water. The latter property, which is related to a higher hydrophobic character combined with the absence of Brønsted acid sites, allows the use of Ti-Beta in gas-phase application at increased reaction temperatures where Brønsted acid-catalysed site reactions would otherwise have become significant.[77]

Ti-Beta was also applied in the selective transformation of α-pinene oxide to camphenolenic aldehyde.[78] A selectivity of over 98% was observed in the gas-phase reaction that was explained as a combination of a Lewis catalysed reaction in the absence of Brønsted acid sites. Furthermore the concentration of α-pinene oxide in the zeolite pores was found to be an important factor not ruling out additional transition state selectivity as well.

6 OXIDATION CHEMISTRY

After the discovery of TS-1 as a high effective and selective oxidation catalyst using aqueous hydrogen peroxide as the oxidant,[79,80] it was soon realized that the pore size of TS-1 was too small for most organic molecules and for the application of *tert*-butyl hydroperoxide as the oxidizing agent. The larger pore apertures and high silica nature of zeolite Beta made it the interesting candidate for titanium incorporation. Two types have been reported sofar, Ti,Al-Beta and Ti-Beta in oxidation chemistry.[29]

The Ti,Al-Beta shows both acidic and oxidative properties which is reflected in unwanted side-reaction. The group of Corma used the bifunctionality in the epoxidation/rearrangement of α-terpineol to cineol alcohol and in the formation of furans from linalool.[81,82] Similarly van Klaveren *et al.* applied Ti,Al-Beta in the one-pot conversion of styrene to phenyl acetaldehyde.[83] Sato *et al.*[84] solved the unwanted acid-catalyzed side reaction by neutralizing the acid site by ion exchange with alkali metals. Nevertheless the bifunctionality restricts the use of this catalyst to a limited number of reactions.

It is therefore that in recent articles the focus has shifted to Ti-Beta as the catalyst. Allthough differences in activity between Ti-Beta and TS-1 exist for small molecules, the latter usually being slightly more active, the real difference between the two types of catalysts is in the epoxidation of bulky molecules. Whereas TS-1 is not capable of epoxidizing cyclohexene, Ti-Beta has no problem with a wide range of cyclic alkenes, bulky olefins or terpenes.[85]

7 CONCLUSION

Altogether zeolite Beta is a versatile catalyst that can be tuned in many ways. As a consequence Beta is already serving several industrial processes and the expectations for further applications are high.

8 REFERENCES

1   R.L. Wadlinger, G.T. Kerr and E.J. Rosinski, *US Pat. Appl.* 3.308.069 (1967).
2   R.L. Wadlinger, N.Y. Niagara, G.T. Kerr and E.J. Rosinski, *US Pat. Appl. Reissue* 28.341 (1975).
3   See for reviews: P.B. Venuto, *Microporous Mater.*, 2 (1994) 297; W.F. Holderich and H. van Bekkum, *Stud. Surf. Sci. Catal.*,**58** (1991) Chp. 16.
4   J.C. van der Waal and H. van Bekkum, *J. Porous Mater.*, 5 (1998) 289
5   J.B. Higgins, R.B. LaPierre, J.L. Schenker, A.C. Rohrman, J.D. Wood, G.T. Kerr and W.J. Rohrbaugh, *Zeolites*, 8 (1988) 446.
6   J.M. Newsam, M.M.J. Treacy, W.T. Koetsier and C.B. de Gruyter, *Proc. R. Soc. Lond. A*, **420** (1988) 375.
7   B. Marler, R. Böhme and H. Gies, in *"Proc. 9th Int. Zeolite Conf."*, (R. von Ballmoos *et al.* Eds.), Butterworth-Heinemann, Montreal, (1992) p. 425.
8   J. Pérez-Pariente, J.A. Martens and P.A. Jacobs, *Appl. Catal.*, **31** (1987) 35.
9   M.A. Camblor and J. Pérez-Pariente, *Zeolites*, 11 (1991) 202.
10   M.J. Eapen, K.S.N. Reddy and V.P. Shiralkar, *Zeolites*, 14 (1994) 294.
11   N.A. Briscoe, J.L. Casci, J.A. Daniels, D.W. Johnson, M.D. Shannon and A. Stewart, in *"Zeolites: Facts, Figures, Future"*, (P.A. Jacobs and R.A. van Santen Eds.), Elsevier Science Publishers B.V., Amsterdam, (1989) p. 151.
12   J.C. van der Waal, M.S. Rigutto and H. van Bekkum, *J. Chem. Soc., Chem. Commun.*, (1994) 1241..
13   S.I. Zones, D.L. Holtermann, L.W. Jossens, P.S. Santilli, A. Rainis and J.N. Ziemer, *PXCT Int. Pat. Appl.* WO 91/00.777 (1991).
14   P. Caullet, J.L. Guth, A.C. Faust. F. Raatz, J.F. Joly and J.M. Deves, *Eur. Pat. Appl* A1 0.419.334 (1990).

[15]   P. Caullet, J. Hazm, J.L. Guth, J.F. Joly, J. Lynch and F. Raatz, *Zeolites*, **12** (1992) 240.
[16]   M.R. Rubin, *Eur. Pat. Appl.* 0.159.847 (1985).
[17]   J.P. van den Berg. P.C. de Jong-Versloot, A.G.T. Kortbeek and M.F.M. Post, *Eur. Pat. Appl.* 0.307.060 (1988).
[18]   LO. Li-Jeu, H. Liang-Yuan, K. Ben-Chang, C. Li, W. Shwu-Tzy and W. Jung Chung, *Appl. Catal.*, **69** (1991) 49.
[19]   R.N. Bhat and R. Kumar, *J. chem.. tech. Biotechnol.*, **48** (1990) 453.
[20]   Y. Sun, T. Song, S. Qiu and W. Pang, *Chem. Research in Chin. Universities*, **10** (1994) 2.
[21]   M.A. Camblor, M. Constantini, A. Corma, L. Gilbert, P. Esteve, A. Martínez and S. Valencia, *J. Chem. Soc., Chem. Commun.*, (1996) 1339.
[22]   M.K. Rubin, *Eur. Pat. Appl.* A2 0.159.846 (1985).
[23]   R. de Ruiter, K. Pamin, A.P.M. Kentgens, J.C. Jansen and H. van Bekkum, *Zeolites*, **13** (1993) 611.
[24]   A. Berreghis, P. Ayrault, E. Fromentin and M. Guisnet, *Catal. Lett.*, **68** (2000) 121.
[25]   R.W. Tschernich, in "Zeolite of the World", Geoscience Press Inc., 1992, p. 509; R.C. Boggs, D.C. Howard, J.V. Smith and G.L. Klein, *Am. Mineral.*, **78**, (1993), 822.
[26]   M. Taramasso, G. Perefo and B. Notari, in "*Proc. 5th Int. Conf. Zeolites*", (L.V.C. Rees Ed.), Heyden, London, (1980) p. 40.
[27]   R. Kumar, A. Thangaraj, B.N. Bhat and P. Ratnasamy, *Zeolites*, **10** (1990) 85.
[28]   R. Fricke, H. Kosslick, G. Lischke and M. Richter, *Chem. Rev.*, **100** (2000) 2303.
[29]   [Ti,Al-β] : M.A. Camblor, A. Corma and J. Pérez-Pariente, *J. Chem. Soc., Chem. Commun.*, (1992) 589; M.A. Camblor, A. Corma and J. Pérez-Pariente, *Zeolites*, **13** (1993) 82; [Ti-β] : R.J. Saxton, J.G. Zajacek and K.S. Wijeseker, *Eur. Pat. Appl.* A1 0.659.685 (1994); J.C. van der Waal, P.J. Kooijman, J.C. Jansen and H. van Bekkum, *Microporous Mesoporous Mater.*, **25** (1998) 43; see also ref. 21.
[30]   N. van der Puil, F.M. Dautzenberg, H. van Bekkum and J.C. Jansen, *Microporous Microporous Mater.*, **27** (1999) 95.
[31]   O.L. Oudshoorn, M. Janissen, W.E.J. van Kooten, J.C. Jansen, H. van Bekkum, C.M. van den Bleek and H.P. Calis, *Chem. Eng. Sci.*, **54** (1999) 1413.
[32]   M.J. Verhoef, R.M. Koster, E. Poels, A. Bliek, J.A. Peters and H. van Bekkum, submitted to 13th IZA conference (2001).
[33]   P.J. Kunkeler, B.J. Zuurdeeg, J.C. van der Waal and H. van Bekkum, *J. Catal.*, **180** (1998) 234.
[34]   Y. Sun, T. Song, S. Qiu and W. Pang, *Chem. Res. Chin. Univ.*, **10** (1994) 2.
[35]   P. Caullet, J. Hazm, J.-L. Guth, J.-F. Joly and F. Raatz, *Zeolites*, **12** (1992) 240.
[36]   M.J. Verhoef, Thesis Delft University of Technology, 2000.
[37]   J. Stelzer, M. Paulus, M. Hunger and J. Weitkamp, *Microporous Mater.*, 1998, 22.
[38]   C. Jia, P. Massiani and D. Barthomeuf, *J. Chem. Soc., Faraday Trans.*, **89** (1993) 3659.
[39]   F. Fajula, in "Zeolites, a refined Tool for designing catalytic Sites", Eds. L. Bonneviot and S. Kaliaguine, 1995, Elsevier, p. 133.
[40]   I. Kiricsi, C. Flego, G. Pazzuconi, W.O. Parker, R. Millini, C. Perego and G. Bellussi, *J. Phys. Chem.*, **98**, 1991, 4627.
[41]   A. Vimont, F. Thibault-Starzyk and J.C. Lavalley, *J. Phys. Chem.*, **104**, (2000) 286.
[42]   J.A. van Bokhoven, Thesis University of Utrecht, 2000.
[43]   E.J. Creyghton, J.A. Elings, R.S. Down ing, R.A. Sheldon and H. van Bekkum, *Microporous Mater.*, **5** (1996) 299.
[44]   P.J. Kunkeler, J.A. Elings, R.A. Sheldon and H. van Bekkum, in "*Proc. 11th IZC*", Baltimore, 1999, p. 1975.
[45]   M.J. Verhoef, J.A. Peters and H. van Bekkum, *Stud. Surf. Sci. Catal.*, **125** (1999) 465.
[46]   P. Botella, A. Corma, J.M. Lopez Nieto, S. Valencia, M.E. Lucas and M. Sergio, *J. Catal.*, (2000) in press.
[47]   K. Hensen, C. Mahaim and W.F. Hölderich, *Stud. Surf. Sci. Catal.*, **105** (1997) 1133.
[48]   K.P. de Jong, C.M.A.M. Mesters, D.G.R. Peferoen, P.T.M. van Brugge and C. de Groot, *Chem. Eng. Sci.*, **51** (1996) 2053.
[49]   M.A. Camblor, A. Corma, S. Iborra, S. Miquel, J. Primo and S. Valencia, *J. Catal.*, **172** (1997) 76.
[50]   M.S. Rigutto, H.J.A. de Vries, S.R. Magill, A.J. Hoefnagel and H. van Bekkum, *Stud. Surf. Sci. Catal.*, **78** (1993) 661.
[51]   J.C. van der Waal, H. van Bekkum and J.M. Vital, *J. Mol. Catal.A: Chem.*, **105** (1996) 185.
[52]   J.S. Beck and W.O. Haag, in "*Handbook of Heterogeneous Catalysis*", Eds. G. Ertl, H. Knözinger and J. Weitkamp, Wiley-VCH Vol 5 (1997) 2123; J.A. Horsley, *Chem. Tech.*, October (1997) 45.

53 G. Bellussi, G. Pazzuconi, C. Pereo, G. Girotti and G. Terzoni, *J. Catal.*, **157** (1995) 227.

54 A. Corma, V. Martinez-Soria and E. Schnoeveld,. *J. Catal.*, **192** (2000) 163.

55 S. Sivanker, A. Thangaray, R.A. Abdulla and P. Ratnasamy, *Stud. Surf. Sci. Catal.*, **75** (1993) 397.

56 C. He, Z. Liu, F. Fajula and P. Moreau, *J. Chem. Soc., Chem. Commun.*, (1998) 1999.

57 P. Moreau, Z. Liu, F. Fajula and J. Joffre, *Catal. Today*, **60** (2000) 235.

58 P.Andy, J. Garcia-Martinez, G. Lee, H. Gonzalez, C.W. Jones and M.E. Davis, *J. Catal.*, **192** (2000) 215.

59 A. Vogt and A. Pfenninger, EP Pat appl. 0701987 A1 (1996) Uetikon AG.

60 A. Corma, M.J. Climent, H. Garcia and J. Primo, *Appl. Catal.*, **49** (1989) 109.

61 M. Spagnol, L. Gilbert and D. Alby, in "The Roots of Organic development", Eds. Desmurs and S. Ratton, Industrial Chemistry Library, Elsevier, Vol 8 (1996) p. 29; M. Spagnol, L. Gilbert, E. Benazzi and C. Marcilly, WO Pat appl. 96/35655 (1996) Rhodia.

62 P. Metivier, in *"Fine Chemicals through Heterogeneous Catalysis"*, Eds. R.A. Sheldon and H. van Bekkum, Wiley-VCH (2000) in press.

63 Jap. Pat. Appl. 81.142.233 (1981), Mitsui Toatsu Chemicals Inc.

64 A. Heidekum, M.A. Harmer and W.F. Hölderich, *J. Catal.*, **176** (1998) 260.

65 A.J. Hoefnagel and H. van Bekkum, *Appl. Catal. A:General*, **97** (1993) 87.

66 H. van Bekkum, A.J. Hoefangel, M.A. van Kooten, E.A. Gunnewegh, A.H.G. Vogt and H.W. Kouwenhoven, in *"Zeolites and Microporous Crystals"*, Eds. T. Hattori and T. Yashima, Elsevier, (1994) p. 379.

67 D. Vassena, D. Malossa, A. Kogelbauer and R. Prins, in *"Proc. 12th Int. Zeolite Conf."* (Eds. M.M.J. Treacy *et al.*), 1999, p. 1471.

68 K. Smith, A. Musson and G.A. DeBoos, *J. Org. Chem.*, **59** (1994) 4939.

69 D. Vassena, A. Kogelbauer and R. Prins, *Stud. Surf. Sci. Catal.*, 130A (2000) 515.

70 E. Bourgeat-Lami, P. Massiani, F. Di Renzo, P. Espiau and F. Fajula, *Appl. Catal.*, **72** (1991) 139.

71 E.J. Creyghton, S.D. Ganeshie, R.S. Downing and H. van Bekkum, *J. Chem. Soc., Chem. Commun.*, (1995) 1859.

72 E.J. Creyghton, S.D. Ganeshie, R.S. Downing and H. van Bekkum, *J. Mol. Catal.A: Chem.*, **115** (1997) 457.

73 P.J. Kunkeler, B.J. Zuurdeeg, J.C. van der Waal , J.A. van Bokhoven, D.C. Koningsberger and H. van Bekkum, *J. Catal.*, **180** (1998) 234.

74 J.C. van der Waal, K. Tan and H. van Bekkum, *Catal. Lett.*, **41** (1996) 63.

75 G. Sastre and A. Corma, *Chem. Phys. Lett.*, **302** (1999) 447.

76 J.C. van der Waal, E.J. Creyghton, P.J. Kunkeler, K. tan and H. van Bekkum, *Topics Catal.*, **4** (1997) 261.

77 J.C. van der Waal, P.J. Kunkeler, K. Tan and H. van Bekkum, *J. Catal.*, **173** (1998) 74.

78 P.J. Kunkeler, J.C. van der Waal, J. Bremmer, B.J. Zuurdeeg and H. van Bekkum, *Catal. Lett.*, 53 (1998) 135.

79 M. Taramasso, G. Perego and B. Notari, *US Pat. Appl.* 4.410.501 (1983).

80 For a review on the oxidation chemistry see G. Bellusi and M.S. Rigutto, *Stud. Surf. Sci Catal.*, **85** (1994) 177.

81 A. Corma, M.T. Navarro, J. Pérez-Pariente and F. Sánchez, *Stud. Surf. Sci. Catal.*, **84** (1994) 69; A. Corma, P. Esteve and A. Martínez, *J. Catal.*, **161** (1996) 11.

82 A. Corma, P. Esteve, A. Martínez and S. Valencia, *J. Catal.*, **152** (1995) 18.

83 M. van Klaveren and R.A. Sheldon, in *"4th Int. Symp. On Heterogeneous Catal. And Fine Chemicals"*, Basel, Switzerland, (1996) p. 83.

84 T. Sato, J. Dakka and R.A. Sheldon, *J. Chem. Soc., Chem. Commun.*, (1994) 1877; T. Sato, J. Dakka and R.A. Sheldon, *Stud. Surf. Sci. Catal.*, **84** (1994) 1853.

85 J.C. van der Waal, M.S. Rigutto and H. van Bekkum, *Appl. Catal. A:General,*. **167** (1998) 331.

CHIRAL MESOPOROUS HYBRID ORGANIC-INORGANIC MATERIALS IN ENANTIOSELECTIVE CATALYSIS

Daniel Brunel

Laboratoire des Matériaux Catalytiques et Catalyse en Chimie Organique-UMR-5618-CNRS -ENSCM, 8 rue de l'Ecole Normale, F-34296 -MONTPELLIER Cédex 05.
Email :brunel@cit.enscm.fr

1 INTRODUCTION

In the past decade, many efforts have been devoted to the preparation of solids displaying chirality in order to develop heterogeneous enantioselective catalysis. Among the possible ways to reach such a goal, the addition of an auxiliary to already efficient heterogeneous catalysts was mostly explored.[1-3] Another interesting route concerns the immobilisation of chiral homogeneous catalysts onto solids. The main attractive features of such a strategy are the high efficiency of supported catalysis due to site isolation, friendliness to the environment and low cost despite the high prize of most chiral products, because they are easily recyclable. In this field, heterogeneisation of chiral catalysts was more developed on polymer supports[4-14] than on mineral support.[15-28]. It is also noteworthy that few works dealt with the covalent anchorage of chiral catalysts to porous metal oxides.

On the other hand, the nineties have seen important advances in the synthesis and the characterisation of new mesoporous micelle-templated mineral oxides such as MCM-41 type silicates.[30,31] The major advantages of these new supports are their large surface area, their regular mesoporous system of pore-monodispersed size and homogeneity of chemical surface properties. Hence, these materials have given us useful opportunities for the preparation of hybrid organic-inorganic mesoporous materials based on the functionalisation of their surface which allows the preparation of well-defined mesoporous hybrid materials.[24] This presentation mainly deals with the design of the more efficient supported catalysts on MCM-41-type silica. This goal is based on the control of the different parameters including the texture of the support, the anchoring mode, the chemical nature of the grafted moieties during the different modification steps. The aim of such a control is i) the improvement of the surface coverage in order to insure preservation of the unique texture of the materials and lowering the possible interaction between the catalytic site and the support surface; ii) the best definition of catalytic centres and their environment. The interest of this approach is highlighted by the preparation of two kinds of chiral supported ligands able to provide higher levels of enantioselectivity in epoxidation and C-C bond formation reactions.

The literature survey concerning enantioselective hydrogenation, amination, epoxidation, and C-C bond reactions using different chiral catalysts supported on silicas, zeolites and micelle-templated materials is firstly presented in order to compare the different methodologies.

2 ENANTIOSELECTIVE HYDROGENATION AND HYDRIDE TRANSFER

The first example of preparation of chiral catalytic sites anchored onto a mineral surface and optically active in hydrogenation reaction was reported by Nagel et al in 1986.[15] The supported catalysts were prepared through the covalent fixation of rhodium complex of 3,4-(R,R)-bis(diphenylphosphino)pyrrolidine onto silica surface with different linkers. (Scheme 1).

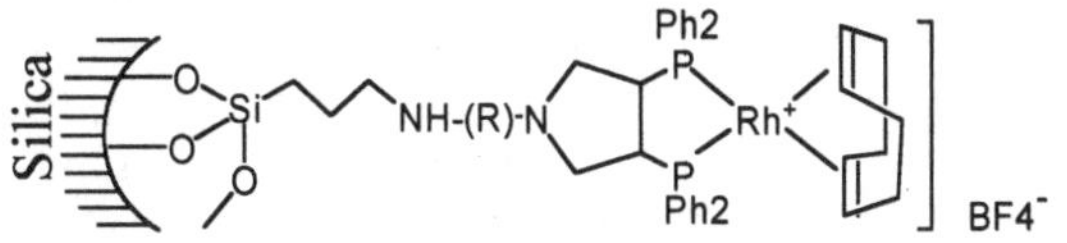

Scheme 1 *Supported chiral Rhodium complex anchored on silica surface with different linkers :* R = C(O)C(O)-, -C(O)C$_6$H$_4$C(O)-, -C(O)[CH$_2$]$_3$C(O)-

Heterogeneisation of chiral rhodium complex of 1,2-diphosphines already known as very efficient catalysts for enantioselective hydrogenation[32] was achieved through amine functionality borne by pyrrolidine molecule. The supported Rh complex revealed as its homogeneous counterpart very high enantioselectivity (<90%) in hydrogenation of α-(acetylamino)cinnamic acid and its methyl ester.

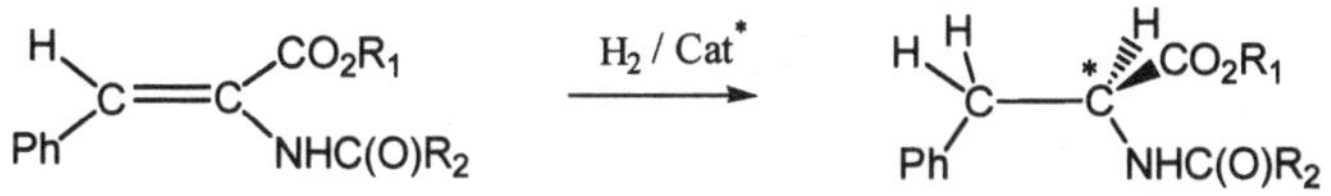

Scheme 2 *Enantioselective hydrogenation α-(acetylamino)cinnamyl derivatives*

The authors have pointed out the possible role of the interaction of the metal complex and the surface leading to a limitation in the reactant accessibility by metal shielding.

More recently Pugin has investigated the effect of the site isolation during the same hydrogenation reaction with anchored rhodium complex of (2S,4S)-4 diphosphino-2-(diphenylphosphino-methyl)pyrrolidine and during hydrogenation of N-(2-methyl-6-ethylphen-1-yl)methoxymethylmethylketimin with neutral Ir complex based on the same ligand. [33] Whereas cationic diphosphine rhodium complexes have no tendency to interact with each other, the neutral catalyst do interact in a way that is detrimental for their catalytic performance due to the formation of inactive dimers. Hence, the activity of the immobilised neutral complex increases with decreasing catalyst loading.

Corma et al have anchored Rh(I), Ru (II), Co(II) and Ni(II) chiral complexes based on β-aminoalcohols such as (L) prolinol onto silica and modified USY-zeolites (scheme 3) to perform enantioselective hydrogenation of the same prochiral alkenes than shown in scheme 2.[20,34]

The authors stressed the higher enantioselectivity and activity of zeolite-supported complexes than on either silica-supported or unsupported complexes.

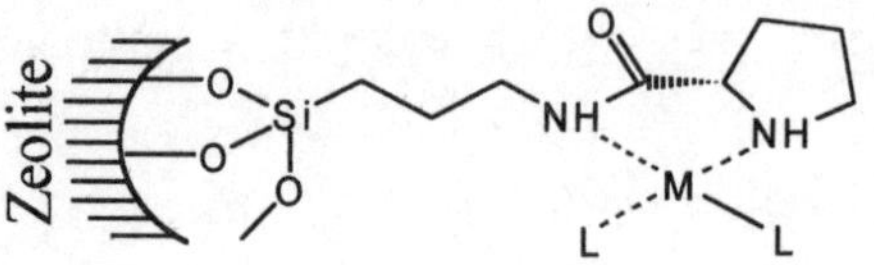

Scheme 3 *Chiral transition metal complexes anchored onto zeolite surface with M =*
Rh(I), Ru (II), Co (II) and Ni (II), and L = COD, Acac and CO according to
refs. 19, 20, 34.

They assigned this beneficial effect to the important role of the concentration effect of the zeolite in addition with steric constraints due to the presence of supermicropores (20-60 Å) into the texture of an ultrastable Y zeolite (USY) after washing with citric acid.

Another interesting strategy to prepare heterogeneous catalysts emerged from sol-gel chemistry. The co-hydrolysis of substituted alkoxysilanes, containing ligand and ethyl silicates followed by co-condensation have been shown to generate supported homogeneous catalysts and heterogeneous silica supported catalysts. [35,36] Moreau et al. have applied this method in order to prepare hybrid organic-inorganic solids bearing chiral organic fragment attached to the inorganic silicate network.[36-38]. The chiral moieties consisted of trans diaminocyclohexane[37] and binaphthyl units.[38] (Scheme 4)

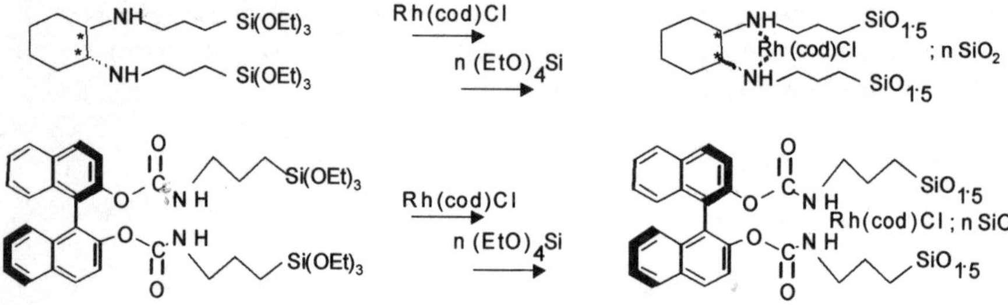

Scheme 4   *Sol-gel preparation of chiral organic-inorganic hybrid materials*
according to ref. 37 and 38.

The catalytic performance of these hybrid materials has been evaluated in the hydrogen-transfer reduction of prochiral ketones (Scheme 5).

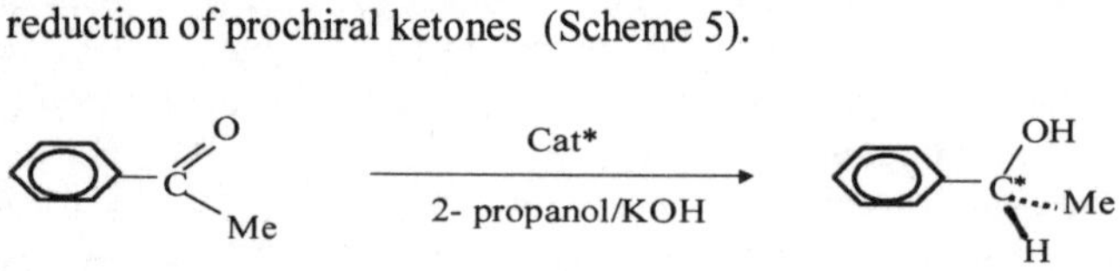

Scheme 5   *Enantioselective hydrogen –transfer reduction of acetophenone.*

In the case of hybrid organic-inorganic materials containing trans 1,2-diaminocyclohexane, higher selectivity were obtained compare to the homogeneous counterpart and the gel prepared with n = 0 gave higher enantioselectivity (58%) than with added TEOS (15% with n = 3). The enantioselectivity rise until 98% when the aromatic nucleus is the hindered naphthyl group. However accessibility to the catalytic site drastically limits the conversion.

The rhodium containing hybrid materials prepared with the BINOL bicarbamate moieties are less active and enantioselective than the previous hybrid catalyst. The observed enantioselectivity was attributed to supramolecular effect of the chiral tridimensional network owing to the weakness of the interaction of the transition metal and the chiral ligand. The control of the texture and morphology of these solids by templating methods firstly reported by Macquarrie[39] and Mann[40] with suitable surfactants would improve the catalytic performance of this new class of chiral materials.

3 ENANTIOSELECTIVE AMINATION

Chiral moieties have been tethered onto MCM-41 surface in order to obtain optically active supported catalysts possessing the intrinsic texture brought by this particular support.[21,23,24,26,41-43] Thomas et al [26,41] have developed a very interesting methodology to anchor a chiral ligand 1,1'-bis(diphenylphosphino)ferrocene to the inner walls of MCM-41 silica (Figure 1).

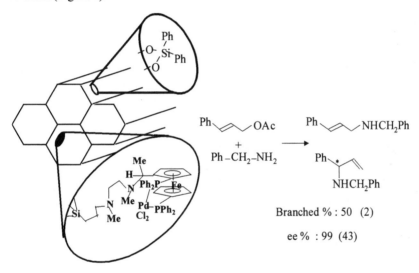

Branched % : 50 (2)

ee % : 99 (43)

Figure 1 *Supported chiral palladium complex anchored onto interior surface of MCM-41 and catalytic benzylamination of cinnamyl acetate (values in brackets obtained with functionalised silica gel counterparts (from reference 26)*

The exterior surface of the mesoporous support was first deactivated by treatment with Ph$_2$SiCl$_2$ under non-diffusive conditions. The inner walls of the materials were then functionalised with 3-{(*S*)-1-[(*R*)-1',2-bisdiphenyl-phosphino)ferrocenyl]ethyl-*N*,*N*'-dimethylethylenediamino}propylsilane chains. The mesoporous chiral catalyst obtained on reaction of with PdCl$_2$, revealed higher regioselectivity (50%) and enantioselectivity (99%) in the Trost-Tsuji reaction than the same catalytic site bound on non porous high-area silica (values in brackets on Fig.1).

These results highlight the emerging potential of the mesoporous dimension in regio-and enantioselective catalysis.

4   ENANTIOSELECTIVE EPOXIDATION

The heterogeneous asymmetric epoxidation of alkenes is central to many of the recent research activities that are focused on the design of highly enantioselective catalysts. The approach consisting in the immobilisation of either Sharpless tartrate-titanium isopropoxide[44,45] or Jacobsen-Katstuki complexes[7-10,46-48] on polymer has received more attention than the immobilisation on mineral supports. To my knowledge, the only example of immobilisation of Sharpless-type system on mineral support concerns the combinaison of a dialkyl tartrate and titanium pillared montmorillonite.[49,50] However this system is rather a modification of an achiral heterogeneous catalyst with a chiral auxiliary than a genuine chiral supported catalyst.

On the contrary, a truly heterogenised Jacobsen's complex has been successfully obtained by a "ship-in-a-bottle" approach reported by both T. Bein et al[51] and A. Corma et al[52] at the same time (Scheme 6a). The first group has encapsulated the chiral salen in EMT zeolite, then metalation with manganese followed by oxidation in presence of LiCl afforded the chloro[*N,N'*-bis(salicylidene)-cyclohexanediamine]manganese (III). The highest ee (88%) in epoxide was achieved with cis-β-methylstyrene using NaOCl in presence of pyridine *N*-oxide as an additional axial ligand. The second group has directly accomplished the preparation of encapsulated Jacobsen's complex by intrazeolite assembling condensation of chiral diaminocyclohexane and salicylaldehyde around Mn(II) metal ion of Mn-exchanged Y zeolite. Finally, oxygen zeolite framework is acting as counteranion for the oxidised Mn(III) complex. The lower performance of this catalyst compared to the homogeneous counterpart was explained by possible diffusion limitation.

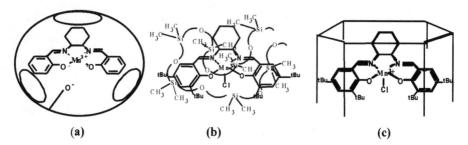

                    (a)   **(b)**   **(c)**

Scheme 6 *Jacobsen's complex entrapment into a) zeolite ; b) polydimethylsiloxane membrane ; c) MCM-41.*

Jacobs et al has also reported immobilisation of Jacobsen's catalyst in a polydimethylsiloxane (PDMS) membrane for the epoxidation of terminal alkene (Scheme 6b).[53] In the case of styrene epoxidation using NaOCl, styrene oxide was obtained with nearly the same activity and enantioselectivity (52 %) than in homogeneous conditions. Interestingly, the catalytic membranes are easily regenerable.

Finally, the same manganese complex has been embedded in the channels of MCM-41 (Scheme 6c).[54,55] When the Jacobsen's complex was supported on mesoporous Al-, Ga- and Fe-substituted MCM-41 silicates by impregnation, the catalyst was thought to be maintained inside the channels by strong guest/host interactions between the aromatic rings of the complex and the internal silanol groups of the mesopore walls.[54] On the other hand, Hutchings et al have performed the Mn-complex catalyst by reacting the chiral salen ligand with Mn-exchanged Al-MCM-41 in order to enhance the hitching by ionic interaction.[55]

Similarly, Che et al have recently immobilised a chromium (III) binaphthyl Schiff base complex through coordination of the chromium ion to the terminal NH_2 group of surface –bound 3-aminopropyl tethering chains (scheme 7).[27]

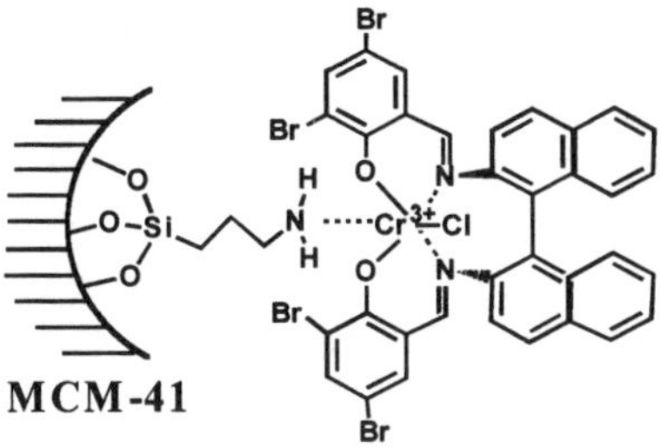

MCM-41

Scheme 7   *Chromium (III) binaphthyl Schiff base complex supported on MCM-41.*

The heterogeneised catalyst exhibits significantly higher enantioselectivity than the same free complex. The increase in the chiral recognition could arise from the enhanced stability of the chromium complex upon immobilisation and from the unique spatial environment as previously suggested by Thomas et al.[26]

Following the Corma group's work on the immobilisation of molybdenum by complexation with prolinol derivatives strongly bonded on zeolite surface (Scheme 8a),[19] Salvadori et al have reported the first example of a silica gel-grafted Jacobsen complex (Scheme 8b)[25] using a stepwise approach analogous to that by Sherrington et al.[10] to prepare polymer containing salen complex and similar to the Song et al's preparation of silica functionalised with bis-cinchona alkaloid.[22] The complex was anchored to the silica surface by coupling reaction under radical conditions of the vinyl group bearing the aromatic nucleus with 3-mercaptopropylsilane moieties.

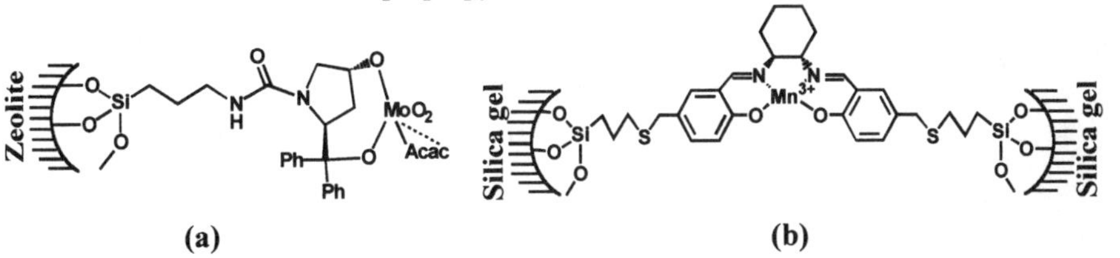

(a)   **(b)**

Scheme 8   *Metal transition complexes grafted on the surface of a) Y dealuminated zeolite ; b) silica gel.*

While the enantioselectivity found with this supported catalyst in the epoxidation of dihydronaphthalene and indene was far lower than in homogeneous conditions or with the same catalyst supported on polymer, a significant 53-58% ee was observed with diphenylcyclohexene.

Recently, we have investigated the MCM-41 surface functionalisation with a new chiral tetrazamacrocycle we prepared in order to synthesise another Mn (II) complex able to perform enantioselective epoxidation (scheme 9)[42].

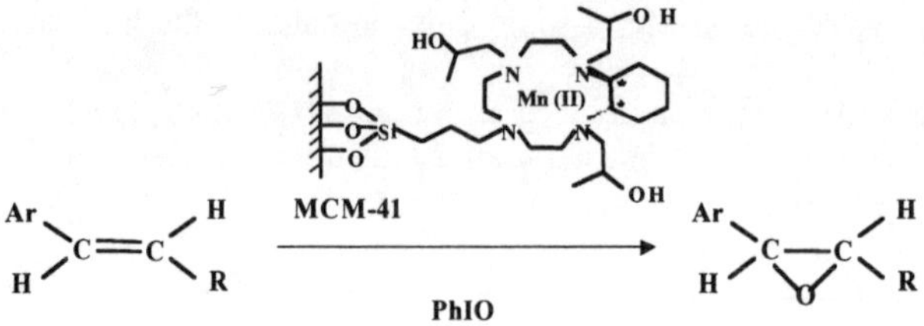

Scheme 9 *Styrene epoxidation with MCM-41-grafted Mn(II) 2R,3R-cyclohexano-*
N,N',N'',N'''-tetrakis(2-hydroxypropyl)1,4,7,10-tetraazacyclododecane

This chiral macrocycle is based on tetraazacyclododecane containing the chiral
diaminocyclohexane unit. The ligand anchorage on MCM-41 surface was achieved by
chlorine substitution of tethered 3-chloropropyl silane. The surface was previously end-
capped before the coupling reaction in order to prevent the further polymerisation of
propene oxide used as *N*-alkylating agent. Moreover this treatment could avoid the
undesirable action of silanol groups as achiral ligand of manganese salts. Iodosylbenzene
was used as oxygen donor to regenerate the Manganese (IV) complex that is the active
site for transferring oxygen to styrene or trans-β-methylstyrene. The trans-β-
methylstyrene oxide was obtained with a relatively high enantioselectivity (76% ee) by
comparison with other immobilised system although lower than that obtained with
homogeneous counterpart (94%). This promising result prompts further investigation to
improve the selectivity in epoxide and the catalyst reusability.

5   ENANTIOSELECTIVE ALKYLATION

Only few reports deal with the use of mineral supports for immobilising chiral auxiliaries
able to activate dialkylzinc addition to aldehyde (Scheme 10).

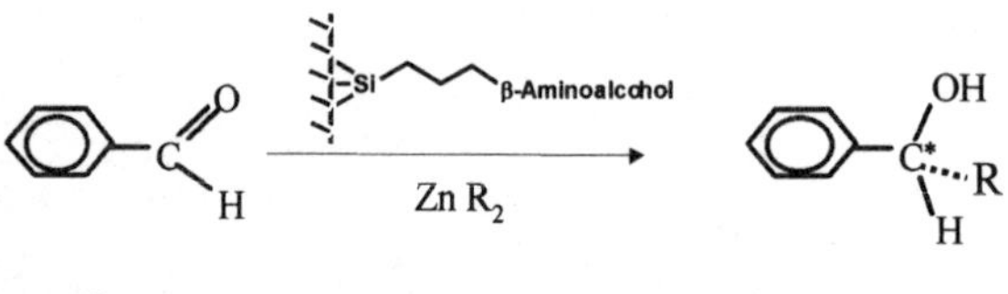

Scheme 10 *Benzaldehyde alkylation with dialkylzinc catalysed by grafted*
β-aminoalcohol.

The pioneering work reported by Soai et al. in this field, deals with anchorage of
ephedrine onto silica and alumina surface (Scheme 11).[16]

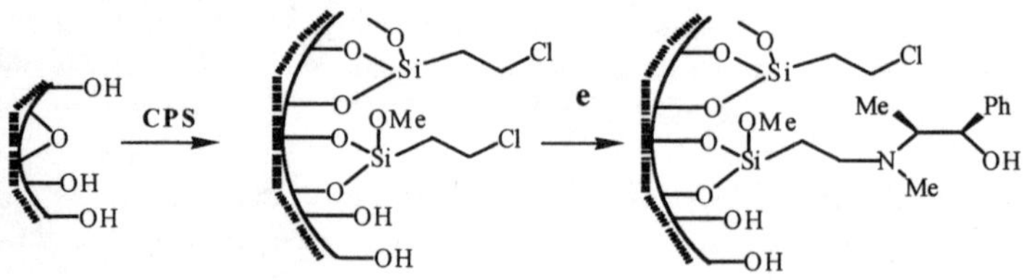

Scheme 11 *Anchorage of ephedrine onto silica surface.*

According the same methodology, the Montpellier's group has investigated the design of functionalised MCM-41 surface in order to improve the catalytic performance.[21,23,56,57]

Table 1. *Comparison of the catalytic performance of various chiral catalysts.*

Mineral Support :	*β-aminoalcohol* :	*Enantiomeric excess :*		*Ref.:*
		Heterogeneous	*Homogeneous*	
SILICA	☞ **Ephedrine**	37%	76%	16
MCM-41	☞ **Ephedrine**	56%	76%	56
MCM-41	☞ **Prolinol**	75%	93%	43

Kim et al have recently reported also the preparation and application of prolinol derivatives on mesoporous silicas[43]. In this case, higher enantioselectivities were obtained mainly due to the higher intrinsic efficiency of the chiral auxiliary and to the addition of butyl lithium as extra-metal reagent.

Actually, these enantiomeric excesses are far lower than the ones obtained with the same β-aminoalcohol in homogeneous conditions (Table 1) or anchored on polymer support.

It should be noted that the catalytic cycle involves the formation of a 1:1 chelate between β-aminoalcohol and alkyl zinc in a first initiation step according to Soai,[4] Noyori[6] and Corey.[58] This monoalkyl chelate would be the actual catalytic site which coordinate benzaldehyde and a second molecule of dialkylzinc. The stereoselection occurs during the alkyl transfer from the complexed ZnR_2 molecule to the coordinated aldehyde due to the geometry of the site environment. We envisaged that the low ee's obtained with ephedrine immobilised on silica support result from the activation of dialkylzinc by the naked surface. Taking into account the mechanism of the silylation of MCM-41 –type surface, the grafted chains are surrounded by hydrophilic silanol groups.[59] End-capping of the silica surface by hexamethyldisilazane proved quite inefficient to totally passivate the uncovered surface.[60] Hence such a treatment failed to deactivate the contribution of these achiral catalytic sites (Scheme 12).

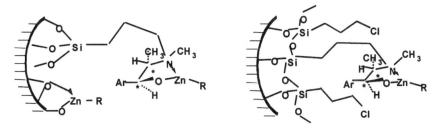

Scheme 12 *Grafted chiral and nonchiral catalytic sites for alkylation reaction.*

In recent works, MCM-41 was functionalised by a new sol-gel method involving prehydrolysis of the alkoxy groups of the silylating agent and control of the condensation reaction between monolayer-packed silylating agent.[61] This method afforded high surface coverage that limits the undesirable activity of the uncovered support surface.[57]

This new generation of catalysts demonstrated remarkable enantioselectivity at the level of the homogeneous catalysis[57] as shown in the Abramson et al 's contribution in this book. After removal of the soluble product, the catalyst featured the same activity and enantioselectivity without further regeneration.

6   CONCLUSION

Some degree of success in supported enantioselective catalysis was accomplished by using functionalisation of mineral support. Due to their unique textural and surface properties, mesoporous micelle-templated silicas are able to bring new interesting properties for the preparation of optically active solids. Many successfully examples have been reported for enantioselective hydrogenation, epoxidation and alkylation. However, the stability of the immobilised catalysts still deserves efforts to allow industrial development of such attractive materials.

References

1.   H.U. Blaser, *Tetrahedron Asymm*. 1991, **2**, 843.
2.   A. Pfaltz, T. Heinz, *Top. Catal*., 1997,**4**, 229.
3.   A. Baiker , *J. Mol.Catal*., 1997, **118**, 473.
4.   K. Soai, S. Niwa and M. Watanabe, *J. Chem. Soc. Perkin Trans 1*, 1989, 109.
5.   S. Itsuno, Y. Sakurai, K. Ito, T. Maruyama, S. Nakahama and J.M.J. Frechet, *J. Org. Chem.*, 1990, **55**, 304.
6.   M. Yamakawa, and R. Noyori, *J. Am. Chem. Soc.*, 1995, **117**, 6327.
7.   B.B. De, B.B. Lohray, S. Sivaram and P.K. Dhal, *Tetrahedron Asymm.*, 1995, **6**, 2105.
8.   F. Minutolo, D. Pini and P. Salvatori, *Tetrahedron Lett.*, 1996, **37**, 3375.
9.   F. Minutolo, D. Pini, A. Petri and P. Salvatori, *Tetrahedron Asymm.*, 1996, **7**, 2293.
10.   L. Canali, E. Cowan, H. Deleuze, C.L. Gibson and D.C. Sherrington, *Chem. Comm.*, 1998, 2561.
11.   E. Breysse, C. Pinel and M. Lemaire, *Tetrahedron Asymm.*, 1998, **9**, 897.
12.   J.K. Karjalainen, O.E.O., Hormi and D.C. Sherrington, *Tetrahedron Asymm.*, 1998, **9**, 2019.
13.   D.W.L. Sung, P. Hodge and P.W. Stratford, *J. Chem. Soc., Perkin Trans 1*, 1999, 1463.
14.   P. Hodge, D.W.L. Sung and P.W. Stratford, *J. Chem. Soc., Perkin Trans 1*, 1999, 2335.
15.   U. Nagel and E. Kinzel, *J. Chem. Soc. Chem Comm.*, 1986, 1098.
16.   K. Soai, M. Watanabe and A. Yamamoto, *J. Org. Chem.*, 1990, **55**, 4832.
17.   A. Corma, M. Iglesias, C. del Pino and F. Sanchez, *J. Organometal. Chem.* 1992, **431**, 233.
18.   B. Pugin and M. Muller, *Stud. Surf. Sci. Catal.*, 1993, **94**, 107.
19.   A. Corma, A. Fuerte, M. Iglesias and F. Sanchez, *J. Mol. Catal.*, 1996, **107**, 225.
20.   A. Corma, M. Iglesias, F. Mohino and F. Sanchez, *J. Organometal. Chem.*, 1997, **544**, 147.
21.   N. Bellocq, D. Brunel, M. Laspéras and P. Moreau, *Stud. Surf. Sci. Catal.*, 1997, **108**, 435.
22.   C.E. Song, J.W. Yang and H-J. Ha, *Tetrahedron Asymm.*, 1997, **8**, 841.
23.   N. Bellocq, S. Abramson, M. Laspéras, D. Brunel and P. Moreau, *Tetrahedron Asymm.*, 1999, **10**, 3229.
24.   D. Brunel, *Microporous Macrop. Mater.*, 1999, **27**, 329.
25.   D. Pini, A. Mandoli, S. Orlandi and P. Salvatori, *Tetrahedron Asymm.*, 1999, **10**, 3883.
26.   B.F.G. Johnson, S.A. Raynor, D.S. Shephard, T. Mashmeyer, J.M. Thomas, G. Sankar, S. Bromley, R. Oldroyd, Lynn Gladden and M.D. Mantle, *Chem . Comm.*, 1999, 1167.
27.   X-G. Zhou, X-Q. Yu, J-S. Huang, S-G. Li, L-S. Li and C-M. Che, *Chem Comm.*, 1999, 1789.
28.   S.J. Bae, S-W. Kim, T. Hyeon and B.M. Kim, *Chem. Comm.*, 2000, 31.

29. *Chiral catalyst Immobilization and Recycling* » Eds D. de Vos, I.F.J. Vankelecom and P.A. Jacobs, Wiley-VCH, 2000.
30. J S Beck, C T-W Chu, I D Johnson, C T Kresge, M E Leonowicz, W J Roth, J C Vartuli, WO 91/11390 , 1991.
31. J S Beck, J C Vartuli, W J Roth, M E Leonowicz, C T Kresge, K D Schmitt, C T-W Chu, D H Olson, E W Sheppard, S B McCullen, J B Higgins and J L Schlenker, *J. Amer. Chem. Soc.,* 1992, **114**,10834.
32. T.P. Dang and H.B. Kagan, *J. Chem. Soc., Chem. Commun.*, 1971, 481.
33. B. Pugin, *J. Mol. Catal. A.*, 1996, **107**, 273.
34. A. Corma, M. Iglesias, C. del Pino and F. Sanchez, *J. Chem. Soc., Chem. Commun.*, 1991, 1253.
35. E. Lindner, M. Kemmler, H.A. Mayer and P. Wegner, *J. Am. Chem. Soc.*, 1994, **116**, 348.
36. A. Adima, J.J.E. Moreau and M.W.C. Man, *J. Mater. Chem.* 1997, **7**, 2331.
37. A. Adima, J.J.E. Moreau and M.W.C. Man,*Chirality*.2000, **12**, 411
38. P. Hesemann and J.J.E. Moreau, *Tetrahedron Asymm.*, 2000, **11**, 2183.
39. D J Macquarrie, *Chem. Commun*, 1996, 1961.
40. S L Burkett, S D Sims and S Mann, *Chem. Commun.*, 1996, 1367;
41. J.M. Thomas, T. Maschmeyer, B.F.G. Johnson and D.S. Shephard, *J. Mol. Catal. A*, 1999, **141**, 139.
42. D. Brunel, P. Sutra and F. Fajula, *Stud. Surf. Sci. Catal.*, 2000, **129**, 773; P. Sutra, A. Blanc, F. Fajula and D. Brunel, *Tetrahedron Asymm.*, submitted.
43. S.J. Bae, S-W. Kim, T. Hyeon and B.M. Kim, *Chem. Commun.*, 2000, 31.
44. L. Canali, J.K. Karjalainen, D.C. Sherrington and O. E.O. Horni, *Chem. Commun.*, 1997, 123.
45. J.K. Karjalainen, O.E.O. Horni and D.C. Sherrington, *Tetrahedron Asymm.*, 1998, **9**, 2019.
46. L. Canali, D.C. Sherrington and H. Deleuze, *React. Funct. Polym.*, 1999, 34, 155.
47. L. Canali and D.C. Sherrington, *Chem. Soc. Rev.*, 1999, **28**, 85.
48. D.C. Anis and E.N. Jacobsen, *J. Am. Chem. Soc.*, 1999, **121**, 4147.
49. M.J. Farral, M. Alexis and M. Tracarten, *Nouv. J. de Chim.*, 1983, **7**, 449.
50. B.M. Choudary, V.L.K. Valli and A.D. Durga Prasad, *J. Chem. Soc., Chem. Comm.*, 1990, 1186.
51. S.B. Ogunwumi and T. Bein, *Chem. Commun.*, 1997, 901.
52. M.J. Sabater, A. Corma, A. Domenech, V. Fornés and H. Garcia, *Chem. Commun.*, 1997, 1285.
53. I.F.J. Vankelecom, D. Tas, R.F. Parton, V. Van de Vyver and P.A. Jacobs, *Angew. Chem. Int. Ed. Engl.*, 1996, **35**, 1346.
54. L. Frunza, H. Kosslick, H. Landmesser, E. Höft and R. Fricke, *J. Mol. Catal.A*, 1997, **123**, 179.
55. P. Piaggio, P. McMorn, C. Langham, D. Bethell, P.C. Buman-Page, F.E. Hancock and G.J. Hutchings, *New J. Chem.*, 1998, 1167.
56. S. Abramson, N. Bellocq and M. Laspéras, *Topics in Catalysis*, 2000, in press.
57. S. Abramson, M. Laspéras, A. Galarneau, D. Desplantier-Giscard and D. Brunel, *Chem. Commun*, 2000, 1773.
58. E.J. Corey and F.J. Hannon, *Tetrahedron Lett.*, 1987, **28**, 5233 and 5237.
59. D. Brunel, A. Cauvel, F. Di Renzo, F. Fajula, B. Fubini, B. Onida and E. Garrone, *New J. Chem.*, 2000, **24**, 807.
60. P. Sutra, F. Fajula, D. Brunel, P. Lentz, G. Daelen and J. B.Nagy, *Colloids and Surfaces A,* 1999, **158**, 21.
61. A. Galarneau, T. Martin, V. Izard, V. Hulea, A. Blanc, S. Abramson, F. Di Renzo, F. Fajula and D. Brunel, *in preparation.*

IMMOBILISED LEWIS ACIDS AND THEIR USE IN ORGANIC CHEMISTRY

James H Clark, Arnold Lambert, Duncan J Macquarrie. David J Nightingale,
Peter M Price, J Katie Shorrock and Karen Wilson

Clean Technology Centre
Department of Chemistry
University of York,
York YO10 5DD

1   INTRODUCTION

Acid catalysis is by far the most important type of catalysis in the chemical and allied process industries.   The range of important organic reactions catalysed by acids is enormous and covers all sectors of chemicals manufacturing. The petrochemical sector is a major user of acid catalysis and its technology has matured into highly efficient continuous vapour phase processes typically based on fixed bed reactors with solid acids that have very long on-bed lifetimes. The fine and speciality chemicals manufacturing industries are also major users of acid catalysis for reactions including Friedel-Crafts acylations and alkylations, halogenations, nitrations, sulphonylations, oligomerisations, isomerisations and Diels-Alder reactions. These reactions are however usually carried in the liquid phase, often in volatile organic solvents and with soluble or liquid acids such as sulfuric acid, hydrogen fluoride, phosphoric acids, aluminium chloride and boron trifluoride. These inexpensive inorganic Bronsted and Lewis acids acids have been used in organic chemistry for many years but their use is increasingly being challenged as a result of process inefficiency and waste[1]. Most organic reactions that use these acids require a water quench to separate the organic components. The resulting aqueous acid stream is neutralised leading to salt waste. Volumes of process waste are commonly several times larger than product volumes and waste disposal costs can easily exceed the cost of the raw materials. It is very important that we seek more environmentally benign processes that can operate in these industries[2]. One approach to this is to design reaction systems based on heterogeneous solid acids that stay in a separate phase to the organic components[3]. This can facilitate separation and reuse, and is amenable to moving the more traditional batch type reactors towards continuous processing. In this paper we will look at the use of three different types of solid acids based on immobilised Lewis acids as replacements for traditional soluble and liquid acids in important organic reactions.   The relative advantages and limitations of the three approaches will be considered alongside illustrations of their value as new catalysts for organic chemistry.

2   TYPES OF IMMOBILISED LEWIS ACIDS

There are numerous possible support materials that can be adapted for so-called solid phase synthesis or for heterogenisation of inorganic reagents and catalysts[3]. We will focus here on mesoporous inorganic oxides including silica, hexagonal mesoporous silica, and zirconia. For these materials there are three possible approaches for the preparation of immobilised Lewis acids:

- Physisorbed Lewis acid

Here the Lewis acid is adsorbed on to the support material typically from a solvent such as an alcohol followed by evaporation of the solvent. A good example of this is zinc chloride adsorbed onto acid-treated clay[4] or onto mesoporous silica[5] and includes a commercial catalyst[6]. The catalyst preparation procedure is simple and assuming the solvent is recovered, relatively green. The major concern over such materials will be stability especially in polar media when the Lewis acid may simply be washed from the support. Experience has taught has however that these materials may not be simple; supported metal salts can prove to be suprisingly resistant to solvent and reaction media leaching which leads us to believe that there are more complex interactions between the support and the Lewis acid. A better understanding of such interactions is important. One example of a solid acid in this category which we have recently studied is supported zinc triflate[7], the reaction chemistry of which is described later.

- Chemisorbed Lewis acid

Here the Lewis acid (either as is or as a complex) is forced to react with the support surface or is introduced by complexation of a Lewis acid centre with a chemically modified solid surface. A good example of the former is supported aluminium chloride which can be produced by reacting $AlCl_3$ or aluminium alkyls with the hydroxyl groups on the surface of various solids including mesoporous silicas[8]. This is also the basis of a commercial catalyst. Surface (e.g. Si) – O – Lewis acid centre bonds should provide a greater degree of stability towards leaching through dissolution in the reaction media or washing solvent though they are also likely to be sensitive to hydrolysis and complexation (poisoning) with Lewis bases (as with all Lewis acids). It can also be difficult to control the surface chemistry so as for example, to avoid formation of $(SiO)_2AlCl$ in competition with the more active $(SiO)AlCl_2$ sites. The effect of the

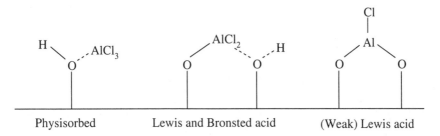

Physisorbed   Lewis and Bronsted acid   (Weak) Lewis acid

Figure 1 *Active sites on silica chemically modified with aluminium chloride*

polarisation of neighbouring hydroxyl groups leading to Bronsted acidity cannot be ignored and may be desirable (Figure 1). We have recently extended this methodology to preparing supported boron trifluoride-organic complexes[9] and supported antimony trifluoride, the latter being a very effective catalyst for Baeyer-Villiger reactions (see later)[10].

• Structural Lewis acid

Structural modification of solids by incorporation of a Lewis acid centre into the bulk structure (e.g. titanium silicates) or the enhancement of structural centres by chemical modification (e.g sulfation of zirconia) is a popular method of making stable catalytically active solids including solid acids. An interesting example of this is mesoporous sulfated zirconia which has been widely studied for gas phase reactions but as we and others have recently shown, can be a very useful catalyst for some liquid phase organic reactions[11,12]. The exact structure of the active sites is uncertain but it seems likely that what are essentially Lewis acid sites can generate Bronsted acid activity through interaction with water molecules thus giving the attractive potential for controllable acidity (Figure 2).

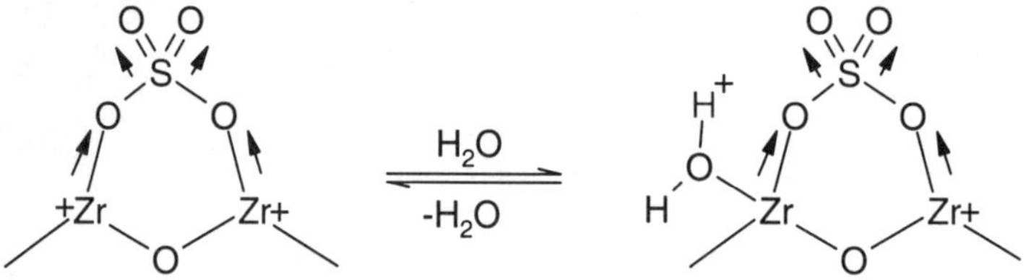

Figure 2 *Acid sites on sulfated zirconia*

The other great attraction of these materials is their stability and we have found that the only limitation to their catalytic activity even in polar reaction media is poisoning by organic products and byproducts. These can however, usually be desorbed by heating (at temperatures as high as ca. 600° C) or solvent treatment (using water and/or organic solvents). Some of the catalytic properties are described later.

Overall there are benefits and limitations to all of the common methods for immobilising Lewis acids. This helps illustrate the apparently inescapable conclusion in heterogeneous acid catalysis that there are likely to be as many "optimum" catalysts as there are acid-catalysed processes.

3   CATALYTIC ACTIVITY OF IMMOBILISED LEWIS ACIDS

Clay- and silica-supported zinc chloride proved to be a useful general purpose mild solid Lewis acid that had at times suprisingly high activity most famously in the benzylation of benzene where a synergistic effect was clearly evident helping to illustrate the comment that these are more than simple physisorbed (i.e. loosely bound) salts[4]. More recently we have found that supported zinc triflate is a very active and quite selective catalyst for the isomerisation of α-pinene oxide to campholenic aldehyde (Figure 2)[7]. This is a reaction where pure Lewis acidity is required as Bronsted acidity leads to other products reducing the yield of the desired aldehyde. Characterisation of the acid sites on Zn triflate-silica reveals essentially only Lewis acidity. On application, the best catalyst proves to be Zn triflate adsorbed on hexagonal mesoporous silica (HMS) at a loading of only 0.001 mmole/g. The catalyst is also completely recyclable (turnover number > 3500) suggesting that this small amount of salt may be somehow strongly bonded to the support. The catalyst is also reasonably active in the benzoylation of anisole (rapid

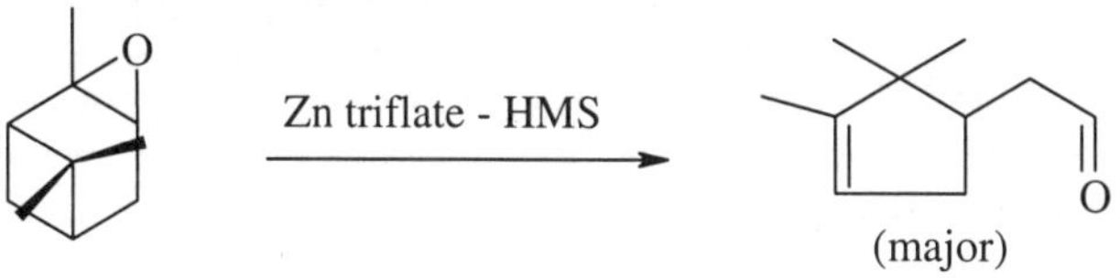

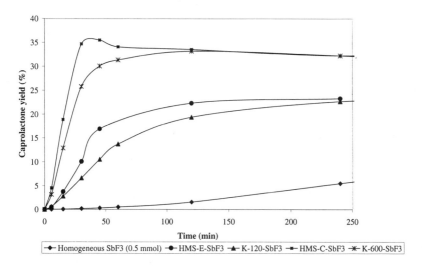

Figure 3 *The rearrangement of α-pinene oxide catalysed by supported zinc triflate*

reaction at 150°C) and may well prove to have wider utility as proved to be the case for supported zinc chloride.

Clay based solid acids have evolved through ion exchanged materials through physisorbed Lewis and Bronsted acids to chemically modified materials. Clay-supported aluminium chloride[8] has proven to be a very active catalyst for the alkylation of benzene and some other aromatics with alkenes. More impressively, the same reaction when catalysed by aluminium chloride supported on HMS leads to added selectivity in the extent of monoalkylation ultimately reaching close to 100%[13]. These materials show both Lewis acid and Bronsted acid sites when characterised by pyridine adsorption and it is probably difficult to control the balance of these. Supported boron trifluoride complexes are essentially Bronsted acids[9] and we have shown these to be capable of various reactions notably esterifications. More recently we have extended this methodology to supported antimony trifluoride[10]. By analogy with supported aluminium chloride we believe that the active sites in this catalyst are based on Lewis acidic Si-O-SbF$_2$ units although polarisation of neighbouring hydroxyl groups to create Bronsted acidity is also very likely. Supported antimony trifluoride is an excellent catalyst for the Baeyer-Villiger oxidation of cyclic ketones such as cyclohexanone to lactones using hydrogen peroxide (Figure 4).

Figure 4 *Effect of the support on the antimony trifluoride-catalysed Baeyer-Villiger oxidation*

The desired oxidation reaction runs in competition with hydrogen peroxide destruction and with polymerisation of the lactone. The route based on the optimally loaded catalyst using HMS as the support is particularly effective in minimising these side-reactions demonstrating how careful control of the active sites in immobilised Lewis acids can be so important. The mechanism of the oxidation reaction is likely to involve Lewis acid-ketone complexation making the ring carbon more susceptible to nucleophilic attack by the hydrogen peroxide. Remarkably the catalyst resists the aqueous reaction conditions although the best reaction rates and catalyst lifetimes are achieved with continuous water removal from the reaction. The catalyst can be easily recovered and is reusable with a small loss in activity.

Sulfated zirconia is a good example of a structural Lewis acid which has been chemically treated to enhance acidity. It has been extensively studied as a solid acid catalyst for vapour phase reactions and we[11,12] and others[14] have found that a mesoporous version of this material is a particularly effective catalyst for liquid phase Friedel-Crafts alkylation reactions and to a lesser extent Friedel-Crafts benzoylations. The commercial (MEL Chemicals Ltd) material SZ999/1 shows a nitrogen isotherm characteristic of a mesoporous solid (surface area 162 m^2g^{-1}, pore volume 0.22 cm^3g^{-1}). Whereas microporous and mesoporous materials are capable of rapidly catalysing the alkylation of benzene with various alkenes (Table 1), on reuse only the mesoporous

Alkene	Composition of product mixture (%)		
	Alkene	Monalkylate	Dialkylate
1-hexene	<0.5	91	8
1-octene	0	94	6
1-dodecene	<0.5	93	6
1-hexadecene	2.5	94	0

Table 1 *Alkylation of benzene catalysed by optimally activated sulfated zirconia*

material gives comparable activity and selectivity (washing and thermal activation of the microporous material also restores activity indicating that it was blocked with involatile organic compounds, e.g. alkene oligomers). The pre-activation of the catalyst is very important in determining its behaviour in catalysis. A temperature of ca. 500°C generally gives good activity in Friedel-Crafts reactions but it is also important to control the cooling of the calcined material and its transport to the reactor. In the benzoylation of benzene for example, cooling in air lowers the optimal activity to form benzophenone (Figure 5). At best the rate of formation of benzophenone at 85°C using a catalyst loading of 1g/5 mmol acylating agent is ca. 5%/h in the early stages of the reaction which compares favourably with other reported examples of Friedel-Crafts acylations catalysed by sulfated zirconias[14]. The catalyst is easily recoverable and as described above it is thermally stable enough to allow rigorous thermal reactivation after washing. While this is not a particularly fast reaction it is quite attractive when one considers that the (very fast) traditional route is based on stoichiometric quantities of aluminium chloride which is not recoverable (being destroyed in an aqueous quench work-up).
Sulfated zirconia may not be the best Friedel-Crafts catalyst but it is generally useful and its structural stability when compared to other sometimes more active but less stable

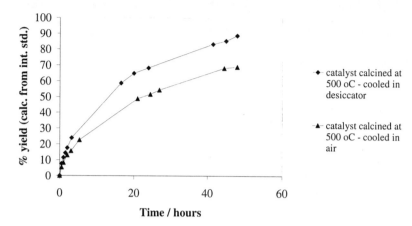

Figure 5 *The benzoylation of benzene (0.5 mol) using benzoyl chloride (0.05 mol) at 85 °C catalysed by sulfated zirconia (10g)*

solid acids, does give it advantages in terms of leaching and reactivation. As more fine chemical reactions move towards continuous processes, long term catalyst stability will become an increasingly important issue.

We gratefully acknowledge the financial support of the EPSRC, Royal Academy of Engineering (Clean Technology Fellowship to JHC), the Royal Society (University Research Fellowship to DJM) and our other industrial sponsors. We are also indebted to our research collaborators notably the Process Intensification Group at Newcastle University.

4   REFERENCES

1. J. H. Clark, *Green Chemistry,* 1999, **1**, 1.
2. P. T. Anastas and J. C. Warner, "Green Chemistry: Theory and Practice", Oxford University Press, Oxford, 1998.
3. J. H. Clark and C. N. Rhodes, "Clean Synthesis using Porous Inorganic Solid Catalysts and Supported Reagents", *RSC Clean Technology Monograph Series*, J. H. Clark ed., RSC, Cambridge, 2000
4. J. H. Clark, S. R. Cullen, S. J. Barlow and T. W. Bastock, *J. Chem. Soc., Perkin Trans. 2,* 1994, 1117.
5. D. R. Brown, H. G. M. Edwards, D. W. Farell and J. Massam, *J. Chem. Soc. Faraday Trans.,* 1996, **92**, 1027.
6. Envirocats™, Contract Catalysts, Knowsley Industrial Park, Prescot, Merseyside, UK.
7. K. Wilson, A. Renson and J. H. Clark, *Catal Lett.,* 1999, **61**, 51.
8. J. H. Clark, K. Martin, A. J. Teasdale and S. J. Barlow, *Chem. Commun.,* 1995, 2037.
9. K. Wilson and J. H. Clark, *Chem. Commun.,* 1998, 2135.
10. A. Lambert, D. J. Macquarrie, G. Carr and J. H. Clark, *New J. Chem.,* 2000, in the press.

11. J. H. Clark, G. Monks, D. J. Nightingale, P. M. Price and J. White, *J. Catal.* 2000, in the press.
12. J. H. Clark, G. Monks, D. J. Nightingale, P. M. Price and J. White, *Proc. Org. React. Catal. Soc.,* 2000.
13. J. H. Clark and D. J. Macquarrie, *Chem Commun.,* 1998, 853.
14. G. D. Yadav and J. J. Nair, *Microporous and Mesoporous Materials,* 1999, **33**, 1.

INFLUENCE OF ZEOLITE COMPOSITION ON CATALYTIC ACTIVITY

M. Guisnet

Laboratoire de Catalyse en Chimie Organique, UMR CNRS 6503
Université de Poitiers, Faculté des Sciences
40 avenue du Recteur Pineau
86022 Poitiers Cedex, France

1 INTRODUCTION

Zeolite catalysts are used in the most important processes of Refining and Petrochemicals[1,2] and could play in the near future a significant role in the synthesis of Fine Chemicals[3,6] and in Depollution[7]. It is firstly because, with these perfectly crystallized solids, reactions occur inside uniform pores of molecular size which can therefore be considered as **nanoreactors**. The consequence is that the selectivity can be oriented in the desired way not only by adjusting the characteristics of the active sites as is the case with classical catalysts but also by adjusting the characteristics of the pores. Another reason to the remarkable development of zeolite catalysts is the ease of tailoring the active sites for the reaction to be catalyzed. Thus, zeolites can be used as acid, base, acid-base, redox and bifunctional catalysts. Moreover, the characteristics of the active sites can be adjusted through well controlled methods[8,9].

2 CHARACTERISTICS OF ACID SITES AND ACTIVITY

The activity, stability and selectivity of acid zeolites are obviously determined to a large extent by the characteristics of the acid sites i.e. chemical structure, strength, density (or proximity) and accessibility.

It is generally admitted that skeletal transformations of hydrocarbons are catalyzed by **protonic sites** only. Indeed good correlations were obtained between the concentration of Brönsted acid sites and the rate of various reactions, e.g. cumene dealkylation, xylene isomerization, toluene and ethylbenzene disproportionation and n-hexane cracking[10-12] On the other hand, it was never demonstrated that isolated Lewis acid sites could be active for these reactions. However, it is well known that Lewis acid sites located in the vicinity of protonic sites can increase the strength (hence the activity) of these latter sites, this effect being comparable to the one observed in the formation of superacid solutions. Protonic sites are also active for non skeletal transformations of hydrocarbons e.g. cis trans and double bond shift isomerization of alkenes and for many transformations of functional compounds e.g. rearrangement of functionalized saturated systems, of arenes, electrophilic substitution of arenes and heteroarenes (alkylation, acylation, nitration, etc.), hydration and dehydration etc. However, many of these transformations are more complex with simultaneously reactions on the acid and on the base sites of the solid catalysts and even with the joint participation of acidic sites (Lewis and/or Brönsted) and basic sites in the reaction mechanism (bifunctional acid base catalysis)[11,13].

The acid **strength** required for catalyzing reactions can be quite different[12,13]. Certain reactions demand very strong acid sites e.g. short chain alkane transformation while others e.g. Beckmann rearrangement can be catalyzed by very weak acid sites. Generally, **the stronger the protonic acid sites the faster the reactions**. This inverse relation between the strength of the acid sites and their activity was demonstrated by determining for various reactions the poisoning effect of pyridine on the activity of a HFAU zeolite which presented protonic sites with very different strengths[11,14]. The strength of the sites was characterized by T_D the temperature up to which they conserved the adsorbed pyridine. Indeed, the stronger the acid sites the more difficult to desorb the basic poison. The minimum strength that an acid site requires to be active for a given reaction was estimated by means of T_{Do}, i.e., the value of T_D above which the activity can be measured. A great difference (300°C) was found between the values of T_{Do} for 3,3-dimethyl-1 butene isomerization which is very facile ($T_{Do} = 220°C$) and n-hexane transformation ($T_{Do} = 520°C$). The difference would be even more pronounced if the reactions could be carried out at the same temperature, instead of 400°C for n-hexane transformation and 200°C for 3,3-dimethyl-1 butene isomerization.

The **density** of the protonic sites (or inversely the spacing between sites) would play a determining role in the rate of bimolecular reactions[11]. This effect of site density, well known in metallic catalysis (the so-called structure sensitive reactions) was shown in acid catalysis by using series of dealuminated zeolites. One of the first examples concerned the bimolecular isomerization of n-butane over a series of dealuminated protonic mordenites[15]. The Si/Al ratio of the samples varied from 7.5 to 110 and the density of protonic sites n_H^+ from 11 to 0.8 10^{20} per gram of zeolite. Practically no difference in strength was found between the acid sites of the various samples However, the activity per acid site (turnover frequency, TOF) depended very much on the sample. The lower n_H^+ the lower the TOF value. This is in agreement with the complex bimolecular mechanism of this reaction (dimerization-isomerization-cracking) which requires several acid sites for the various successive steps[16]. The effect of the acid site density was also shown in catalytic cracking (FCC)[17,18]. The research octane number of the gasoline obtained over catalysts containing HFAU zeolites with different Si/Al ratios increases when the percentage of **paired sites** (two aluminium atoms attached at the same silicon) in the protonic sites decreases[18].

3   TAILORING OF THE CATALYTIC PROPERTIES OF ACID ZEOLITES

3.1 Protonic sites of zeolites

As indicated above, most of the hydrocarbon transformations as well as many transformations of functional compounds are catalyzed by protonic sites only. In zeolites, these sites are associated to hydroxyl groups attached to framework oxygens bridging Si and Al tetrahedrally coordinated (Al(-OH)Si). **Their maximum number is equal to the number of framework aluminium atoms**, the actual number being smaller due to cation exchange and to dehydroxylation during activation at high temperatures. The number (and the density) of protonic sites can therefore be adjusted either during the synthesis or during post synthesis treatments of the zeolite : dealumination, ion exchange etc. However, as aluminium atoms cannot be adjacent (Lowenstein's rule), the maximum number of protonic sites is obtained for a framework Si/Al ratio of 1 (8.3 mmol H^+ g^{-1}

zeolite). Moreover no purely protonic zeolite can be prepared with this low framework Si/Al ratio, hence this maximum number can never be achieved.

In order to design zeolite catalysts, the parameters which control the other characteristics of the protonic sites and particularly their strength have also to be known. A first point to be emphasized is **the stronger acidity of H zeolites compared to amorphous aluminosilicates**. This stronger acidity is demonstrated in particular by the higher values of the heats of adsorption for nitrogen bases. To explain this stronger acidity, Mortier[19] proposed the existence of an enhanced donor-acceptor interaction in zeolites. This interaction was extended by Rabo and Gajda[20] into a resonance model of

the (Al(OH)Si) bond structure with bridging hydroxyls $\left(\begin{array}{c} \text{H} \\ | \\ \text{O} \\ \text{Al} \quad \text{Si} \end{array} \right)$ and terminal silanols

$\left(\begin{array}{c} \text{H} \\ | \\ \text{O} \\ \text{Al} \quad \text{Si} \end{array} \right)$ as extreme limits. In agreement with NMR data, the OH groups of amorphous

aluminosilicates are primarily terminal whereas those of zeolites are primarily bridging, the interaction of O with Al weakening the OH bond and increasing the acid strength.

A relation exists between the **T-O-T bond angles and the acid strength of zeolites**[20]. Thus, the protonic sites of HMOR (bond angle range of 143-180°) and HMFI (133-177°) zeolites are stronger than those of HFAU (133-147°). This explains why HMOR is active for butane and n-hexane isomerizations at 200-250°C which require very strong acid sites whereas it is not the case for HFAU.

The synthesis of **metallosilicates** containing trivalent elements in the framework other than Al (B, Ga, In, Fe) is of interest for the design of acid strength. The order in strength drawn from theoretical calculations is in good agreement with acidity measurements[21]. Thus, according to IR spectroscopy (wavenumber of the OH groups) and ammonia TPD over MFI samples the order of strength was the following :

$$B(OH)Si < In(OH)Si \ll Fe(-OH) \, Si < Ga(OH)Si < Al(OH)Si$$

The weaker acidity of gallosilicates (compared to aluminosilicates) was confirmed by the lower frequency shift caused by adsorption of weak bases (e.g. CO) and by the lower activity for model reactions.

The **accessibility** of the protonic sites plays also a significant role in the zeolite activity. Obviously this accessibility depends on the site location (Figure 1) but also on the size of the reactant molecules. Thus, one part of the protonic sites of HFAU zeolites are located in the supercages, hence are accessible to many organic molecules whereas the other part located in the hexagonal prisms are inaccessible.HMOR has also protonic sites accessible to organic molecules (in the large channels) and inaccessible sites (in the side pockets). With HOFF, the acid sites located in the gmelinite cages are accessible to linear organic molecules only whereas those located in the large channels are also accessible to bulkier molecules. With HMFI, all the protonic sites being located at the channel intersections are equally accessible (or inaccessible) to reactant molecules. The same can be observed with HERI, the protonic sites of which are located in large cages with small apertures, hence accessible to linear organic molecules only.

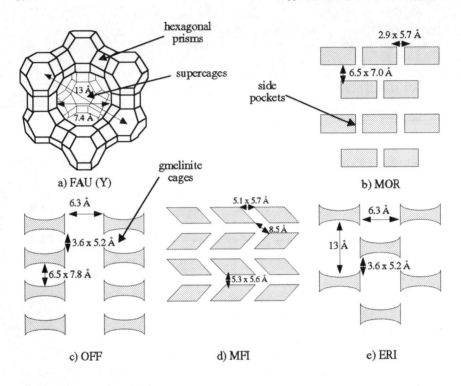

Figure 1 *Schematic representation of the pore structure of various zeolites with large pore apertures: FAU (Y), MOR, OFF, with average pore apertures: MFI (ZSM5), with small apertures: ERI.*

3.2 Na-H zeolites : influence of the degree of sodium exchange

Protonic sites of zeolites are created by removal of basic organic templates by calcination at high temperatures and/or by exchange of the alkaline cations (often Na^+) present in the synthesized samples. Generally protonic sites are introduced in two steps : exchange with ammonium ions followed by thermal decomposition.

$$NaZ \xrightarrow{\ NH_4^+\ } NH_4Z \xrightarrow{\ heat\ } HZ + NH_3 \nearrow$$

The exchange in one step by acid treatment can only be used with high silica zeolites, low silica zeolites undergoing significant dealumination and degradation of the framework. It should also be emphasized that, as was demonstrated by IR spectroscopy, protonic sites are also created by exchange with multivalent cations (e.g. rare earth cations in the FAU zeolite of FCC catalysts)[22]. This is due to a partial hydrolysis of the cations, the positive charge of which is not well neutralized because of their distance from the negative charges of the framework :

$$3Na^+Z^- + RE^{3+} \rightarrow RE^{3+}\,3Z^- \xrightarrow{\ H_2O\ } [RE(OH)]^{2+}2Z^- + HZ \xrightarrow{\ H_2O\ } [RE(OH)_2]^+Z^- + 2HZ$$

The acid properties of NaH zeolite (NaHY, NaHMOR, NaHMFI) series with different Na contents were investigated with adsorption microcalorimetry[23]. Whatever the zeolite, the initial heat of base adsorption (NH_3, pyridine, etc.) as well as the total acidity increases with the exchange of Na^+ by H^+. Thus for a series of NaHMOR zeolites[24], the initial heat of ammonia adsorption passes from 145 to 184 kJ mol^{-1} when the exchange rate passes from 33 to 98 % and the number of acid sites with adsorption heat greater than 110 kJ mol^{-1} passes from 0.4 to 9.3 10^{20} g^{-1} i.e. from 7% of the theoretical number of protonic sites to 56 %.

This very significant increase in the acid strength caused by the exchange of the last sodium cations by protons was also demonstrated by pyridine adsorption followed by IR spectroscopy over HY zeolites[25]. For a decrease of Na content from 80 ppm to 10 ppm in a HFAU zeolite with a Si/Al ratio of 10 the number of protonic sites able to retain pyridine adsorbed at 150°C passes from 3.5 to 5.6 10^{20} g^{-1} (x 1.6) ; the effect is larger (x 3) if only the strongest acid sites, able to retain pyridine adsorbed at 350°C, are considered. With this example it is clear that at high exchange rate very strong protonic sites are created by removal of Na^+ ions but also that there is a significant increase in the strength of the protonic sites already present. To explain this observation, two hypotheses were considered :

- Extraframework aluminium species could be formed on the very exchanged samples during ammonia elimination[26] ; the interaction of these species with the framework protonic sites would result in a significant increase of their acid strength.

- The very few residual Na^+ ions would cause a remote perturbation of the acid sites structure by modifying significantly the TOT bond angles[26].

Obviously, the increase in acid strength caused by Na^+ exchange by protons causes an increase of the zeolite activity and of the turnover frequency (calculated per potential acid site). The more difficult the reaction (hence the more demanding in acid strength) the greater the significance of the activity increase. Thus, for n-octane cracking, the rate constant is multiplied by 60 when the exchange rate of a NaHY zeolite passes from 56 to 90 % whereas for the easier isooctane cracking it is multiplied by 20[28].

In summary, the exchange of Na^+ by H^+ leads to an increase in their strength. **At high exchange rate, there is creation of very strong protonic sites but also an increase in the strength of the protonic sites already present in the zeolite.** This has for consequence a very significant increase in the rate of the desired reaction but also in the rate of coking hence of deactivation.

3.3 Influence of framework composition

3.3.1. Framework Si/Al ratio Both theoretical and experimental approaches of the effect of the framework Si/Al ratio on the acidity led to the conclusion that the strength of the protonic sites of zeolites is influenced by the presence of neighboring sites[20,21]. Each framework Al atom has 4 Si atoms (Lowenstein's rule) in the first surrounding layer (nearest neighbors) and, depending on the zeolite, 9-12 Al or Si atoms in the second layer (next-nearest-neighbors or NNN). According to the NNN concept the acid strength of a protonic site depends on the number of Al atoms in the NNN position[17]; the strength is maximum at 0 Al and minimum at full occupancy of the NNN sites with Al. Through statistical calculations, Wachter[29] determined the Si/Al value when all Al atoms were isolated, to be 7 for all zeolites with 9 Al or SiNNN (e.g. FAU, LTA). Barthomeuf[30] brought improvement to this idea by using topological densities to include the effects of layers 1 to 5 surrounding the Al atom. The values of Si/Al for which **the protonic sites are isolated, hence have the maximum strength**, calculated for 33 different zeolites were found between 5.8 (FAU zeolite) and 10.5 (Bikitaite – BIK).

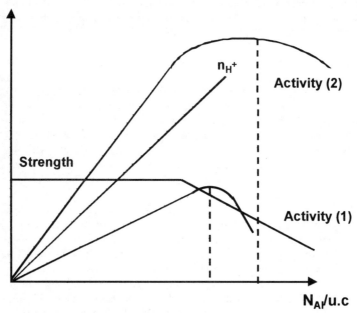

Figure 2 *Change with N_{Al}, the number of framework Aluminium atoms per unit cell in the number of potential protonic sites per unit cell $nH+$ $(=N_{Al})$, in their acid strength, in the zeolite activity for reactions demanding strong acid sites (1) or catalyzed even by weak acid sites (2).*

The acid strength of zeolite samples differing only by their framework Si/Al ratio should therefore change with x_{Al} $(x_{Al} = N_{Al}/(N_{Al} + N_{Si}))$, the molar fraction of aluminium. As the number of acid sites is proportional to x_{Al}, the activity should pass through a maximum, the position of which depends on the acid strength required for the reaction considered. The more demanding the reaction (hence the more difficult it is) the lower x_{Al} corresponding to the maximum in activity (Figure 2). For instance, with HFAU zeolites the value of x_{Al} at which all the protonic sites are isolated was found to be equal to 0.146. According to Barthomeuf[30], for reactions demanding very strong acid sites the maximum in activity should be located at slightly higher x_{Al} values whereas a flat maximum should be observed at much higher x_{Al} values for reactions demanding weak acid sites. In agreement with Figure 2, a maximum in activity was found for many reactions carried out over zeolites. In certain cases, the agreement between the experimental effect of x_{Al} on the activity and the effects expected from the NNN model : maximum in activity for expected values of x_{Al} and proportionality between activity and x_{Al} for values of x_{Al} corresponding to isolated sites (e.g. 0.096 for MOR zeolites). This was found in particular for n-hexane isomerization over PtHMOR catalysts[31] and for o-dichlorobenzene isomerization[32] over HMOR catalysts (Figure 3A). However, in many other examples the location of the maximum is far from the expected value : e.g. for x_{Al}

= 0.029 for liquid phase alkylation of toluene with 1 heptene over HFAU zeolites[33] whereas the x_{Al} value corresponding to isolated sites is equal to 0.147[30]. Furthermore the location of the maximum depends often very much on the mode of preparation of the zeolite samples[34] and even in certain cases e.g. m-xylene transformation over a series of commercial HFAU samples[35], there is no maximum in activity. This effect of the mode of preparation of the zeolite samples suggests that these samples differ not only by the framework composition (Si/Al or x_{Al}) but also by other characteristics which influence the activity.

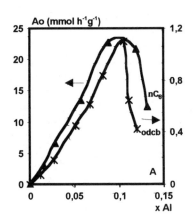
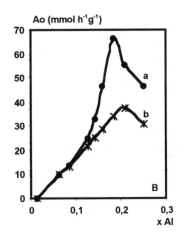

Figure 3 *A : Isomerization of n-hexane over PtHMOR catalysts[31] and of dichlorobenzene over HMOR catalysts[32]. B : n-heptane cracking over HFAU samples[41] dealuminated by steaming (i.e. with a large amount of extraframework Al species (a) followed by acid treatment, i.e. with a low amount of these species (b). Initial activity Ao vs.x_{Al} .*

It can therefore be concluded that changing the framework Si/Al ratio of zeolites is a good way to adjust the number (density) and the strength of their acid sites hence the activity. The effect on the activity should be different to the effect of Na^+ exchange by H^+ ; indeed strength and number of acid sites change in the opposite way whereas both were affected in the same way by Na^+ exchange.

3.3.2. Dealumination of zeolites   Dealumination is the most important commercial method for obtaining the desired x_{Al} value. This is furthermore the only way to increase the low framework Si/Al ratio of zeolites (e.g. HFAU) synthesized by crystallization of inorganic gels. This method is also used with zeolites synthesized with organic structuring agents although these zeolites (e.g. MFI) can be prepared with selected Si/Al ratios (higher than 10).

Table 1   *Dealumination of zeolites*

1   Hydrothermal treatment (steaming)
2   Chemical treatment
      A) Dealumination with Si enrichment
         -   $(NH_4)_2SiF_6$ in solution
         -   $SiCl_4$ in vapor phase
      B) Dealumination without Si enrichment
         -   with acids (HCl, HNO_3...)
         -   with chelating agents (e.g. EDTA)
         -   with volatile, non siliceous halides (e.g. $COCl_2$)
3   Hydrothermal and chemical treatment

Table 1 summarizes the variety of methods which can be used for dealuminating zeolites. Two categories of methods can be distinguished according as there is Si enrichment of the zeolite (methods 2A) or not, which is often the case : hydrothermal treatment, acid treatment, etc. Hydrothermal treatment (self-steaming or steaming) often followed by acid leaching is the most employed : e.g. for the preparation of ultrastable Y (USHY) zeolites used in cracking (FCC) and hydrocracking. Substitution of Al by Si of silicon compounds ($(NH_4)_2$ SiF_6, $SiCl_4$) can also be used for preparing USHY zeolites. Furthermore, acid treatment is well adapted to the dealumination of high silica zeolites (HBEA, HMOR, etc.).

Dealumination by hydrothermal treatment of NH_4NaY zeolites was extensively investigated. The main steps involved in this dealumination[36] are presented in Figure 4. After deammoniation (not indicated in Figure 4) there is successively hydrolysis of the four Si-O-Al bonds with formation of non framework Al species and of defect sites (hydroxyl nests). Only the aluminium atoms with which NH_4^+ or H^+ (and not Na^+) are associated can be extracted from the framework. Parts of the zeolite framework, having undergone a pronounced dealumination (these parts are probably the richest in aluminium atoms) have numerous defects sites and therefore collapse with formation of silicic and aluminosilicic spcies. The following steps are the migration of the non framework hydroxylated species, the filling in by Si of the defect sites with as a consequence stabilization of the framework. Two periods can be distinguished[37], namely an initial one in which dealumination is fast, and a second one in which it is much slower. The first period would correspond to the dealumination of the protonic zeolites, the second one to that of the zeolite exchanged by cationic non framework aluminium species (e.g. $(Al(OH)_2)^+$ resulting from protonation of $Al(OH)_3$).

This simplified description of dealumination by steaming shows that the increase of the framework Si/Al ratio is accompanied by other changes in the zeolite characteristics. The partial collapse of the zeolite leads to the **creation of secondary pores** (supermicropores and mesopores) at the expanse of the initial micropores. The size and number of secondary pores increase with the degree of framework dealumination[38]. These secondary pores were shown to have a pronounced positive effect in liquid phase reactions which are often limited by product desorption. For these reactions (e.g. in alkylation of toluene with 1-heptene)[32], the maximum in activity is significantly displaced to high values of the framework Si/Al ratio. These secondary pores were also

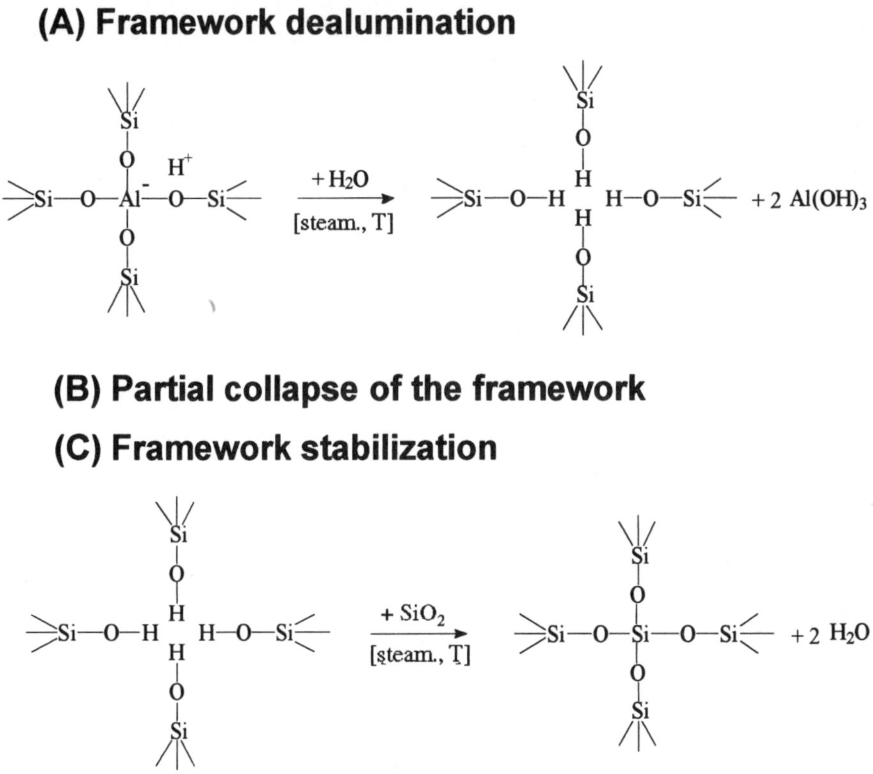

(A) Framework dealumination

$$\ce{>Si-O-Al(-O-Si<)(-O-Si) ->[+H2O][steam., T][H+] >Si-O-H \quad H-O-Si< + 2 Al(OH)3}$$

(B) Partial collapse of the framework

(C) Framework stabilization

$$\ce{>Si-O-H \quad H-O-Si< ->[+SiO2][steam., T] >Si-O-Si-O-Si< + 2 H2O}$$

Figure 4   *Main steps involved in dealumination of zeolites by hydrothermal treatment.*

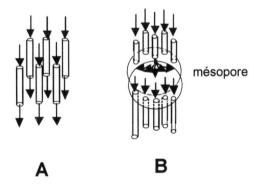

A   **B**

Figure 5   *Pore system of mordenites A : Before dealumination B : After dealumination by hydrothermal or/and acid treatment.*

shown to improve significantly the catalytic stability of monodimensional zeolites such as mordenites. This can be ascribed to a significant decrease in the deactivating effect of coke molecules owing to the change with dealumination of the diffusion of organic reactant molecules from a monodirectional to a quasi tridirectional mode[39] (Figure 5). Moreover, the changes in porosity can allow the accessibility to protonic sites that were not accessible in the parent zeolites. Thus, pyridine molecules were shown to interact with protonic sites located in the hexagonal prisms of dealuminated HFAU zeolites and in the side pockets of dealuminated HMOR samples.

A large variety of extraframework species, cationic or neutral, monomeric to polymeric were identified in dealuminated samples[p] : Al^{3+}, AlO^+, $Al(OH)^{2+}$, $Al(OH)_2^+$,

$[Al-OAl]^{4+}$, $\left[Al \begin{smallmatrix} O \\ O \end{smallmatrix} Al \right]^{2+}$, $AlO(OH)$, $Al(OH)_3$, Al_2O_3, silicates and aluminosilicates.

The smallest can be located within the zeolite micropores, the bulkiest being on the outer surface of the crystallites or in the mesopores. Extraframework species can influence significantly the zeolite activity. They can block or limit the diffusion of the molecules in the micropores, causing then a decrease in activity and favoring the retention of coke deposits hence the deactivation. Cationic species cause a decrease in the number of protonic sites, hence in activity. On the other hand, extraframework aluminium species created by mild steaming were shown to increase the activity of zeolites. This increase in activity was ascribed to the creation of **"superacid"** sites through interaction of bridging hydroxyl groups and neighboring Lewis acid sites[40], in particular small extraframework aluminium species.

$$\begin{array}{c} H \\ | \\ O \longrightarrow (AlO^+)p \\ Al \quad Si \end{array}$$

In agreement with this proposal the removal of these EFAL species by acid treatment[41] or with ammonium hexafluorosilicate causes a significant decrease in activity (Figure 3B). The creation by mild steaming of "superacid" sites resulting from interaction of protonic sites and EFAL species is demonstrated in the case of HFAU[42] and HEMT[43] zeolites by the appearence of two additional bands at 3600 and 3525 cm^{-1}. These bands result from a bathochromic shift of the HF and LF framework hydroxyl bands of the parent zeolite (OH located in supercages and in hexagonal prisms, respectively).This bathochromic shift is in agreement with an electron withdrawal from the bridging hydroxyls by EFAL species, with consequently a weakening of the OH bond hence an increase in acid strength. According to Lunsford and coworkers[44,45], the EFAL species would be

$Al(OH)^{2+}$ or $\left[Al \begin{smallmatrix} O \\ O \end{smallmatrix} Al \right]^{2+}$ located in sodalite cages. The presence of "superacid" sites

is particularly desired in the catalysis of reactions demanding very strong acid sites such as the commercial isomerization of light gasoline ($nC_5 + nC_6$). Very active catalysts for the process were tailored by optimizing the amount and nature of the EFAL species created during HMOR dealumination as well as the framework Si/Al ratio[46].

Dealumination with ammonium hexafluorosilicate, developed by Skeels and Breck[47], was mainly applied to HFAU zeolites. NH_4 zeolites are treated with solutions of $(NH_4)_2SiF_6$ under controlled pH conditions at $\approx 80°C$. Aluminium is removed from the zeolite in the form of soluble hexafluoroaluminate salts whereas most of the created framework vacancies are filled in by silicon.

$$\text{(Si–O–Al(NH}_4^+\text{)–O–Si)} + (NH_4)_2SiF_6 \longrightarrow \text{(Si–O–Si–O–Si)} + (NH_4)_3AlF_6$$

This method of dealumination seems therefore ideal for preparing samples differing only by their Si/Al ratio. Actually, it is not the case as was shown by several teams[47-49] which have determined the effect of operating parameters : reaction time, temperature, pH, $(NH_4)_2SiF_6$/zeolite ratio, on the characteristics of dealuminated HY samples.

The optimal conditions were found to be : relatively short reaction times (3 hours after addition of $(NH_4)_2SiF_6$), pH close to 6, temperature between 75 and 95°C. The degree of dealumination increases with temperature and with X_M the ratio between the number of Si atoms in the solution and the number of framework Al atoms (i.e. the maximum degree of dealumination). However, a plateau in X_R, the actual degree of dealumination (≈ 0.6 at 75°C) is obtained for high values of X_M. Moreover, a large decrease in crystallinity is then observed, which is certainly due to a rate of aluminium removal much higher than the rate of silicon insertion. Therefore this method is limited to the preparation of not very dealuminated samples (e.g. HFAU samples with a maximum Si/Al rato of 7). The effect of washing on the thermal and hydrothermal stability of the dealuminated samples is particularly significant[48,49]. At least 4 washings are necessary to obtain the maximum stability. Traces of hexafluorosilicate, difficult to eliminate completely by washing were shown to be responsible for zeolite degradation[49].

A combined XPS-SIMS study of dealuminated HFAU samples shows a significant gradient in the Si/Al ratio in the crystallites[50]. This gradient was ascribed both to a **gradient of dealumination** (i.e. dealumination is diffusion controlled) and to a selective **deposit of silica on the outer surface of the crystallites**. Silica was proposed to be formed from hydrolysis of SiF_4 that results from the decomposition of hexafluorosilicate.

In summary, dealumination with $(NH_4)_2SiF_6$ leads to samples differing not only by their framework Si/Al ratio but also by other characteristics : gradient of Si/Al in the crystallites, SiO_2 deposited on their outer surface, etc. It should however be noted that the differences between the parent and dealuminated samples are here much less pronounced than with dealumination by steaming. For identical values of the framework Si/Al values, large differences are furthermore found between the acid and catalytic properties of samples dealuminated by these two methods.

4 CONCLUSION

The acid properties of zeolites can be modified by various treatments : ion exchange, dealumination etc. which can be carried out in different ways. As was shown for the most classical methods, the effect of these treatments on the characteristics of the acid sites is generally complex. Indeed they provoke also modifications of the pore system which can favor or limit the diffusion of reactant and product molecules hence influence the catalytic properties. All these effects being well-known, it is relatively easy to tailor zeolites for obtaining active, stable and selective catalysts for desired reactions.

References

1. J.A. Rabo, *Zeolite Chemistry and Catalysis*, ACS Monograph 171, American Chemical Society, Washington , 1976.
2. I.E. Maxwell and W.H.J. Stork, *Stud. Surf. Sci. Catal.*, 1991, **58**, 571.
3. P.B. Venuto, *Microp. Mater.*, 1994, **2**, 69.
4. W.F. Hoelderich, *Stud. Surf. Sci. Catal.*, 1989, **49**, 69.
5. G. Pérot and M. Guisnet, *J. Mol. Catal.*, 1990, **61**, 173.
6. H.W. Kouwenhoven and H; van Bekkum, in *Handbook of Heterogeneous Catalysis*, G. Ertl, H. Knözinger and J. Weitkamp, (Eds.), Wiley, 1997, **5**, 2358.
7. B.K. Marcus and W.E. Cormier, *Chem. Engineering Progress*, 1999,47.
8. *Guidelines for mastering the properties of molecular sieves*, D. Barthomeuf et al. (Eds.), NATO ASI Series B : Physics, 221, Plenum Press, New-York, 1990.
9. R. Szostak, Stud. Surf. Sci. Catal., 1991, **58**, 153.
10. B.C. Gates, J.R. Katzer, G.C.A. Schuit, *Chemistry of Catalytic Processes*, Mac Graw Hill, New-York, Chapter 1, p. 1, 1979.
11. M. Guisnet, *Acc. Chem. Res.*, 1990, **23**, 392.
12. M. Guisnet, *Stud. Surf. Sci. Catal.*, 1985, **20**, 283.
13. A. Corma, H. Garcia, *Catalysis Today*, 1997, **38**, 257.
14. G. Bourdillon, C. Gueguen, M. Guisnet, *Appl. Catal.*, 1990, **61**, 123.
15. M. Guisnet, F. Avendano, C. Bearez, F. Chevalier, *J. Chem. Soc., Chem. Commun.*, 1985, 336.
16. C. Bearez, F. Avendano, F. Chevalier, M. Guisnet, Bull. Soc. Chem. Fr., 1985, 346.
17. L.A. Pines, P.J. Maher and W.A. Wachter, *J. Catal.*, 1984, **85**, 466.
18. A.W. Peters, W.C. Cheng, M. Shatlock, R.F. Wormsbecker and E.T. Haib, Jr., in *Guidelines for mastering the properties of molecular sieves*, D. Barthomeuf et al. (Eds.), NATO ASI Series B : Physics, Plenum Press, New-York, 1990, **221**, 365.
19. W.J. Mortier, *Proceedings 6th Int. Zeolite Conference*, D. Olson and A. Bisio, Eds., Butterworth, Guildford, 1984, 734.
20. J. Rabo and G.J. Gajda, *Guidelines for mastering the properties of molecular sieves*, D. Barthomeuf et al. (Eds.), NATO ASI Series B : Physics, Plenum Press, New-York, 1990, **221**, 273.
21. J.A. Martens, W. Souverijns, W. van Rhyn and P.A. Jacobs, in *Handbook of Heterogeneous Catalysis*, G. Ertl, H. Knözinger and J. Weitkamp, (Eds.), Wiley, 1997, **1**, 324.
22. J.W. Ward, in *Zeolite Chemistry and Catalysis*, J. Rabo, Ed., ACS Monograph, 1976, **171**, 118.
23. N. Cardona-Martinez and J.A. Dumesic, *Adv. Catal.* 1992, **38**, 149.
24. I. Bankós, J. Valyon, G.I. Kapustin, D. Kalló, A. Klyachko, T.R. Brueva, *Zeolites*, 1988, **8**, 189.
25. J. Chupin, N.S. Gnep, S. Lacombe, M. Guisnet, to be published.
26. J. Datka, B. Gil and A. Kubacka, *Zeolites*, 1995, **15**, 501.
27. J. Muscas, J.F. Dutel, V. Solinas, A. Auroux, Y. Ben Taarit, *J. Mol. Catal. A : Chemical*, 1996, **106**, 179.
28. I.V. Mishin, A.L. Klyachko, G. I. Kapustin and H.G. Karge, *Kinetics and Catalysis*, 1993, **34**, 828.
29. W.A. Wachter, *Proceedings 6th Int. Zeolite Conference*, D. Olson and A. Bisio, Eds., Butterworth, Guildford, 1984, 141.

30. D. Barthomeuf, *Materials Chemistry and Physics*, 1987, **17**, 49.
31. M. Guisnet, V. Fouché, M. Belloum, J.P. Bournonville, C. Travers, *Appl. Catal.*, 1991, **108**, 107.
32. B. Coq, R. Durand, F. Fajula, C. Moreau, A. Finiels, B. Chiche, F. Figueras and P. Geneste, *Stud. Surf. Sci. Catal.*, 1988, **71**, 282.
33. P. Magnoux, A. Mourran, S. Bernard, M. Guisnet, *Stud. Surf. Sci. Catal.*, 1997, **108**, 107.
34. A. Corma, in *Zeolite Microporous Solids : Synthesis Structure and Reactivity*, E.G.Derouane et al., Eds., NATO ASI Series C, 1992, **352**, 373.
35. S. Morin, P. Ayrault, N.S. Gnep, M. Guisnet, *Appl. Catal. A : General*, 1998, **166**, 281.
36. G.T Kerr, J. Phys. Chem., 1967, **71**, 4155.
37. Q.L. Wang, G. Giannetto, M. Torrealba, G. Pérot, C. Kappenstein and M Guisnet, *J. Catal.*, 1991, **130**, 459.
38. A. Berreghis, S Morin, P. Magnoux, M. Guisnet, V. Le Chanu, H. Kessler, *J. Chim. Phys.*, 1996, **93**, 1525.
39. N.S. Gnep, P. Roger, P. Cartraud, M. Guisnet, B. Juguin, C. Hamon, *C.R. Acad. Sci.* Paris, 1989, **309**, 1743.
40. C. Mirodatos and D. Barthomeuf, *J. Chem. Soc. Chem. Commun.*, 1981, 39.
41. Q.L. Wang, G Giannetto and M. Guisnet, *J. Catal.*, 1991, **130**, 471.
42. S. Khabtou, T. Chevreau, J.C. Lavalley, *Microp. Mater.*, 1994, **3**, 133.
43. S. Morin, A. Berreghis, P. Ayrault, N.S. Gnep, M. Guisnet, *J. Chem. Soc.*, Faraday Trans., 1997, **93**, 3269.
44. P.O. Fritz and J.H. Lunsford, *J. Catal.*, 1989, **118**, 85.
45. F. Lonyi and J.H. Lunsford, *J. Catal.*, 1992, **136**, 566.
46. A. Corma and A. Martinez, in *Catalytic Activation and Functionalization of Light Alkanes*, E.G. Derouane et al., Eds., NATO ASI Series, 1998, **44**, 35.
47. G.W. Skeels and D.W. Breck, *Proceedings 6th Int. Zeolite Conference*, D. Olson and A. Bisio, Eds., Butterworth, Guildford, 1984, 157.
48. G. Garralon, V. Fornes, A. Corma, *Zeolites*, 1988, **8**, 268.
49. Q.L. Wang, G. Giannetto and M. Guisnet, *Zeolites*, 1990, **10**, 301.
50. Q.L. Wang, M. Torrealba, G. Giannetto, M. Guisnet, G. Pérot, M. Cahoreau, J. Caisso, *Zeolites*, 1990, **10**, 703.

SYNTHESIS OF SOLUBLE LIBRARIES OF MACROCYCLES FROM POLYMERS: INVESTIGATIONS OF SOME POSSIBLE SCREENING METHODS USING POLYMERS

P. Hodge, C. L. Ruddick, A. Ben-Haida, I. Goodbody and R. T. Williams

Chemistry Department,
University of Manchester,
Oxford Road,
Manchester, M13 9 Pl, UK

1 INTRODUCTION

Macrocycles can be synthesised in several ways. The classical method is to use an appropriate bifunctional monomer and to react the functional groups together to form new linkages under high dilution conditions: Reaction 1 in Scheme 1. Under the high dilution conditions *intramolecular* reactions to give cyclics are greatly favoured over *intermolecular* reactions. In many cases the linkages are formed under kinetic control and, depending on the relative rates of the *intra-* and *inter-molecular* reactions, a family of cyclic oligomers is formed. In other cases the reactions between the functional groups are reversible. Then, providing the solution is dilute, a family of cyclic oligomers is again formed but this time under thermodynamic control. Whether the macrocyclisation is kinetically or thermodynamically controlled the proportions of the cyclic oligomers formed are often, but not always, for most practical purposes very similar. Essentially, the larger a cyclic oligomer is the smaller the amount of it that is likely to be formed. A recent example relevant to the present work is the preparation of cyclic trimer **1** and higher cyclic oligomers by treating a dilute solution of methyl ester **2**, a derivative of quinine, in toluene with potassium methoxide in the presence of 18-crown-6.[1]

An alternative, and closely related, route to cyclic oligomers starts with the corresponding step-growth polymers : Reaction 2 in Scheme 1. Here, using an appropriate catalyst, the linkages between the monomer moieties of the polymer are broken reversibly at high dilution. When the linkages reform they do so under similar conditions to those existing in the high dilution methods considered above, and the result is the formation of a similar mixture of cyclic oligomers. This type of reaction is called cyclo-depolymerisation (CDP). Clearly, since CDP depends on reversible reactions, the cyclic oligomers formed will be in equilibrium with each other (Reactions 3 in Scheme 1) and so their proportions will be under thermodynamic control. On the assumption that the cyclics formed were all strainless, Jacobson and Stockmayer developed a theory to explain the distibution of the cyclic species formed in such ring-chain equilibria.[2-4] A recent example pertinent to the present work is the preparation of the cyclic oligomers **3** in 93% yield by treating a dilute solution of poly(ethylene naphthalene-2,6-dicarboxylate) **4** in *o*-dichlorobenzene at reflux temperature in the presence of a catalytic amount of dibutyltin oxide.[5] This gave from the cyclic dimer up to the cyclic heptamer in percentage yields of 2, 68, 14, 6, 3, and 2 respectively.

An interesting property of cyclic oligomers is their potential for use as recognition systems. In this role they serve as a cavity (the "host") into which a moiety to be recognised (the "guest") fits. For good recognition the sizes of the "host" and "guest" should match well and the "host" cavity should be lined with moieties with which the

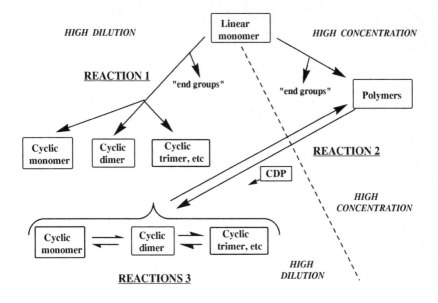

Scheme 1

2 $\longrightarrow$ 1

(Plus higher oligomers and methanol)

4 $\rightleftharpoons$

3 : n = 2 to >9

"guest" can favourably interact. The interactions may, for example, be electrostatic interactions, hydrogen bonding and/or π,π-interactions. If there is an excellent match the binding constant can be substantial.

In recent years a new method, combinatorial chemistry, has been developed for optimising properties of various kinds.[6,7] The object is to synthesise rapidly a large number of compounds (a "library") all of which potentially have a given desired property and then to rapidly screen the library to identify which is the best one. Combinatorial chemistry has been developed and used most to understand structure-activity relationships in protein chemistry[7,8] and pharmaceutical chemistry.[6,7,9] Usually the compounds of interest are prepared on insoluble polymer beads using procedures which can be automated and from hundreds to millions of compounds may be prepared. Most commonly the compounds are screened for biological activity. Such screens are often rapid and require very little material, thereby helping to make it feasible to screen very large libraries. Occasionally soluble libraries have been prepared and screened.[10]

It is clear that the ideas of combinatorial chemistry could potentially be applied to identifying excellent matches of "hosts" and "guests", and that to achieve this requires the synthesis of libraries of macrocycles. As noted above, when cyclic oligomers are prepared under thermodynamic conditions the family of oligomers formed can equilibrate, Reactions 3 in Scheme 1, i.e. dimers can interconvert with trimers and tetramers etc. Use can be made of this to establish equilibria between cyclic oligomers of *different families* , i.e. to "mix" the different families, and in this way a soluble macrocyclic library may be prepared.

The aim of this article is to briefly review our work on the preparation of soluble libraries of macrocycles using the various reactions outlined above and then to consider possible ways of screening such soluble libraries using functional polymers.

2 SYNTHESIS OF SOLUBLE MACROCYCLIC LIBRARIES

As noted above, combinatorial families of macrocycles may be generated by preparing several families of cyclic oligomers, either by macrocyclisation or CDP, and then "mixing" the macrocycles, or, more directly by carrying out the CDP of several polymers at the same time. This general approach is most likely to be successful when each of the families of oligomers contains just one or two major cyclics and the families being used have sufficient similarities that they "mix" well (see later).

Polymers **5 - 8** were prepared using classical polymerisation techniques.[11] Separate CDPs of these polymers, catalysed by dibutyltin oxide, gave the corresponding families of cyclic oligomers **9 - 12** in yields of 54 - 92 %. In each family there was little or no monomer present, and the amounts of the cyclic dimers (18-membered rings) were 61 - 80% (by weight) and cyclic trimers (27-membered rings) 14 - 18 %. These similarities in the compositions of the four families of cyclic oligomers suggested they would "mix" well and that the dimers and trimers would be the only cyclics present in the libraries in significant amounts. Clearly, these macrocyclic esters closely resemble crown ethers.

A soluble library (Library 1), see Scheme 2, was prepared by "mixing" the three familes of cyclic oligomers **9 - 11** by transesterifications. By mass spectrometry and ^{13}C NMR spectroscopy, the library contained all six combinations of the 18- membered ring macrocycles **13** and all ten combinations of the 27-membered ring macrocycles **14**. Morever, the ^{13}C NMR spectroscopy indicated that the proportions of the 18-membered rings were the same within a factor of two, i.e. the families "mixed" well. A second library (Library 2) was prepared directly from the four polymers **5 - 8** by carrying out the CDPs and the "mixing" at the same time. This library contained all the expected ten 18-membered macrocycles **13** plus all the twenty combinations of the 27-membered macrocycles **14**. The presence of these library members was established by mass spectrometry. The libraries just described only varied the diacid component. A further library (Library 3), where both the diacid and diol components were varied, was prepared

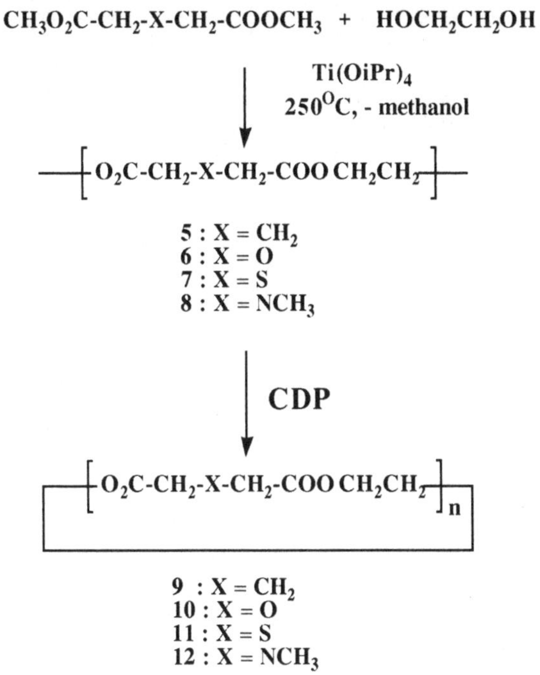

$$CH_3O_2C\text{-}CH_2\text{-}X\text{-}CH_2\text{-}COOCH_3 \quad + \quad HOCH_2CH_2OH$$

$$\downarrow \begin{array}{c} Ti(OiPr)_4 \\ 250^{\circ}C, \text{ - methanol} \end{array}$$

$$-\!\!\left[O_2C\text{-}CH_2\text{-}X\text{-}CH_2\text{-}COO\,CH_2CH_2\right]\!\!-$$

5 : X = CH$_2$
6 : X = O
7 : X = S
8 : X = NCH$_3$

$$\downarrow \quad \textbf{CDP}$$

$$\left[O_2C\text{-}CH_2\text{-}X\text{-}CH_2\text{-}COO\,CH_2CH_2\right]_n$$

9 : X = CH$_2$
10 : X = O
11 : X = S
12 : X = NCH$_3$

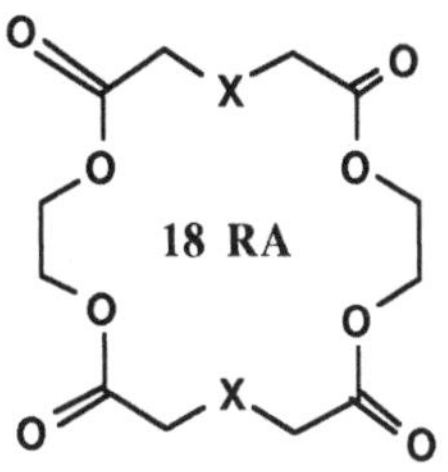

18 RA

13

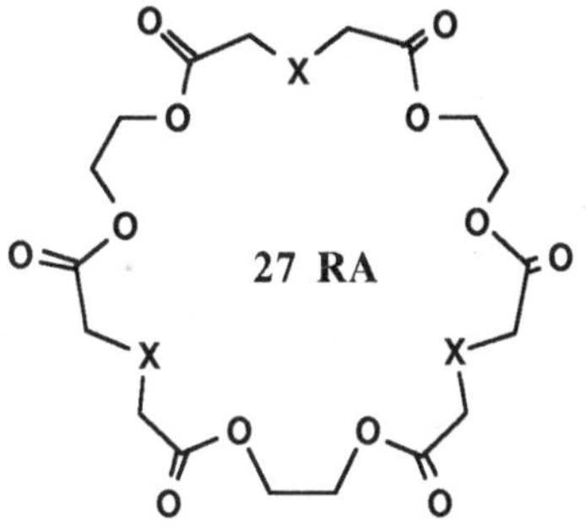

27 RA

14

from macrocycles **15** and **16**. This library consisted of nine 24- and twenty 36-membered rings.[11]

It might appear from the above that preparing such soluble libraries is extremely simple but there are limitations and the work of Sanders' group sheds light on the requirements for the successful "mixing" of the macrocycles.[12,13] Thus, a macrocycle is likely to be a better "host", i.e. exhibit larger binding constants, if the macrocycle is preorganised into the shape (conformation) which best accommodates the "guest". This usually requires the presence in the macrocycle of some moieties which result in it having some rigidity. This rigidity means, however, that the distribution of cyclic oligomer sizes can deviate substantially from that expected for strainless rings as outlined by Jacobson and Stockmeyer.[2] Thus, different families with different moieties to introduce some

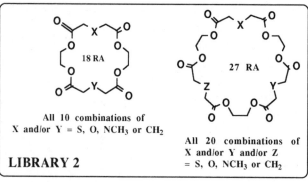

LIBRARY 1

All 6 combinations of
X and/or Y = S, O, or CH$_2$

All 10 combinations of
X and/or Y and/or Z
= S, O, or CH$_2$

LIBRARY 2

All 10 combinations of
X and/or Y = S, O, NCH$_3$ or CH$_2$

All 20 combinations of
X and/or Y and/or Z
= S, O, NCH$_3$ or CH$_2$

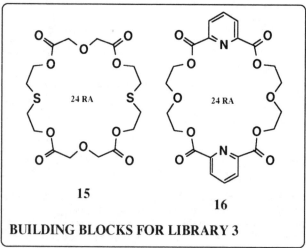

15

16

BUILDING BLOCKS FOR LIBRARY 3

Scheme 2

preorganisation do not necessarily "mix" well. Sanders' group clearly demonstrated this with several examples involving derivatives of cinchona alkaloids such as the cyclic oligomers of the type **1**.[12,13]

As a further development of these ideas, macrocylic libraries have been prepared recently which, whilst they have some preorganisational properties, still mix well. Libraries 4, see Scheme 3, are based on macrocycles of the type **17**,[14] and were "mixed" using transesterification, and Libraries 5 on macrocycles of the type **18**,[15,16] were "mixed" using nucleophilic aromatic substitution reactions catalysed by cesium fluoride.

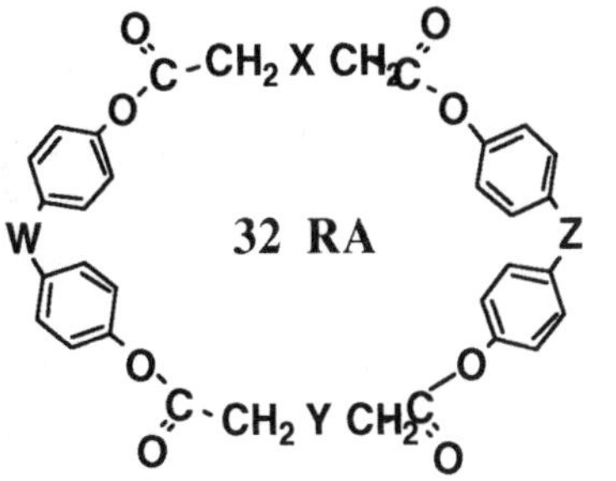

W and/or Z = cyclohexylidene, -O-, -SO$_2$- or various other moieties

X and/or Y = CH$_2$, -O-, -S- or various other moieties

17

BUILDING BLOCKS FOR LIBRARIES 4

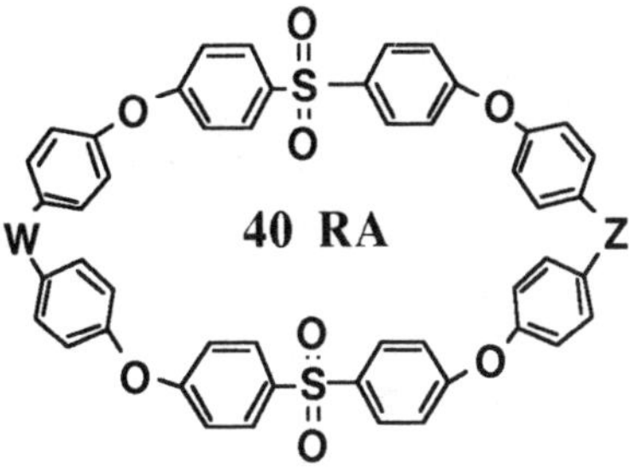

W and/or Z = cyclohexylidene, -O-, -SO$_2$- or other moieties

18

BUILDING BLOCKS FOR LIBRARIES 5

Scheme 3

3 INVESTIGATIONS OF SOME POSSIBLE SCREENING METHODS USING POLYMERS

The other major problem to be addressed in the combinatorial chemistry search to optimise recognition, is how the soluble libraries of macrocycles might be screened. An obvious way is to attach the "guest" to be recognised to a polymer, allow the most appropriate "host" in the library to bind, and then exploit the major differences in physical properties between the polymers and macrocycles to facilitate the separation and identification of the "host" which chose to bind to the supported "guest".[17] This is akin to "fishing" in the pool of soluble macrocycles with the polymer as the "fishing rod" and the guest as the "bait" The same basic idea has recently been used in other contexts.[18]

The first type of polymer we studied was macroporous polystyrene beads functionalised with potassium sulphonate groups.[17] Macroporous beads were used because they are particularly easily to separate. The functionalised beads were used to "fish" in a pool of commercial crown ethers. Whilst some of the crown ethers bound to the beads, unfortunately it appeared that many were just simply adsorbed. The second type of polymer studied, in an otherwise similar system, was gel-type beads. These have no inner surfaces for adsorbtion. However, even here adsorbtion appeared to be a problem.[17] It was concluded from these early studies that it would be better to use a soluble polymer as the "fishing rod".

It was anticipated that to facilitate separation of the polymer bound "host-guest complexes" the binding would need to be very strong and that it may even be necessary to use rotaxane formation. This would clearly be very limiting. Nevertheless, some main chain[19] and side chain[20] pseudopolyrotaxanes were investigated. There were several problems, not the least being the need to cap the chain ends after the formation of the pseudorotaxanes.

At this stage another related, simpler and more general method was tried and thus far it has been found to be much more successful.[20] It is based on High Resolution Diffusion Ordered Spectroscopy (HR-DOSY).[21,22] Here a form of two-dimensional ^{1}H NMR spectroscopy is used where one domain is the usual spectrum and the second the speed of diffusion on the particular species: see Figure 1. Small molecules, such as members of the soluble libraries, diffuse relatively rapidly, but polymers diffuse slowly. A soluble polymer is, therefore, again used as the "fishing rod" with a sidechain bearing the "bait". Then any member of the soluble library which binds to the bait will on average now diffuse more slowly than in the absence of the polymer and the magnitude of the change in the speed of diffusion will reflect the magnitude of the binding constant. Preliminary trials of this method, using weak acid - weak base interactions, have shown that it can easily screen a library of eleven compounds and that with some further development it could possibly be used to screen significantly larger libraries.[20]

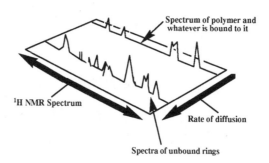

Figure 1

4 CONCLUSIONS

We have shown how soluble libraries of macrocycles can be prepared from appropriate macrocycles or polymers by transesterifications or nucleophilic aromatic substitutions. The rules for successfully achieving good "mixing" of the various macrocycles are becoming clearer, in part due to the work of Sanders' group.[12,13] A novel method for screening the soluble libraries is being developed. It is expected that these novel combinatorial chemistry approaches to "host-guest" chemistry will soon have reached the point where they can be used to specifically identify or optimise excellent "host-guest" recognition.

Acknowledgements

We thank the EPSRC for financial support for CLR and AB-H and the Thai Government for a Scholarship for PM.

References

1. S. J. Rowan and J. K. M. Sanders, *J. Org. Chem.*, 1998, **63**, 1536.
2. H. Jacobson and W. H. Stockmeyer, *J. Chem. Phys.* , 1950, **18**, 1600.
3. J. A. Semlyen, *Adv. Pol. Sci.*, 1976, **21**, 41.
4. U.W. Suter in *Comprehensive Polymer Science*, ed. G. Allen and J.C. Bevington, Pergamon, Oxford, 1989, vol 5, p91.
5. P. Hodge, Z. Yang, A. Ben-Haida and C. S. McGrail, *J. Mat. Chem.*, 2000, **10**, 1533.
6. *Combinatorial Chemistry, Synthesis and Application*, ed. S.R. Wilson and A. W. Czarnik , Wiley, New York, 1997.
7. L. A. Thompson and J. A. Ellman, *Chem. Rev.*, 1996, **96**, 555.
8. G. Lowe, *Chem. Soc. Rev.* , 1996, **25**, 309.
9. N. K. Terrett, *Tetrahedron*, 1995, **51**, 8135.
10. E.A. Wintner and J. Rebek, in ref 6, Chapter 5, pp.95-117.
11. P. Monvisade, P. Hodge and C.L.Ruddick, *Chem. Comm* , 1999, 1987.
12. S. J. Rowan and J. K. M. Sanders, *Chem. Comm* , 1997, 1407.
13. See, for example, S. J. Rowan, P. S. Lukeman, D. J. Reynolds and J. K. M. Sanders, *New J. Chem.*, 1998, 1015 and references sited therein.
14. P. Hodge, A. Ben-Haida, R. Williams and I. Goodbody, unpublished results.
15. I. Baxter, A. Ben-Haida, H. M. Colquhoun, P. Hodge, F. H. Kohnke and D. J. Williams, *Chem. Comm.*, 1998, 2213.
16. P. Hodge, M. Nisar and A. Ben-Haida, unpublished results.
17. C. L. Ruddick and P. Hodge, unpublished results.
18. See, for example: K. Lewandowski, P. Murer, S. Svec and J. M. J. Frechet, *Chem..Comm.*, 1998, 2237; and R. Boyce, G. Li, P. Nestler, T. Suenga and W. C. Still, *J. Am. Chem. Soc.*, 1994, **116**, 7955.
19. P. Hodge, P. Monvisade, G. J. Owen, and Pan Yang, *New J. Chem.*, 2000, accepted for publication.
20. P. Hodge, G. A. Morris and P. Monvisade, unpublished results.
21. M. D. Pelta, H. Barjat, G. A. Morris, A. L. Davis and S. J. Hammond, *Magn. Reson. Chem.*, 1998, **36**, 706.
22. C. S. Johnson, in *Progress in NMR Spectroscopy*, 1999, **34**, 204.

IMMOBILISED CATALYSTS AND THEIR USE IN THE SYNTHESIS OF FINE AND INTERMEDIATE CHEMICALS

Wolfgang F. Hölderich, Hans H. Wagner and Michael H. Valkenberg

Department of Chemical Technology and Heterogeneous Catalysis
University of Technology RWTH-Aachen
Worringerweg 1, 52074 Aachen
Tel.:+49-241-806560; Fax: +49-241-888291
E-mail: hoelderich@rwth-aachen.de

1 ABSTRACT

Catalytical processes in the production of fine and intermediate chemicals have found great interest in industry which is well aware of commercial as well as environmentally sustainable aspects. Generally applicable methods for the transformation of major homogeneous catalysts to give recyclable heterogeneous catalysts will be presented in some examples. The immobilisation will be demonstrated in the following cases:

- homogeneous transition metal complexes on special zeolites containing mesopores completely surrounded by micropores
- homogeneous transition metal complexes on mesoporous MCM-41 type material
- ionic liquids on various carriers.

The catalysts obtained thereby are tested in different reactions such as oxidation, hydrogenation and carbon-carbon linkage.

2 INTRODUCTION

In recent years increasingly stringent environmental constraints have led to a great interest in the application of new catalytic methods in the synthesis of fine and intermediate chemicals. Although the manufacture of bulk and fine/intermediate chemicals have common features, there are some typical differences which have important drawbacks on the environmental assessment of these processes.

Due to their thermal instability fine chemicals often must be produced in the liquid phase at moderate temperatures. They are generally complex and multifunctional as well as chemo-, regio- and stereoselectivity play an important role. The reactor system of the choice are batch or semi batchwise operated multipurpose units. Bulk chemicals, mostly consisting of relatively small, thermostable molecules, can be produced in continuously operated fixed bed or fluidised bed reactors in the gas-phase allowing much higher space time yields, thus minimising investment costs[1]. These procedures can hardly be applied to the synthesis of e.g. fine and intermediate chemicals because of the relatively small scale of the production (<1000-10000 t/a), the complex synthesis routes and the short development time to meet the market. For this reason these processes often lead to tremendously higher E-factors (25->100) compared to manufacturing bulk chemicals (<1-5)[2]. This is still tolerable from an economic point of view because of the high value added, but it is getting more and more difficult to carry out industrial scale reactions in such a manner. There is currently a general trend to develop clean and eco-efficient catalytic processes which minimise the generation of unwanted and harmful by-products. In contrast to homogeneous catalysts, heterogeneous catalysts bear specific advantages concerning especially the workup procedures. The separation of the catalyst from reactants and products can be facilitated by using simple mechanical techniques as filtration or centrifugation. From the technical point of view, a fixed bed reactor equipped with a heterogeneous catalyst might be preferable to a liquid phase reaction. An other significant advantage is the possibility of recovery and reuse of the catalyst. This will be even more important where expensive noble metals or chiral ligands are involved.

Advantages of heterogeneous catalysts have been lined out. However, for most applications involving valuable reactants and thus demanding outstanding selectivity of the reaction homogeneous catalysis is state of the art. Consequently the immobilisation of homogeneous catalysts is a reasonable idea, combining the positive effects of catalytical performance and practical use.

Generally applicable methods for the transformation of major homogeneous catalysts to give recyclable heterogeneous catalysts will be presented in some examples. There are three attempts which have been examined

- Steric hindrance – occlusion in porous systems
  The "ship-in-a-bottle" catalyst's main feature is the host-guest interaction which is neither covalent nor ionic. The guest is retained in the zeolite matrix by restrictive pore openings and will, in principle, keep all properties of the homogeneous complex in addition to the advantages offered by the heterogeneous system.

- Impregnation
  Most of the heterogeneous catalyst which are in practical use consist of one or more catalytically active compounds which are impregnated on supporting carrier materials. This method can be chosen to immobilise acids and bases as well as salts, oxides or complexes. The major drawback is leaching of one or more component which leads to irreversible deactivation of the catalyst. Physisorption can be enhanced by choosing the appropriate porous, chemical and electronical properties. This leads to catalysts with sufficient long term stability due to e.g. ionic linkages.

- Grafting
  The immobilisation of organic molecules on inorganic supports by the creation of a covalent bond is called grafting. This stable bond prevents leaching of the grafted moiety. Established homogeneous catalysts have to be chemically modified by the introduction of a suitable linker for the immobilisation by this method. This can also affect the catalytic properties of the active component.

Other fundamental strategies are grafting of homogeneous catalysts on organic supports or the use of supported liquid phases[3]. Recent reviews reveal the potential of heterogeneous chiral catalysts created by the methods mentioned[4-6].

Some examples for the application of such strategies will be presented in the next chapters. Special emphasis has been laid on typical drawbacks of the methods mentioned and the adequate solution for these problems.

3 RESULTS AND DISCUSSION

3.1 The "Ship In A Bottle" Approach; Immobilised Salen Complexes for the Stereoselective Epoxidation of Olefins

Recent developments in the area of stereoselective oxidation catalysis resulted in a growing interest in organometallic compounds encapsulated in zeolites. Thereby the zeolite-entrapped metal complex is free to move within the confined cavities of its crystalline host but prevented from leaching by restrictive pore openings. This can be achieved by the physical occlusion of metal complexes in zeolitic micro- or mesoporous space via the "ship-in-a-bottle"-synthesis method."Ship-in-a-bottle" catalysts provide many advantages in catalysis that cannot be achieved by homogeneous or conventional heterogeneous catalytic systems where the metal complex is attached to a solid surface by a covalent or ionic bond. Regarding conventional heterogeneous catalysts the main advantage of "ship-in-a-bottle" catalysts is the enhanced accessibility of the catalytical active metal complex due to its ability to move freely within the zeolite's cavity. In addition, because of the nature of their immobilisation, i. e. the physical entrapment in zeolite pores, "ship-in-a-bottle" catalysts are unlikely to leach. Romanovsky et al. reported in 1977 on the feasibility of this method[7]. The superiority of these catalysts to homogeneous systems is based on their easy separation from the reaction mixture and, thus, their recyclability and environmental compatibility[8]. Furthermore, it is likely that the zeolitic host bestows size and shape selectivity to the catalyst as well as a stabilising effect on the organometallic complex. Due to the site isolation of the single complexes multimolecular deactivation pathways such as formation of μ-oxo- or peroxo-bridged species will be rendered impossible[9-11].

However, even a large-pore zeolite such as zeolite Y, whose structure consists of almost spherical 13 Å supercages interconnected tetrahedrally through smaller apertures of 7.4 Å in diameter, is limited regarding the size of guest molecules by the space available in these. Indeed, organometallic complexes as they are widely used for stereo controlled reactions in the production of fine chemical production are often too spacious to enter those pores. This was previously demonstrated for Rhodium diphosphine or Cobalt salen complexes[12, 13].

Our approach was to enlarge the intrazeolitic cavities in order to generate superior hosts for bulky homogeneous chiral catalysts. Mesopores created this way are completely surrounded by micropores and offer additional advantages. The entrapped metal complex can move freely and is more accessible during catalysis and even sterically demanding transition states can be formed within the individual pores.

The (salen)manganese-catalysed oxidation of olefins is currently being investigated by various groups, Jacobsen and co-workers being only one of them[14]. Immobilisation of the Jacobsen catalyst is an attracting target. However, its outstanding activity, selectivity and chiral induction are accompanied by several disadvantages, such as quick deactivation, difficult separation and salt formation due to the use of sodium hypochlorite as oxidant. It has been attempted to immobilise the Jacobsen catalyst. However, its entrapment in zeolitic space has presented a problem the average faujasite supercage being a cavity too small for such a spacious complex[13]. So far only complexes of the less bulky Bissalicylidene-1,2-cyclohexanediamine have been successfully occluded in faujasites.

In the following a new approach is described to create mesopores surrounded by micropores in a faujasite matrix. These are spacious enough to accommodate the Jacobsen ligand, (R, R')-(N, N')-Bis(3,5-di-*tert*-butylsalicylidene)-1,2-cyclohexanediamine, but narrowed by the pore openings to the external surface to prevent leaching. In catalytic tests this catalyst showed promising results, e.g. in the epoxidation of α-pinene and (R)-(+)-limonene concerning activity, selectivity and chiral induction.

3.1.1 Synthesis of New Hosts by the Dealumination of Zeolites

For a new method to the generation of mesopores in the zeolitic framework, NaX and NaY zeolites were subjected to the combined dealumination steps with $SiCl_4$ and steam[15].

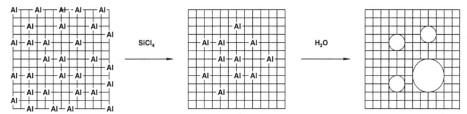

Figure 1 *Scheme of dealumination mechanism*

The scheme of dealumination mechanism is presented in **Figure 1**. During treatment with silicon tetrachloride, a dealumination method first reported by Beyer et al., the faujasite's framework aluminium is isomorphously replaced by silicon maintaining the microporous structure[16]. The reaction is self-terminating due to the precipitation of $NaAlCl_4$ in the outer pores of the zeolite crystal. The temperature to conduct the reaction (523 K) is relatively mild. Therefore it can be assumed that only the outer layers of the zeolite crystal are dealuminated by the treatment with $SiCl_4$. During the successive ion-exchange the chloro aluminium complexes are extracted and the zeolite is converted into the ammonium form. The zeolite is then steam-dealuminated. After this hydrothermal treatment the amount of framework aluminium has considerably decreased to a high degree of dealumination with a SiO_2/Al_2O_3 ratio ranging from 125 to 190 for both zeolites X and Y. Using the described dealumination procedure over all steps highly dealuminated faujasite zeolites are obtained regardless of the parent material.

Final treatment of the host material with hydrochloric acid removes some of the extra-framework aluminium in order to make room for guest molecules. The overall procedure is described elsewhere in detail[11, 15].

3.1.2 Synthesis of the Complex within the Host

To introduce the desired transition metal cation the zeolitic host was ion-exchanged. The synthesis of the salen ligands in the mesopores was conducted at room temperature under an inert atmosphere. To the ion-exchanged zeolitic material was added an amount of the optically pure diamine in dichloromethane in slight excess (1.2 equiv.) to the metal content. The appropriate amount of aldehyde was added to the slurry The final material was Soxhlet-extracted with dichloromethane and toluene, respectively, until the solvent remained colourless. This procedure is described elsewhere in detail[11, 15].

Figure 2 shows the salen complexes that have been entrapped in the mesopores of the new host materials and identified with FTIR measurements, thermogravity, XRD, and wet analysis.

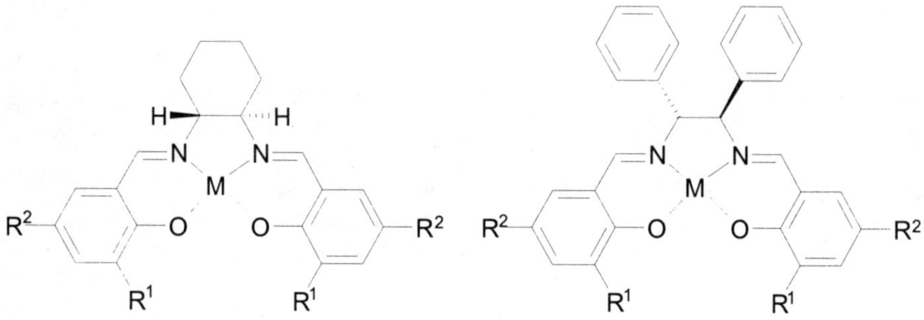

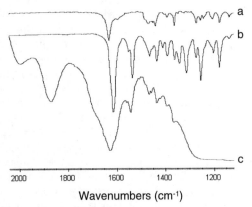

salen-**1**: R[1] = H, R[2] = H   salen-**5**: R[1] = t-Bu, R[2] = t-Bu

salen-**2**: R[1] = t-Bu, R[2] = t-Bu   salen-**6**: R[1] = t-Bu, R[2] = Me

salen-**3**: R[1] = t-Bu, R[2] = ME   salen-**7**: R[1] = t-Bu, R[2] = H

salen-**4**: R[1] = t-Bu, R[2] = H

Figure 2   *Salen complexes based on (R,R)-cyclohexanediamin and (R,R)-diphenyl-ethylendiamin*

Wavenumbers (cm⁻¹)

Figure 3   *FTIR-spectra of the Jacobsen ligand (A), the Jacobsen catalyst (B) and our catalyst salen 2 (C).*

In **Figure 3** the FTIR-spectra of the Jacobsen ligand (A), the Jacobsen catalyst (B) and our immobilised catalyst salen 2 (C) are compared. While the spectra A and B have been done using standard KBr technique, the "ship-in-a-bottle" catalyst (C) has been prepared as a self-supported wafer. The bands at wavenumbers 1466 cm^{-1}, 1434 cm^{-1}, 1399 cm^{-1}

and 1365 cm^{-1} in spectrum C can be assigned to the salen ligand because they also appear in spectra A and B. The band at 1542 cm^{-1} in spectrum C can only be assigned to the (salen)manganese complex as it does appear in the spectrum of the Jacobsen catalyst (A) at 1535 cm^{-1}, too, whereas in the spectrum of the salen ligand (B) there is no band in the region between 1470 cm^{-1} and 1600 cm^{-1}. This can be considered as a hint of the presence of the Jacobsen catalyst in our "ship-in-a-bottle" catalyst.

3.1.3 Stereoselective Epoxidation with Immobilised Salen Complexes

These catalysts have been tested in the stereoselective epoxidation of R-(+)-limonene and (-)-α-pinene. Here only the epoxidation of (-)-α-pinene as depicted in **Figure 4** is considered. The oxidant applied in the reaction is somewhat similar to the one introduced by Mukaiyama et al. and was favored over the system used by Jacobsen and coworkers[14, 17]. The major benefit of this system is that undesirable salt formation can be avoided by the use of environmentally benign molecular oxygen at RT instead of NaOCl as oxidant at 0°C.

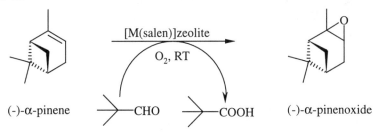

Figure 4 *Oxidation of (-)-α-pinene*

Figure 5 shows that the (salen-2) complexes of V and the Co(salen-5) complex retained their catalytic properties upon entrapment in the host materials. In contrast to the Mn(salen-2) complex which loses only some of its epoxide selectivity upon immobilisation, the corresponding Co and Cr complexes show an additional decrease in stereoselectivity as well. Strikingly, the immobilised Co(salen-5) complex achieved with 100 % conversion, 96 % selectivity and 91 % de even better results in the epoxidation of (-)-α-pinene than its homogeneous counterpart. However, it is worth notifying that among the (salen-2) complexes neither the homogeneous nor the occluded Jacobsen complex catalysed the epoxidation of (-)-α-pinene best.

We can conclude that a comparison of the respective catalytic results of these new heterogeneous catalysts and their homogeneous counterparts showed that the entrapment of the organometallic complex was achieved without considerable loss of activity and selectivity. The immobilised catalysts are reusable and do not leach. The oxidation system applies only O$_2$ at RT instead of sodium hypochloride at 0°C. A disadvantage is the use of pivalic aldehyde for oxygen transformation via the corresponding peracid. This results in the formation of pivalic acid which has to be separated from the reaction mixture. The best results so far – 100 % conversion, 96 % selectivity and 91 % de – were achieved with the immobilised Cobalt(salen-5) complex in the epoxidation of (-)-α-pinene.

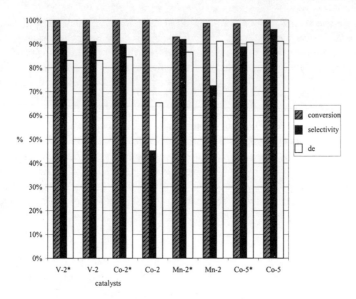

Figure 5   *Epoxidation of (-)-α-pinene over various homogeneous and immobilized transition metal (salen) complexes, * = homogeneous complex. Reaction Conditions: 10ml C₆H₆F, 4.6 mmol pivalic aldehyde, 1.85 mmol (-)-α-pinene, 25 mg Catalyst, 30 bar O₂, RT*

3.2 Immobilised Rh Diphosphine Complexes as catalysts for Enantioselective Hydrogenations

Homogeneous Rh diphosphine complexes belong to the most important catalysts in enantioselective hydrogenation reactions. Various attempts towards the immobilisation of organometallic complexes have been investigated previously[18] such as attachment to supporting materials by grafting[19, 20] or in form of supported liquid phase catalyst[3].

Nanosized channels of ordered mesoporous materials related to the M41S family can be tailored to pore sizes ranging from 1.5 nm to ca. 100 nm[21]. MCM-41 type molecular sieves are suitable as carriers for transition metal complexes offering new opportunities for the encapsulation of large catalyst species and for the catalytic conversion of substrates much larger than in common zeolites[22]. So far, the application of such host/guest compounds as catalysts has only been successfully tested for oxidative reactions[23-28].

We previously prepared surface-bonded rhodium phosphine complexes in Al-MCM-41. In a solution of dichloromethane, [Rh(acac)(chiraphos)] and Al-MCM-41 react to a surface bonded [(Os)x-Rh(chiraphos)] complex due to an exchange reaction of the acetylacetonato ligand and surface oxygens of the acidic support[12]. Here we present a heterogeneous Rhodium diphosphine catalyst and its application in the enantioselective hydrogenation of dimethylitaconate. The results indicate the localisation of the complex inside of the mesoporous channel system.

3.2.1   Preparation of the Immobilised Rh Phospine Complex

Principally Al-MCM-41 is impregnated with chiral rhodium diphosphine complexes. This leads to a strong adsorption of the complex in the mesoporous system of the carrier material (**Figure 6**). The preparations of the Rh complexes are carried out carefully under argon by standard Schlenk techniques. Due to high air sensitivity of the organometallic complexes combined with the high surface area of the heterogeneous catalysts best results are achieved with freshly prepared materials.

Al-MCM-41 was synthesised according to a slightly altered method described in literature[21, 29]. The Si/Al ratio is set to 40. The complexes are prepared from [CODRhCl]$_2$ and 1.1 equivalents of chiral diphosphine in dichloromethane as solvent. The carrier material, Al-MCM-41 is then added. After stirring the mixture for 24 h the solid is filtered, washed with dichloromethane until the solvent shows no colour in order remove the excess of the Rh complex, not fixed to the Al-MCM-41. The catalysts are dried in vacuum and extracted with methanol in a soxhlet apparatus for 24 h. ICP-AES analysis as well as FTIR-spectra of the remaining solvent indicate no content of homogeneous complex. The resulting catalyst has the pale yellow colour similar to the colour of the homogeneous complex.

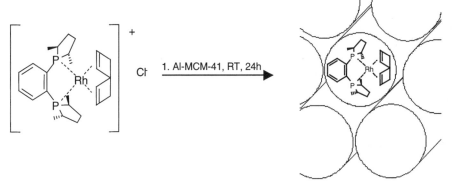

Figure 6   *Impregnation of Al-MCM-41 with Rhodium diphosphane complex results in the supported catalyst*

Rhodium complex contents vary between 0.02 and 0.07 mmol per g Al-MCM-41 depending on the amount of complex offered and on the batch of Al-MCM-41 used.

X-ray powder diffraction shows a strong peak of the (100) reflex, indicating that the structure of the carrier material remains throughout the immobilisation treatment.

The graphical illustration calculated with MSI-software of mmol [(1,5-Cyclooctadiene)Rh(1,2-bis((2S,5S)-2,5-dimethylphospholano)benzene)]$^+$Cl$^-$ (CODRhDuphos) entrapped in Al-MCM-41 shows that the complex fits easily in a tube of 2.2 nm.

N$_2$-adsorption isotherms of the supported catalyst compared to the corresponding carrier material shows a decrease of ca. 10 % of mesoporous pore volume (**Table 1, Figure 7**). These results indicate that the complex is deposited on the inner surface of the Al-MCM-41. The specific surface of the catalyst is also reduced by loading it with the complex.

Table 1   *Surface areas and pore volumes of Al-MCM-41 (carrier material) and immobilised CODRhDuphos*

	surface area* / $m^2 g^{-1}$	pore volume** / $cm^3 g^{-1}$
Al-MCM-41	1310	1.04
CODRhDuphos immobilised on Al-MCM-41	1250	0.93

* calculated according to the BET-method ** according to the BJH-theory

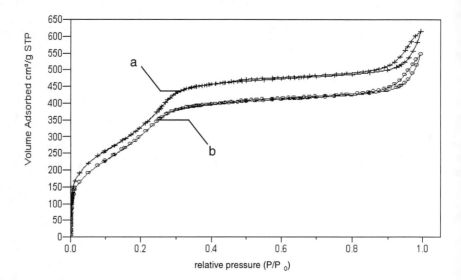

Figure 7   *N_2 adsorption and desorption isotherms of a) Al-MCM-41 (carrier material) and b) immobilised CODRhDuphos*

Thermogravimetric analysis was performed to investigate diffusion or even stabilising effects of the incorporation of the organometallic complex within the mesoporous host. Thermogravimetric and differential scanning calorimetric (DSC) measurements show that the immobilised complex is stable up to 250 °C. Oxidative decomposition of the complex takes place in 2 steps at 295 °C and 420 °C whereas a mechanical mixture of the homogeneous complex with the carrier material exhibits no distinct peaks. This also proves that due to the mesoporous properties some interaction takes place. However this cannot be specified by this method. The loss of weight of ca. 4.5 wt% caused by the burning of the complex is consistent with the content determined by chemical analysis.

Temperature programmed desorption (TPD) of ammonia shows weak acidic centers on the Al-MCM-41 carrier material. On the immobilised complex the amount of desorbed ammonia slightly decreases compared to the pure carrier material. This indicates an interaction of the immobilised complex with acidic sites on the Al-MCM-41. This result is also supported by IR-spectroscopy. The infrared spectra show no change of wavenumber but a slight decrease of intensity for the signal at ca. 3740 cm-1 which is assigned to the streching vibration of terminal silanol groups.

The Rhodium complex presented here is coordinated by a chiral diphosphine and a cyclooctadiene ligand. Various forces can cause the bonding of the complex on Al-MCM-41. So far the spectroscopic data suggest that a combination of various adsorptive forces and ionic bonding is responsible for the stability of the whole system. Electrostatic interaction of the cationic complexes occurs with the anionic framework of the Al-MCM-41 structure. This mechanism was also described for the immobilisation of manganese complexes on Al-MCM-41[24]. Direct bridging of the Rhodium to surface oxygen of the mesoporous walls has also been observed and could occur after cleavage of the diene complex during the hydrogenation reaction[12].

3.2.2 Enantioselective Hydrogenation with Immobilised Rh Phosphine Complexes

Several diphosphine ligands have been tested for the immobilisation and as catalysts in the enantioselective hydrogenation (**Figure 8**).

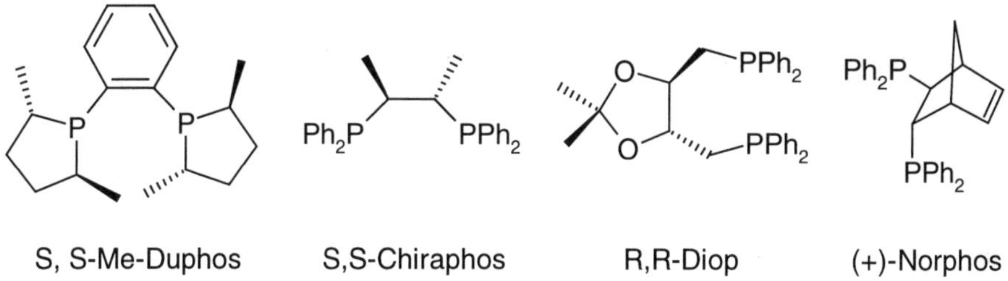

| S, S-Me-Duphos | S,S-Chiraphos | R,R-Diop | (+)-Norphos |

Figure 8   *Immobilised diphosphine ligands*

As a test reaction for the catalytic activity the hydrogenation of dimethylitaconate was employed. No reaction takes place in the blank test, when the carrier Al-MCM-41 itself is used as catalyst. The immobilised rhodium complexes give enantioselectivities up to 92 % ee of dimethyl-(R)-methylsuccinate with a turn over number of 4000 for the S,S-Me-Duphos ligand. The corresponding supported catalysts with R,R-Diop and S,S-Chiraphos ligands lead to enantioselectivities of 34 % ee and 47 % ee with lower activities. With the (+)-Norphos ligand the favoured enantiomer is dimethyl-(S)-succinate which is formed with 47 % ee.

Table 2   *The catalytic activity and enantioselectivity of immobilised rhodium diphosphine complexes in the hydrogenation of dimethylitaconate*

immobilised ligand	Conversion / %	TON	e.e. / %
S,S-Me-Duphos	100	> 4000	92
R,R-Diop	57	2280	28
S,S-Chiraphos	8	320	47
(+)-Norphos	8	320	47[1]

[1] *S* enantiomer

The catalyst can easily be recovered and reused without further treatment. The supported Rh-Me-Duphos catalyst was recycled four times. As the experiment shows, the conversion and the enantioselectivity of the catalyst remains high after several runs (**Figure 9**).

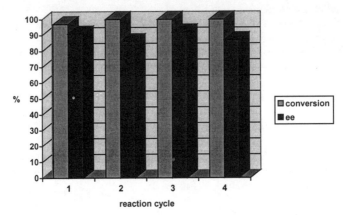

reaction cycle

Figure 9 *The catalytic activity and enantioselectivity of immobilized rhodium Me-Duphos complex during the recycling in the hydrogenation of dimethylitaconate*

In order to prove that the reaction is catalysed heterogeneously and to exclude the possibility of leaching the reaction solution was filtered before complete conversion. Hydrogenation of the reaction solution following filtration after 3.5 h does not give any further reaction. After 24 h the conversion of the filtered sample remains at 14 % whereas the original batch with catalyst goes to complete conversion of 100 %.

In conclusion, chiral heterogeneous catalysts are prepared from chiral Rhodium diphosphine complexes and Al-MCM-41. The bonding supposedly occurs via an ionic interaction of the cationic complex with the host. Also a slight reduction of weak acidic sites of Al-MCM-41 has been observed. These catalysts are suitable for the hydrogenation of functionalised olefins. The organometallic complexes remain stable within the mesopores of the carrier at reaction conditions. The catalyst can be recycled by filtration or centrifugation.

3.3 Immobilised ionic liquids as catalysts for Friedel-Crafts reactions

Ionic liquids (ILs), previously known as molten salts, were mainly used in electrochemistry studies due to their ionic nature. The most important step in the chemistry of the ILs occurred when Osteryoung described a mixture of 1-(1-butylpyridinium)-chloride and aluminium chloride which was liquid at room temperature[30]. Later on, Wilkes discovered other ionic liquids based on dialkylimidazolium salts that featured even more convenient physical and electrochemical properties than the butylpyridinium salts[31].

The use of immobilised ILs as catalysts would result in the easy separation of the catalyst from the reaction mixture, allowing its fast reuse and avoiding the generation of contaminated waste and its subsequent treatment. Ionic liquids have already been proposed as catalysts for Friedel-Crafts alkylation of benzene with olefins in order to produce LABs[32-35]. They show Lewis acidic properties when a Lewis acid (e.g.

aluminium trichloride – AlCl$_3$), forming the counteranion, is used in excess. The acidity of the resulting IL can be controlled by varying the amount of the Lewis acid in the final ionic liquid[36]. The possibility of modifying the acidity of the ILs through a wide range of values is very interesting from the catalytic point of view since the selectivity in some reactions depends on the acid strength of the catalyst. A clear example is the alkylation of phenol, where higher acid strength of the catalyst favours *C*-alkylation rather than *O*-alkylation.

Nevertheless, the use of ILs in their liquid form presents some inconveniences in an industrial continuous system. Immobilising these ILs on inert supports on the other hand brings many advantages for the system, e.g. the easier separation of the catalyst from the reaction media and the possible utilisation of the catalyst in a continuous system. The evaluation of the catalytic properties of acidic ionic liquids, immobilised on known supports, and research of the possible advantages of these materials was the driving force of this work.

Within parts of the scientific community researching ionic liquids, often a notion of different "generations" of ionic liquids is expressed. The chloroaluminate ionic liquids can be viewed as the first generation of ILs. They show new and fascinating properties but their sensitivity towards water limits the possible uses for them. Tetrafluoroborates and Hexafluorophosphates can be considered ILs of the second generation. They have a lower or no sensitivity towards moisture and can be handled under air. Today, research scientists are working on the third and even fourth generation of ionic liquids.

This development has been taken into account in the preparation of immobilised ionic liquids. The first generation, immobilised via the inorganic anion, was easy to prepare but limited to highly acidic ILs which form a stable metal/oxygen bond.

The second generation, immobilised via the organic cation is – at least in theory – not limited to any special anion. The resulting materials are far more versatile. Their main drawback is the fact that the bond between cation and support can be hydrolysed. Supports, containing the organic cation within the structure of the support material, are stable even towards boiling acids. The resulting catalysts have been introduced as Novel Lewis-Acidic Catalysts (NLACs, see **Figure 10**).

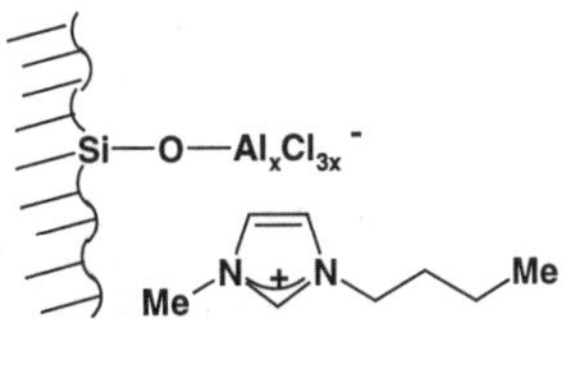

Incipient wetness	Grafting	Sol Gel
NLAC I	**NLAC II**	**NLAC III**

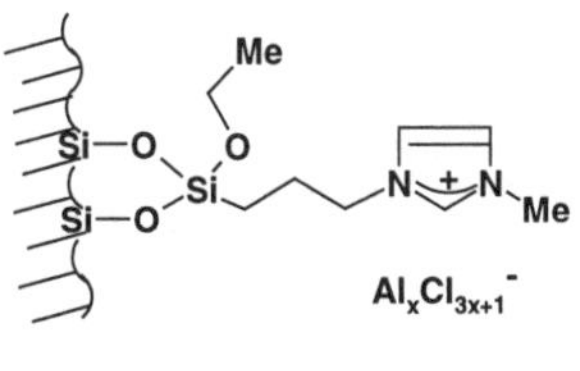

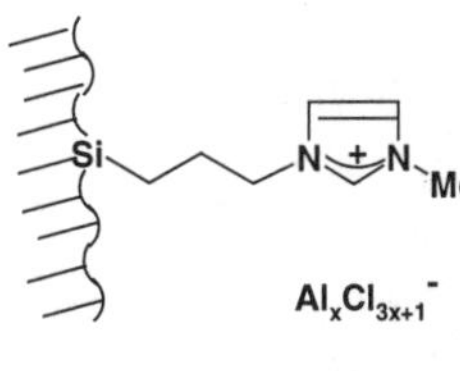

Figure 10 *Three different methods for the preparation of immobilised ionic liquids*

The results for the catalysts NLAC II and NLAC III are presented in a separate talk[37].

3.3.1 Impregnation of Various Supports with Ionic Liquids

The easiest impregnation technique for the immobilisation of ionic liquids is called the method of "incipient wetness". This method basically consists of the slow addition of the ionic liquid to the support material, a period of stirring to guarantee homogenisation, and a following extraction of excess IL with boiling dichloromethane. Immobilisation in solvents is possible as well. The main reason that the method of incipient wetness was chosen, is the fact that here the excess ionic liquid and the dichloromethane can be completely recovered and reused. After the extraction, the IL containing solvent can be separated in a Rotavap; experiments proved the extracted ionic liquid to be still catalytically active. The new class of catalysts obtained this way is summarised as NLAC I.

After extensive testing of different carrier materials silica proved to be a very suitable support, since here the highest amount of immobilised ionic liquid could be found. Other metal oxides were suitable for immobilisation as well, but usually the amount of ionic liquid immobilised was far lower (**Figure 11**). The only exception was Al_2O_3, where the amount of IL found on the carrier after immobilisation was as high as for silica but the product proved to be inactive as a catalyst.

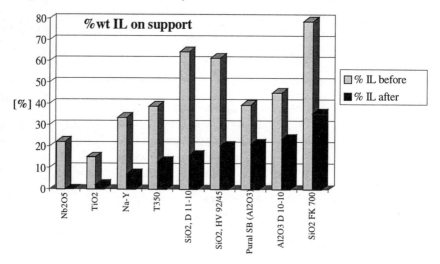

Figure 11 *Comparison of different supports: Content of IL before and after extraction*

The immobilised ionic liquids have been extensively analysed. FT-IR and MAS-NMR spectroscopy show the disappearance of the hydroxyl groups on the surface of the supports. In FT-IR spectra this can be seen by a band at 3741 cm^{-1} assigned to terminal OH-groups which disappears after addition of an ionic liquid (spectra not shown). In crosspolarised MAS-NMR spectra the signals at -91 ppm and -101 ppm assigned to $(SiO)_2Si(OH)_2$ and $(SiO)_3Si$-OH groups respectively cannot be detected after immobilisation (**Figure 12**).

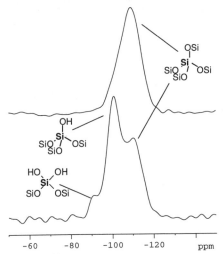

Figure 12 *^{29}Si crosspolarised spectra of SiO_2 without (bottom) and with (top) immobilised ionic liquid*

We consider this to be proof, that the ionic liquids supported on a metal oxide form a covalent bond between the aluminum of the chloroaluminate species and the oxygen of the hydroxyl-groups on the surface of the carrier (**Figure 13**). This is important for the stability of the immobilised ionic liquids since a leaching of the ionic liquid is prevented. As shown in the case of immobilised ILs based on iron chloride, the lack of strong bond between anion and surface leads to leaching.

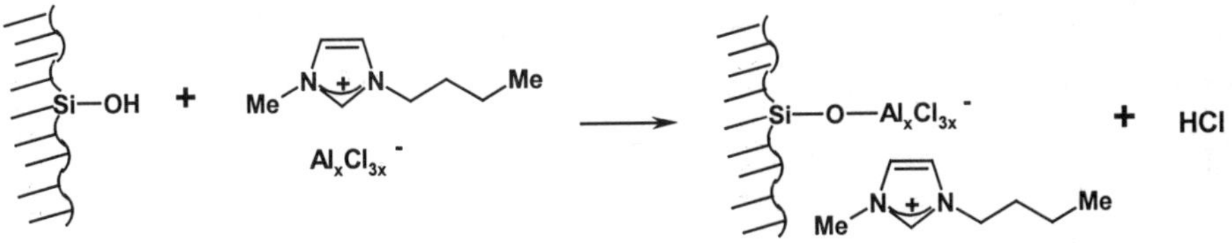

Figure 13 *Immobilisation via the anion: Incipient wetness method*

The method of incipient wetness has one major drawback, which is the destruction of structured support materials. Immobilisation on zeolites or MCM-41 leads to the decomposition of the carrier when a highly acidic ionic liquid is added directly. The reason for this seem to be superacidic properties of HCl in ionic liquids. Especially the MCM-41 materials, which are mesoporous aluminosilicates with a very high surface area, are interesting carrier materials. Alternatives to the method of incipient wetness have been developed and will be presented elsewhere.

ICP-AES and CHN analysis allowed an exact description of the composition of the immobilised ILs. Furthermore, a comparison of the results before and after use of the catalysts allowed statements about leaching.

Thermogravimetric analysis (DTG) confirmed the results of the CHN analysis regarding the content of organic substances and additionally gave information about the thermo-stability of the catalysts. The immobilised ILs are be stable up to temperatures of about 350 °C.

3.3.2   Friedel Crafts Alkylation Reactions With Immobilised Ionic Liquids

As a test reaction to illustrate the potential of such immobilised ionic liquids the alkylation of benzene with olefins is presented. This reaction is economically important, since the linear alkyl benzene compound is one intermediate in the synthesis of biodegradable detergents[38],[39]. Conventional methods for alkylation reactions use HF or AlCl₃ as acid catalysts. These processes feature serious drawbacks: HF is highly toxic, volatile and corrosive, AlCl₃ is also toxic, corrosive and is consumed in large amounts. This generates a large amount of contaminated wastes, mainly salts, that require treatment before their final disposal[40]. Data recently published by UOP discloses Detal technology as a new alkylation process for the alkylation of benzene with heavy olefins, using solid acids as catalysts in the liquid phase reaction[41].

The alkylation reactions of toluene and benzene with dodecene were used as test reactions in order to evaluate the catalyst activity. It could be observed that the IL is catalytically activity after being immobilised. Moreover, due to the better dispersion of the solid catalyst in the reaction media, even the conversions obtained for the supported IL were better than for the pure IL. In the alkylation of toluene conversions reached at standard conditions were twice as high for the immobilised ILs, for benzene the difference was even bigger.

The comparatively low conversions reached with pure ionic liquids are due to the low solubility of the aromatic compounds in the IL. To gain access to its' complete catalytic activity a far higher experimental expenditure is necessary, e.g. an optimised reactor and stirrer design.

When alkylating benzene with dodecene, standard reaction conditions were a reaction temperature of 80°C and a catalyst ratio of 6 wt% catalyst over the total amount of reactants. Dodecene conversion reached 95 mol % with a selectivity of 98% towards monoalkylated products in the first 30 minutes of the reaction. In this case, benzene was used as solvent (molar ratio of benzene to dodecene = 10:1).

Catalyst recycling is an important subject, especially when dealing with supported catalyst. Heterogeneous catalysts tend to lose activity when used in consecutive reactions. In case of the immobilised Ils, the reason can be either leaching of the active phase, deposits on the surface (adsorption on the acid sites) or the deterioration of the catalyst due to water adsorption. Leaching of the IL from the surface of the support was studied for the catalysts mentioned and it was found to be negligible (**Table 3**).

Table 3   *ICP-AES analysis of catalysts before and after reactions (leaching test)*

Catalyst	Al (mg/g)	Si (mg/g)	Al/Si	Procedure
T-350/NLAC I	12,9	377,6	0,0341	Before reaction
T-350/NLAC I	13,0	380,4	0,0342	After 2nd run [a]

[a] batch reaction toluene/$C_{12}^{=}$

In the beginning catalysts were not very active in a second run. This was attributed to the method of recovering the catalyst. Indeed extreme care must be taken to ensure completely dry conditions. Filtration of the catalyst led to its' deactivation, probably because of residues of moisture in the filter. Recycling proved to be possible when the catalyst was left in the flask, the reaction mixture decanted of and the catalyst washed with dichloromethane.

The use of the catalyst in continuous liquid phase reactions avoids such handling problems. Here the advantages of the heterogeneous system is obvious compared to homogeneous discontinuous systems. The reactions had to be carried out at lower temperatures than in the batch reactors to stay below the boiling point of the starting materials. Nevertheless, even at room temperature conversions of about 20 % and a selectivity towards the monoalkylated product of more than 95 % could be achieved (**Figure 14**).

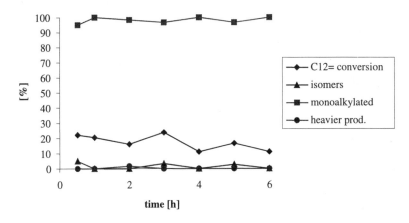

T = 20°C, WHSV=7.8 h^{-1}, cat.:T350/IL, R = 10 : 1

Figure 14 *Alkylation of benzene with dodecene: Continuous liquid phase*

Still deactivation of the catalyst with time on stream (TOS) can be observed. This deactivation is mainly due to the adsorption of dodecene and heavy products on the catalyst which block its active sites for further reaction. Elementary analysis of the catalyst after the reaction has been carried out, and a significant increase of the carbon content was found. This confirms that the deactivation of the catalyst is due to strong adsorption of organic materials on the surface of the catalyst.

4 CONCLUSION

We demonstrated the feasibility of immobilisation reactions on three novel examples. In all of the cases typical catalytic properties of homogeneous catalysts could be transferred to a heterogeneous system. This facilitated recovery of the catalyst and the work in continuous phase. The recyclability of the catalysts was demonstrated.

We prepared a new carrier material, suitable as a host for"ship in a bottle". Zeolites X and Y have been highly dealuminated by a succession of different dealumination methods. This generated mesopores completely surrounded by microporous space. The catalytic behavior of bulky transition metal complexes entrapped in these new was investigated in stereoselective epoxidation. The organometallic complex is entrapped without considerable loss of activity and selectivity

In the second example we immobilised chiral Rhodium diphosphine complexes on Al-MCM-41 was demonstrated. The bonding supposedly occurs via an ionic interaction of the cationic complex with the host. Also a slight reduction of weak acidic sites of Al-MCM-41 has been observed. These catalysts are suitable for the hydrogenation of functionalised olefins. The organometallic complexes remain stable within the mesopores of the carrier at reaction conditions.

The third case shows the immobilisation of Lewis-acidic ionic liquids. The resulting catalysts, named Novel Lewis-Acidic Catalysts (NLACs), are highly active in the Friedel-Crafts alkylation of aromatic compounds with dodecene. Conversions and selectivities to the desired monoalkylated products were excellent. No leaching of the catalytically active component could be observed. The isomer distribution of the monoalkyated products is very similar to that obtained over pure aluminum(III)chloride. The main drawback of the NLACs is that thy are very sensitive towards water, which leads to irreversible deactivation. A second problem is the deactivation after long reaction times. The most likely cause is olefin oligomerisation.

5 REFERENCES

1.   H.-U. Blaser, M. Studer, *Applied Catalysis A: General,* 1999, **189**, 191.
2.   R. A. Sheldon, *Chem. Ind.,* 1997, 12.
3.   K. T. Wan, M. E. Davis, *Nature,* 1994, **370**, 449.
4.   A. Baiker, *Current Opinion in Solid State & Materials Science,* 1998, **3**, 86.
5.   T. Bein, *Current Opinion in Solid State & Materials Science,* 1999, **4**, 85.
6.   J. M. Thomas, T. Maschmeyer, B. F. G. Johnson, D. S. Shephard, *Journal of Molecular Catalysis A: Chemical,* 1999, **141**, 139.
7.   V. Y. Zakharov, B. V. Romanovsky, *Vestn., Mosk. Univ., Khim.,,* 1977, **18**, 142.
8.   G. Schulz-Ekloff, G. Meyer, D. Wöhrle, M. Mohl, *Zeolites,* 1984, **4**, 30.
9.   R. Parton, D. De Vos, P. A. Jacobs, in E. G. e. a. Derouane (Ed.): *Zeolite Microporous Solids : Synthesis, Structure and Reactivity,* Kluwer Academic Publishers 1992, p. 555.
10. K. J. Balkus Jr., A. K. Khanmamedova, K. M. Dixon, F. Bedioui, *Applied Catalysis A: General,* 1996, **143**, 159.
11. C. Heinrichs, W. F. Hölderich, *Catalysis Letters,* 1999, **58**, 75.
12. A. Janssen, J. P. M. Niederer, W. F. Hölderich, *Catalysis Letters,* 1997, **48**, 165.
13. W. Kahlen, H. H. Wagner, W. F. Hölderich, *Catalysis Letters,* 1998, **54**, 58.
14. E. N. Jacobsen, A. R. Muci, N. H. Lee, *Tetrahedron Letters,* 1991, **32**, 5055.
15. C. Schuster, W. F. Hölderich, *Catalysis Today,* 2000, in press.
16. H. K. Beyer, I. M. Belenykaja, in I. e. al. (Ed.): *Studies in Surface Science & Catalysis, Vol. 5,* Elsevier, Amsterdam 1980, p. 203.
17. T. Mukaiyama, T. Yamada, K. Imagawa, T. Nagata, *Chemistry Letters,* 1992, 2231.
18. A. Baiker, H. U. Blaser, in G. Ertl, H. Knözinger, J. Weitkamp (Eds.): *Handbook of Heterogeneous Catalysis, Vol. 5,* VCH, Weinheim 1997, p. 2422.
19. A. Corma, M. Iglesias, F. Mohino, F. Sanchez, *Journal of Organometallic Chemistry,* 1997, **544**, 147.
20. B. Pugin, *Journal of Molecular Catalysis A: Chemical,* 1996, **107**, 273.
21. J. S. Beck, J. C. Vartuli, W. J. Roth, M. E. Leonowicz, C. T. Kresge, K. D. Schmitt, C. T.-W. Chu, D. H. Olson, E. W. Sheppard, S. B. McCullen, J. B. Higgins, J. L. Schlenker, *Journal of the American Chemical Society,* 1992, **114**, 10834.

22. C. Huber, K. Moller, T. Bein, *Journal of the Chemical Society, Chemical Communications,* 1994, 2619.
23. E. Armengol, A. Corma, F. Vicente, H. García, J. Primo, *Applied Catalysis A: General,* 1999, **181**, 305.
24. L. Frunza, H. Kosslick, H. Landesser, E. Höft, R. Fricke, *Journal of Molecular Catalysis A: Chemical,* 1997, **123**, 179.
25. S.-S. Kim, W. Zhang, T. J. Pinnavaia, *Catalysis Letters,* 1997, **43**, 149.
26. G.-L. Kim, S.-H. Kim, *Catalysis Letters,* 1999, **57**, 139.
27. P. Piaggio, P. McMorn, C. Langham, D. Bethell, P. C. Bulman-Page, F. E. Hancock, G. J. Hutchings, *New J. Chem.,* 1998, 1167.
28. S. Ernst, M. Selle, *Microporous and Mesoporous Materials,* 1999, **27**, 355.
29. Z. Luan, C.-F. Cheng, W. Zhou, J. Klinowski, *J. Phys. Chem.,* 1995, **99**, 1018.
30. R. A. Osteryoung, G. Cheek, H. Linga, *The Third International Symposium on Molten Salts* (Hollywood, Florida, USA), Studies in room temperature chloroaluminates, pp. 221 (1980).
31. J. S. Wilkes, in G. Mamantov, R. Marassi (Eds.): *NATO ASI Series C: Mathematical and Physical Sciences, Vol. 202,* D. Reichel Publishing Company 1987, p. 405 .
32. J. A. Boon, J. A. Levisky, J. L. Pflug, J. S. Wilkes, *J. Org. Chem.,* 1986, **51**, 480 .
33. E. Benazzi, Y. Chauvin, A. Hirschauer, N. Ferrer, H. Olivier, J. Y. Berhard, , Institut Français du Petrole, USA 1996.
34. C. J. Adams, M. J. Earle, G. Roberts, K. R. Seddon, *Chem. Commun.,* 1998, 2097 .
35. E. Benazzi, H. Olivier, Y. Chauvin, J. F. Joly, A. Hirschauer, *Abstracts of Papers of the American Chemical Society,* 1996, **212**, 717.
36. T. Welton, *Chem. Rev.,* 1999, **99**, 2071 .
37. M. H. Valkenberg, C. deCastro, W. F. Hoelderich, *4th Int. Symp. on Supported Reagents* (St Andrews), Immobilised ionic liquids on silica materials (2000).
38. A. H. Turner, J. H. Houston, D. R. Karsa, (University of Salford), Proceedings of the Symposium of the Industrial Division of the Royal Society of Chemistry, pp. 3 (1991).
39. J. L. G. d. Almeida, M. Dufault, Y. B. Taarit, C. Naccache, *J. Am. Oil Chem. Soc.,* 1994, **71**, 675.
40. J. H. Clark, *Green Chemistry,* 1999, **1**, 1.
41. K. Tanabe, W. F. Hoelderich, *Apply. Catal. A: General,* 1999, 399.

CATALYTIC AZIRIDINATION AND EPOXIDATION OF ALKENES USING MODIFIED MICROPOROUS AND MESOPOROUS MATERIALS

Graham J. Hutchings,[a] Christopher Langham,[a] Paola Piaggio,[a] Sophia Taylor,[a] Paul McMorn,[a] David J. Willock,[a] Donald Bethell,[b] Philip C. Bulman Page,[c] Chris Sly,[d] Fred Hancock[e] and Frank King[e]

[a]Department of Chemistry, Cardiff University, P.O. Box 912, Cardiff CF1 3TB
[b]Leverhulme Centre for Innovative Catalysis, Department of Chemistry, University of Liverpool, Liverpool L69 3BX
[c]Department of Chemistry, Loughborough University, Loughborough, Leics. LE11 3TU.
[d]Robinson Brothers Ltd., West Bromwich, West Midlands B70 0AH.
[e]Synetix, R&T Division, P.O. Box 1, Billingham, Teeside TS23 1LB

1 SUMMARY

The immobilisation of enantioselective homogeneous catalysts is described using an approach in which cations are immobilised by ion-exchange within a microporous (zeolite Y) or a mesoporous material (Al-MCM-41). The catalysts are used under non-aqueous conditions so that cation leaching is minimised during the catalytic reaction. The cations can be modified using chiral ligands to form an enantioselective catalyst that is wholly heterogeneous and can be readily reused. Using this approach it is shown that copper-exchanged zeolite Y is a highly active catalyst for the aziridination of alkenes. Modification using bis(oxazolines) leads to the formation of an enantioselective aziridination catalyst. Using a similar approach, manganese-exchanged Al-MCM-41 modified with a chiral salen ligand is found to be an effective enantioselective heterogeneous epoxidation catalyst for *cis*-stilbene. The approach is also used to show that immobilisation of cobalt-exchanged Al-MCM-41 can also give some activity for the enantioselective epoxidation of *cis*-stilbene. However, in this case, although only low enantioselection is observed, the immobilised catalyst gives higher enantioselection than the non-immobilised homogeneous catalyst.

2 INTRODUCTION

The design of asymmetric catalysts is of intense current interest and procedures describing the use of chiral transition metal complexes as homogeneous catalysts have been described for epoxidation, cyclopropanation, aziridination and hydrogenation of alkenes. There is an increasing awareness that heterogeneous catalysts can offer significant advantages over homogeneous catalysts and this has prompted research activity in this field. To date, three experimental approaches have been used in the design of enantioselective heterogeneous catalysts: (i) the use of a chiral support for an achiral metal catalyst; (ii) the immobilisation of an asymmetric homogeneous catalyst onto an achiral support such as a zeolite or a polymer ; and (iii) modification of an achiral heterogeneous catalyst using a chiral cofactor. These approaches have been described in detail by Davies.[1] The first approach was pioneered by Schwab in 1932.[2] Using Cu, Ni,

Pd and Pt supported on enantiomers of quartz he demonstrated low enantioselection in the dehydration of butan-2-ol. A number of other chiral supports have been examined, e.g. natural fibres and chiral polymers. Most recently, attention has focused on using zeolite β, since it is possible that a chiral form of this zeolite could be synthesised.[3] The second approach has been particularly effective for enantioselective hydrogenation reactions using zeolites as supports for asymmetric Ru and Rh catalysts,[4] as well as the immobilisation of chiral manganese-salen catalysts for epoxidation as described by Sherrington.[5] The third approach, involving the creation of a chiral catalyst surface by the adsorption of a chiral modifier onto an achiral catalyst, has been successful in a number of studies, again, in this case, particularly for enantioselective hydrogenation. For example, the modification of platinum catalysts with cinchona alkaloids for the hydrogenation of prochiral α-ketoesters[6] have been extensively studied.

We have also used the third approach, and we have concentrated our design efforts on zeolite Y since we consider that, for the optimal catalyst design, the achiral catalyst should have a well defined structure. The central aspect of our design strategy is that microporous zeolites and mesoporous Al-MCM-41 can immobilise cations by ion-exchange.[7] The ion-exchange is carried out in aqueous solution and when the material is used as a catalyst in non-aqueous media the cation remains immobilised . In our initial proof of concept studies we studied the modification of zeolite H-Y with chiral dithiane 1-oxides and we have shown that this approach can give catalysts that are capable of enantioselection for the dehydration of butan-2-ol in the temperature range 110-150 °C.[8] Although the enantioselection was observed for relatively few catalyst turnovers (*ca.* <20) the surprising observation was the promotional effect observed by the modification of the proton with the dithiane 1-oxide which led to a dramatic increase in the rate of the dehydration reaction being observed.[9] We have now extended this generic approach and have designed catalysts for the enantioselective aziridination[10,11] and epoxidation of alkenes[12,13] and in this paper describe aspects of these catalytic processes.

3. EXPERIMENTAL

The zeolite HY used in this study was supplied by Union Carbide (LZY 82). Al-MCM-41 was synthesised according to literature methods,[14] and prior to use was calcined at 550 °C for 4 h in nitrogen, followed by 16 h in air. Cu-exchanged zeolite (CuHY) was prepared using a conventional ion-exchange method in which zeolite H-Y was treated with aqueous $Cu(OAc)_2$ solution, the concentration of which was chosen so as to obtain the required exchange level (ca. 50-60%).[10,11] The solids were then washed with distilled water until all unbound Cu^{2+} was removed and dried at 110 °C in air. An identical method was used for the preparation of Cu-Al-MCM-41. The aziridination of alkenes was carried out in a batch reactor using (*N*-(*p*-tolylsulfonyl)imino)phenyliodinane (PhI=NTs) and (*N*-(*p*-nitrobenzylsulfonyl)imino)phenyliodinane (PhI=NNs) as the nitrene donors, prepared using standard methods.[11] The alkene, nitrene donor and solvent were stirred together in a flask under controlled temperature in the presence of the Cu catalyst. The nitrene donor reagents are relatively insoluble under the chosen conditions and the reaction was monitored by observing the dissolution of this reagent; when the dissolution was complete the reaction was considered to be complete. In a typical experiment, styrene (500 μl) was reacted with PhI=NTs (0.3g) in acetonitrile together with CuHY catalyst (0.3g) at 25 °C for 2 h. The products were isolated by column chromatography and product identification was confirmed by NMR, elemental analysis, GCMS and infrared spectroscopy. When enantioselective aziridinations were carried out, the CuHY catalyst was pretreated with a chiral modifier prior to reaction, and the products were analysed using chiral HPLC.

The method described above was used to prepare MnHY and Mn-Al-MCM-41 *via* ion exchange with aqueous manganese acetate (0.2M, 25 °C, 24 h), followed by filtration, washed with water and vacuum dried. This procedure was repeated twice and the material was calcined (550 °C, 24 h). The calcined Mn-exchanged materials were refluxed with the chiral salen ligand, (R,R)-$(-)$-N,N'-bis(3,5-di-*tert*-butylsalicylidene)-1,2-cyclohexanediamine, in CH_2Cl_2 (24 h, metal:salen = 1:1), cooled to 0 °C and washed with CH_2Cl_2. In the case of Mn-Al-MCM-41 this procedure resulted in 10% of the salen ligand being incorporated (determined by TGA and solution analysis). The Mn-exchanged material:salen catalyst was then investigated for the epoxidation of *cis* - stilbene using iodosylbenzene as oxidant in a batch reactor[13]. A Co-Al-MCM-41 (4% Co) was prepared using three ion-exchange steps with aqueous cobalt nitrate (0.2M) for 24 h. This material was refluxed with the chiral salen ligand as described above.

4 RESULTS AND DISCUSSION

4.1 Heterogeneous asymmetric aziridination of alkenes with CuHY

CuHY is found to be an effective catalyst for the aziridination of a range of alkenes when using PhI=NTs and PhI=NNs as the nitrene donors (Tables 1 and 2). To confirm the process was wholly heterogeneous, following the reaction the zeolite catalyst was recovered by filtration and another aliquot of reactants was added to the recovered filtrate and no reaction was observed. Further, the removed catalyst was reused with fresh reagents and solvent and the catalyst gave the same reactivity as that observed initially. It is observed that the catalyst gives best results with phenyl-substituted alkenes, and although lower yields are observed with cyclohexene and *trans*-hex-2-ene the reaction is still observed. Interestingly, for the aziridination of *trans*-stilbene no product could be observed. We consider this to be due to the relatively bulky aziridine product being too large to be accommodated within the small pores of CuHY and is further evidence that the reaction proceeds inside the pores of the zeolite. This interesting result illustrates the potential for a heterogeneous catalyst to possess size-specificity to a precise degree. Such a property could be exploited by making use of zeolites with a range of pore sizes, and could also be developed to achieve regioselectivity in a reagent containing two or more double bonds. Modification of the CuHY with chiral bis oxazoline ligands (e.g. 2,2-bis-[2-((4R)-(1-phenyl)-1,3-oxazolenyl)]propane) leads to the aziridine being obtained with up to 75% enantiomeric excess in these initial experiments (Table 2). We have found that a temperatures in the range of -10 to -20 °C provide the highest combination of enantioselectivity, yield and reaction time when using acetonitrile solvent. In these experiments, the chiral modifier is simply added to the reaction mixture, and no special pretreatment of the catalyst system is required.

To demonstrate that alternative types of silicate framework can be used for this reaction, experiments were carried out with copper-exchanged MCM-41. Yields of up to 87% of the aziridine with e.e. of 37% were obtained using PhI=NTs as the nitrene donor. Using

Table 1   *CuHY-Catalysed Aziridination of Representative Alkenes*

alkene [a]	Yield /% [b]
styrene	90 (92)
styrene [c]	87 (35)
α-methylstyrene	33
p-chlorostyrene	76
p-methyl styrene	66
cyclohexene	50 (60)
methyl cinnamate	84 (73)
rans-stilbene	0 (52)
rans-2-hexene	44

[a] Unless otherwise specified reaction conditions were: solvent MeCN, 25°C, styrene: PhI=NTs = 5:1 molar ratio.

[b] Isolated yield of aziridine based on PhI=NTs.

[c] styrene:PhI=NTs = 1:1 molar ratio.
Values in parentheses indicate yields obtained from homogeneous reaction under the same reaction conditions.

this type of mesoporous catalyst system greatly enhances the versatility of the heterogeneous aziridination reaction since larger substrate molecules can be used.

The major advantage of the use of CuHY as a catalyst for this reaction is the ease with which it can be recovered from the reaction mixture by simple filtration if used in a batch reactor (alternatively it can be used in a continuous flow fixed bed reactor). We have carried out the heterogeneous asymmetric aziridination of styrene until completion, filtered and washed the zeolite then added fresh styrene, PhI=NTs and solvent, without further addition of chiral bis(oxazoline), for several consecutive experiments. The yield and the enantioselectivity decline slightly on reuse; we have found that adsorbed water can build up within the pores of the zeolite on continued use and we believe that this is the cause of loss of activity and enantioselection. However, full enantioselectivity and yield can be recovered if the catalyst is simply dried in air prior to reuse, or alternatively the catalyst can be recalcined and fresh oxazoline ligand added.

4.2 Asymmetric epoxidation using modified Mn- and Co-exchanged materials.

To demonstrate the flexibility of the approach to catalyst design that we set out in this paper, the epoxidation of alkenes using iodosylbenzene has also been studied. Initial studies focused on MnHY:salen catalysts for the epoxidation of styrene, however, the reaction was slow, and low yields of styrene oxide were observed. Analysis of the reaction mixture revealed the breakdown of the salen ligand within a few turnovers. Subsequently Mn-Al-MCM-41 was used with iodosylbenzene as the oxygen donor and *cis* -stilbene was used as substrate, and the results, together with those of control experiments, are shown in Table 3.

Table 2   *CuHY-Catalysed Aziridination of Representative Alkenes.*

Bis-oxazoline	Alkene[a]	Temp °C	PhINTs Yield[b] %	PhINTs e.e.[c] %	PhINNs Yield[b] %	PhINNs e.e.[c] %
None	Sytrene	25	90 (92)	-	93(97)	-
	Methyl cinnamate	25	84 (73)	-		
	trans-Stilbene	25	0 (52)	-		
				-		
(1)	Sytrene	25	87	29	69	52
	Styrene	-10	82	44	100[e]	75[e]
	Trans-β-Methylstyrene	-10	74	36		
	trans-β-Methylcinnamate	-10	8 (21)	61 (70)	30[f]	59[f]
(2)	Styrene	25			87	64
	Styrene	-20	64	0		
	Styrene[d]	-20	15 (89)	18		
	Styrene[d]	0		(63)	100	34
(3)	Styrene	25	78 (75)	10 (10)		
(4)	Styrene	25	73 (74)	0 (15)		
(5)	Styrene	-10	4	61		

[a]Solvent CH_3CN, styrene: $PhI=NTs$ = 5:1 molar ratio; [b]Isolated yield of aziridine based on $PhI=NTs$. Values in parentheses indicate yields obtained from homogeneous reactions; [c]Enantioselectivity determined by chiral HPLC; [d]styrene was used as solvent, [e]0 °C, [f]25°C, Absolute configurations of major products, determined by optical rotation, are (S) for trans-β-methylstyrene and trans-β-methylcinnamate, (R) for styrene.

$Mn(OAc)_2$ in the absence of Al-MCM-41 or salen ligand is not particularly reactive, and only 1.5% yield of the epoxide was formed after reaction for 24 h at 25 °C. Modification of Mn^{2+} in solution by the salen ligand, as expected, leads to a significant rate enhancement, and both the *cis*-epoxide and the *trans* -epoxide were formed, the latter with 78% e.e. Interestingly, immobilisation of the Mn^{2+} within Al-MCM-41 leads to an increase in reactivity, and the epoxide is formed with an enhanced *cis/trans* ratio to the homogeneously catalysed Mn:salen catalyst. This effect suggests that the Al-MCM-41 is occupying part of the Mn^{2+} coordination sphere, restricting the *cis* →*trans* transformation. Further modification of the Mn-exchanged Al-MCM-41 with salen leads to a further enhancement in reactivity, and the *trans* epoxide is formed with an 70% e.e., very similar to that observed for the equivalent homogeneous reactions. *trans* -Stilbene is found to be a significantly less reactive substrate, and the e.e. of the resultant *trans*-epoxide is significantly decreased with the salen modified Mn-exchanged Al-MCM-41. The use of Mn-exchanged Al-MCM-41:salen catalyst for this epoxidation does not result in the formation of significant levels of by-products as has been observed when manganese bypiridyls have been used as catalysts, and typically only deoxybenzoin is observed at low levels (*ca.* 5-10%). A further set of experiments was carried out to examine the reusability of the Mn-exchanged Al-MCM-41:salen catalyst. Following the reaction, the Mn-exchanged Al-MCM-41:salen catalyst was recovered by filtration and the solid was reused in a new catalytic reaction; although the reactivity and enantioselectivity had declined, epoxide was still formed and the *cis/trans* ratio was unchanged. Recalcination of the recovered material and addition of new salen ligand essentially restored both the reactivity and the enantioselection. Use of the solution following the filtration did not give any activity, and furthermore this solution contained no Mn^{2+}.

Table 3 *Epoxidation of cis-stilbene at 25°C using Mn-exchanged Al-MCM-41*

Catalyst	Time[a] h	Conv.[b] %	Epoxide Yield (%)[f]	Selectivity (%) cis	Selectivity (%) trans	e.e. *trans* (%)[g]
None	25	0	0	0	0	0
Mn(OAc)2[c,d]	24	100	1.5	0	100	0
Mn(salen) complex[c]	1	100	86	29	71	78
Al-MCM-41[c]	24	0	0	0	0	0
Mn-MCM-41[c]	2	45	3	0	100	0
Mn-MCM-41+salen[c]	2	100	69	58	42	70
Mn-MCM-41+salen[c,e]	26	100	35	0	100	25
solution[c]	2	0	0	0	0	0
Mn-MCM-41 reused[c]	2	37	18	61	39	30
recalcined +salen[c,h]	2	100	52	63	37	54

[a] reaction time, [b] as determined by decomposition of iodosylbenzene to iodobenzene, using HPLC, [c]solvent CH_2Cl_2 with molar ratio of cis-stilbene:catalyst:iodosylbenzene=7:1:0.13, [d] reaction conducted in CH_3OH, [e] *trans*-stilbene used as substrate, [f] Conversions, yields and selectivity determined by HPLC, [g] enantiomeric excess determined by chiral HPLC., [h]Mn-Al-MCM-41 from entry 6, recalcined and refluxed with salen ligand.

These experiments demonstrate that the reaction occurring with Mn-exchanged Al-MCM-41:salen is wholly heterogeneously catalysed. At this stage we have made no attempt to optimise the catalytic performance, but we anticipate that appropriate modification of the chiral salen ligand and the reaction conditions will lead to enhanced reactivity and enantioselection.

Recently, there has been great interest in polymer-supported chiral salen complexes[15], and, in particular, the use of chiral cobalt salen complexes for the hydrolytic kinetic resolution of terminal epoxides has attracted much interest.[16] We were prompted to investigate the use of chiral cobalt salen complexes immobilised on Al-MCM-41 since we observed that cobalt cations catalysed the decomposition of iodosyl benzene at a similar rate to manganese cations, and at a much higher rate than other transition metal cations.[17] The results for the epoxidation of *cis*-stilbene are given in Table 4. Cobalt acetate dissolved in methanol or dichloromethane is not a particularly active catalyst under our conditions and even after 48 h epoxide yields are low (19% in dichloromethane, 24% in methanol) with the *cis*-epoxide being the preferred product (experiments 2-5). The use of cobalt-exchanged Al-MCM-41 significantly increases the rate of reaction whereas non-exchanged Al-MCM-41 demonstrates no catalytic activity (experiments 6,7). This indicates that the immobilisation of cobalt ions within the pores of Al-MCM-41 has a rate enhancing effect in a similar way to ligand accelerating effects observed in homogeneous catalysis[18]. Use of cobalt-exchanged Al-MCM-41 modified with the chiral salen ligand as catalyst shows a similar enhancement in rate, and the reaction is complete after 6h, based on the consumption of iodosyl benzene, but the yield of the epoxide formed is decreased and the *trans*-epoxide is formed with an e.e. of 20% (experiment 8). In contrast, the reaction with the homogeneous cobalt (II) chiral salen complex gives the *trans*-epoxide with a much decreased e.e. (9%) (experiment 10). This is a significant observation, since in the case of chiral manganese salen catalysts, immobilisation of the catalyst, e.g. in Al-MCM-41, always leads to a decrease in enantioselectivity[13]. No leaching of cobalt (II) ions into solution was observed with the cobalt-exchanged Al-MCM-41 catalysts. Filtration of the reaction mixture following reaction to remove the solid catalyst gave a solution which was then re-used with fresh reagent and no epoxide was formed. This confirms that the reaction is catalysed wholly heterogeneously.

A cobalt exchanged Al-MCM-41 catalyst was refluxed with the salen ligand in dichloromethane for a longer time (72h) to determine if improved reactivity can be induced through enhanced incorporation of the salen ligand. When used as a catalyst, this material showed a much slower reaction rate and gave very little *trans*-epoxide with a much lower enantioselectivity (experiment 9). The higher enantioselection for the salen modified cobalt exchanged Al-MCM-41 catalyst, compared with the homogeneous complex, may be due to dilution of the salen ligand within the mesopores and, in this case, dimerisation of the cobalt complexes is restricted. Increasing the concentration of salen within the mesopores, as demonstrated by the 72h reflux preparation, decreases the yield and enantioselection, and this could be caused by the salen ligand interacting with more than one Co^{2+} cation, an effect that could be expected to be pronounced for the homogeneous complex.

A further set of experiments were carried out to demonstrate the effect of temperature on both the heterogeneous and homogenous catalysts and the results are given in Table 5 for reactions between 0-35°C. Both catalysts show maximum enantioselection at about 25°C with complete loss of enantioselection at 35°C. At the lower temperature, the reaction is slower and more *cis*-epoxide is formed because the radical pathway that leads to the *trans* product is less favoured at a lower temperature. Above 25°C, the decomposition of the epoxide possibly contributes to the loss of yield and enantioselection. The results obtained with the immobilised chiral Co-salen catalyst are not promising with respect to the enantioselection achieved.

Table 4 *Epoxidation of cis-stilbene at 25°C using Co-exchanged Al-MCM-41*

Expt.	Catalyst	Solvent	time/h	Conversion of PhIO/%	Epoxide Yield %[b]			e.e.
					Total	cis[c]	trans[c]	trans[d]/%
1	None	CH_2Cl_2	48	0	0	-	-	-
2	$Co(CH_3CO_2)_2$	CH_2Cl_2	2	20	3.2	76	24	-
3	$Co(CH_3CO_2)_2$	CH_2Cl_2	48	60	19	79	21	-
4	$Co(CH_3CO_2)_2$	CH_3OH	2	11	1.2	83	17	-
5	$Co(CH_3CO_2)_2$	CH_3OH	48	63	24	85	15	-
6	Al-MCM-41	CH_2Cl_2	24	0	0	-	-	-
7	Co-Al-MCM-41	CH_2Cl_2	4	100	56	84	16	-
8	Co-Al-MCM-41 + salen ligand (24)[e]	CH_2Cl_2	6	100	8	33	66	20
9	Co-Al-MCM-41 + salen ligand (72)[f]	CH_2Cl_2	24	100	1	46	54	4
10	Co-salen complex	CH_2Cl_2	2	94	42	24	76	9

[a]Reaction conditions : cis-stilbene (1.75 mmol), PhIO (0.25 mmol), catalyst (0.032 mmol) reacted in solvent (4 ml) at 25°C
[b]Isolated yields determined using HPLC (APEX ODS 5μ)
[c]Normalised to 100% epoxide
[d](S,S) configuration, determined using chiral HPLC (Pirkle Covalen (R,R) Whelk-O column)
[e]Co-exchanged Al-MCM-41 refluxed with salen ligand for 24h
[f]Co-exchanged Al-MCM-41 refluxed with salen ligand for 72h.

Table 5   *Effect of Temperature on Epoxidation of cis-stilbene[a]*

Catalyst	Temp/°C	Time/h	Conversion of PhIO/%	Epoxide yield[b]			e.e.
				Total	cis[c]	trans[c]	trans[d]/%
Co-Al-MCM-41	35	24	87	4.4	57	43	0
+ salen ligand	25	6	100	8	34	66	20
	15	24	88	2.4	47	53	3
	0	24	99	2.4	53	47	2
Co-salen complex	35	4	100	7.7	48	52	0
	25	2	97	41.8	24	76	9
	15	4	100	24.7	54	46	6
	0	24	99	2.0	76	24	3

[a]Reaction conditions : cis-stilbene (1.75 mmol), PhIO (0.25 mmol), catalyst (0.032 mmol) reacted in solvent (4 ml) at 25°C
[b]Isolated yields determined using HPLC (APEX ODS 5μ)
[c]Normalised to 100% epoxide
[d](S,S) configuration, determined using chiral HPLC (Pirkle Covalen (R,R) Whelk-O column)

The enantioselection is considerably lower than that achieved with the Mn-salen complex under the same conditions. However, this is not considered to be the significant aspect of the results with the immobilised Co-salen system. Rather this catalyst provides an interesting example of a system where immobilisation within the mesopores of Al-MCM-41 enhances the enantioselection when compared with the non-immobilised catalyst. For the Mn-salen complex and for the Cu-oxazoline complexes the enantioselection achieved with the immobilised catalysts tends to be lower than the corresponding homogeneous complex. Recently, a similar, and much more pronounced, effect has been observed by Johnson *et al.*[19] for 1,1'-bis(diphenylphosphino)ferrocene anchored to the inner walls of MCM-41 and coordinated to Pd^{2+} which is far superior to the homogeneous analogue for the enantioselective amination of cinnamyl acetate. Hence it is concluded that immobilisation within inorganic materials such as zeolites and MCM-41 can provide a route for the design of improved enantioselective catalysts.

5 CONCLUSIONS

In this paper we have described a design approach for heterogeneous enantioselective catalysts. The approach is based upon modification of the counter-cation of zeolites or mesoporous alumino-silicates with a suitable chiral ligand.[7] We have demonstrated the approach with two examples: (a) enantioselective aziridination of alkenes using Cu^{2+}-exchanged zeolite Y modified with chiral oxazolines and (b) the modification of manganese- and cobalt-exchanged Al-MCM-41 by a chiral salen ligand for the enantioselective epoxidation of alkenes. Since there is a broad range of zeolites and mesoporous materials available as catalytic materials, it is anticipated that the approach described in this paper could form the basis of a generic design of new enantioselective catalysts.

Acknowledgements

We thank Synetix, Robinson Brothers, the DTI/LINK Programme and EPSRC for financial support.

References

1. M.E. Davies, *Microporous Mesoporous Mater.*, 1998, **21**, 21.
2. G.M. Schwab and L. Rudolph, *Naturwiss.*, 1932, **20**, 362.
3. M.E. Davis and R.L. Lobo, *Chem. Mater.*, 1992, **4**, 756.
4. A. Corma, M. Iglesis, C. del Pino and F. Sanchez, *Stud. Surf. Sci. Catal.*, 1993, **75**, 2293.
5. D.C. Sherrington, *Catal. Today*, 2000, **57**, 87.
6. G. Webb and P.B. Wells, *Catal. Today*, 1992, **12**, 319.
7. G.J. Hutchings, *Chem. Commun.*, 1999, 301.
8. S. Feast, D. Bethell, P.C.B. Page, M.R.H. Siddiqui, D.J. Willock, F. King, C.H. Rochester and G.J. Hutchings, *J.Chem.Soc., Chem. Comm.*, 1995, 2409.
9. S. Feast, M.R.H. Siddiqui, R.P.K. Wells, D.J. Willock, F. King, C.H. Rochester, D. Bethell, P.C.B. Page and G.J. Hutchings, *J. Catal.*, 1997, **167**, 533.
10. C. Langham, P.Piaggio, D. Bethell, D.F. Lee, P. McMorn, P.C.B. Page, D.J. Willock, C. Sly, F.E. Hancock, F. King and G.J. Hutchings, *Chem. Commun.*, 1998, 1601.
11. C. Langham, S. Taylor, D. Bethell, P. McMorn, P.C.B. Page, D.J. Willock, C. Sly, F.E. Hancock, F. King and G.J. Hutchings, *J. Chem. Soc., Perkin Trans. 2.*, 1999,1043.
12. P.Piaggio, P. McMorn, C. Langham, D. Bethell, P.C.B. Page, F.E. Hancock, and G.J. Hutchings, *New J. Chem.*, 1998, 1167.
13. P.Piaggio, C. Langham, P. McMorn, D. Bethell, P.C.B. Page, F.E. Hancock, C. Sly and G.J. Hutchings, *J. Chem. Soc., Perkin Trans. 2.*, 2000,143.
14. J.S. Beck, J.C. Vartuli, W.J. Roth, M.E. Leonowicz, C.T. Kresge, K.D. Schmitt, C.T.-W. Chu, D.H. Olson, E.W. Shepherd, S.B. McCullen, J.B. Higgins and J.L. Schlenker, *J. Am. Chem. Soc.*, 1992, **114**, 10834.
15. L. Canali and D.C. Sherrington, *Chem. Soc. Rev.*, 1999, **28**, 85.
16. D.A. Annis and E.N. Jacobsen, *J. Am. Chem. Soc.*, 1999, **121**, 4147.
17. P. Piaggio, PhD Thesis, Cardiff University, 1999.
18. D.J. Berrisford, C. Bolm and B. Sharpless, *Angew. Chem. Int. Ed. Engl.*, 1995, **34**, 1059.
19. B.F.G. Johnson, S.A. Raynor, D.S. Shephard, T. Maschmeyer, J.M. Thomas, G. Sankar, S. Bromley, R. Oldroyd, L. Gladden and M.D. Mantle, *Chem. Commun.*, 1999, 1167.

ENANTIOSELECTIVE ALKYLATION OF BENZALDEHYDE BY DIETHYLZINC WITH (–)-EPHEDRINE SUPPORTED ON MTS. A NEW CLASS OF MORE EFFICIENT CATALYSTS

S. Abramson, M. Laspéras and D. Brunel

Laboratoire des matériaux catalytiques et catalyse en chimie organique.
U.M.R. 5618. CNRS – ENSCM, 8 rue de l'école Normale,
34 296 Montpellier Cedex 5

1 INTRODUCTION

In the field of C-C bond formation, enantioselective addition of dialkylzincs to aldehydes is one of the most promising way in order to obtain enantiomerically pure alcohols. Since the precursor work by Ogumi and Omi in 1984,[1] numerous chiral auxiliaries were used as catalysts in this reaction.[2] Among them, β-aminoalcohols led to high enantiomeric excesses. Chiral auxiliaries anchored on a support would be preferable owing to a variety of practical advantages such as recovery and reuse of the catalyst. However, the efficiency of the heterogenised chiral auxiliaries is different compared with their homogeneous counterparts.

Immobilisation of the chiral auxiliary on a polymeric support leads to ee's comparable to that obtained in homogeneous conditions.[3-5] However, rates are lower and limitations reliable to the mechanical stability and stirring difficulties are described.[6] Inorganic oxides may be preferred as supports because of their rigid structure and their physical stability.[7-10]

In our previous studies, we supported a chiral aminoalcohol, (–)-ephedrine, on Mesoporous Templated Silicas (MTS).[8-10] The enantioselective alkylation of benzaldehyde by diethylzinc (Model reaction) was performed using catalytic amounts of the supported aminoalcohol. Our first results[8] were in good agreement with precedent results by Soai and al. using silica gel and alumina supported (–)-éphédrine.[7] Lower rates, selectivities and enantioselectivities were obtained compared with homogeneous catalysis.

Investigation involving changes with the pore size,[9] the composition of the support,[10] end-capping of the surface[9] and the dilution of the catalytic sites[9-10] failed to improve ee's, significantly. However, we demonstrated that ee is limited by the racemic 1-phenylpropan-1-ol formation catalysed by the unfunctionalised mineral surface.[9] Thus, our aim was to decrease the negative effect of the uncovered mineral surface by increasing the coverage by organics on the inner surface of the pores.

In this work, the synthesis of high surface densities of chlororopropyl groups covalently grafted on mesoporous micelle templated aluminosilicates (Al-MTS) of various initial pore diameters is presented. The hybrid chiral materials resulting from halogen substitution are applied in the enantioselective addition of diethylzinc to benzaldehyde.

2 SYNTHESIS AND CHARACTERISATION OF THE HYBRID MATERIALS

2.1 The Supports

Two mesoporous aluminosilicates of the same composition (Si/Al=27), Al-MTS **1** and Al-MTS **2** with pore diameters of 3.6 nm (S=833 m^2/g, V= 0.76 ml/g) and 8.3 nm (S=822 m^2/g, V= 1.71 ml/g), respectively were used as supports. Al-MTS **1** is synthesised according to the litterature procedure[11] by hydrothermal condensation of silicates and aluminates around micellar rodes, folllowed by calcination at 550°C for 12h under a flow of synthetic air. Synthesis of Al-MTS **2** is close[12] excepted that the micellar rodes are « swelled » with trimethylbenzene (TMB). (Scheme 1)

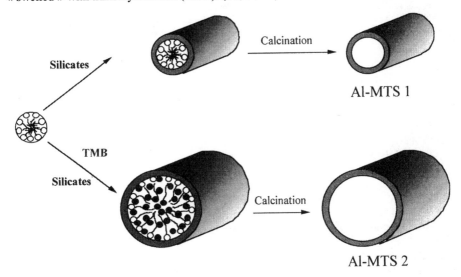

Scheme 1 *Synthesis of the supports*

2.2 Synthesis of the Hybrid Materials

Up to now, covalent anchorages have been performed in anhydrous conditions by silylation (method **a**) with 3-choropropyltrimethoxysilane (CPTMS) (Al-MTS-Cl **1a**) followed by the nucleophilic substitution of the chlorine by the amino group of (-)-ephedrine (Al-MTS-Cl-E **1a**). Alternatively, the surface coating of the MTS surface with CPTMS was achevied by a sol-gel method[13-14] (method **b**, Al-MTS-Cl **1b**, Al-MTS-Cl **2b**) and reaction with (–)-ephedrine led to new hybride chiral materials (Al-MTS-Cl-E **1b**, Al-MTS-Cl-E **2b**) (Scheme 2).

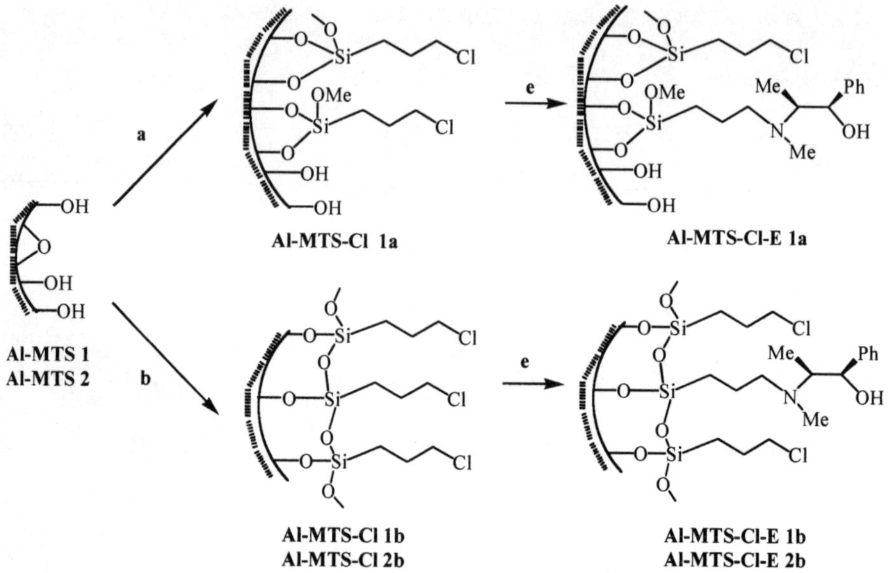

a : CPTMS - Toluene - anhydrous conditions -120°C, 4h
b : CPTMS -Toluene - H_2O - NH_4F / pTsOH - 60°C, 6h
e : *(1R,2S)*-(–)-ephedrine - Xylene - 140°C, 6h

Scheme 2 *Synthesis of the hybrid materials*

2.3 Characterisation of the Hybrid Materials

2.3.1 Loading of the Solids by Elemental and Thermogravimetric Analyses : Loading of grafted chloropropylsiloxanes moieties (before modification by ephedrine $N_{Cl\ i}$, after modification by ephedrine $N_{Cl\ R}$) or ephedrine moieties (N_N) were calculated from elemental and thermogravimetric analyses (see Table 1). Loadings were calculated relatively to the weight of dry initial silica content. They depend mainly on the grafting method used, the method **b** leading to notably higher amounts of chloroalkylsiloxanes moities on the surface, before or after substitition by (–)-ephedrine.

Table 1 *Loading of the hybrid materials*

	$N_{Cl\ i}$ (mmol/g)	N_N (mmol/g)	$N_{Cl\ R}$ (mmol /g)	Modification ratio (%)
Al-MTS-Cl-E 1a	2.1	1.3	0.8	62
Al-MTS-Cl-E 1b	4.4	1.9	2.5	43
Al-MTS-Cl-E 2b	4.3	1.9	2.4	45

2.3.2 Textural Properties of the Materials :   Textural properties were determined by nitrogen volumetry at 77°K on a micromeretics ASAP 2000 apparatus. Figures 1 and 2 show nitrogen adsorption-desorption isotherms of the hybrid solids obtained from the Al-MTS **1** and Al-MTS **2** supports, respectively.

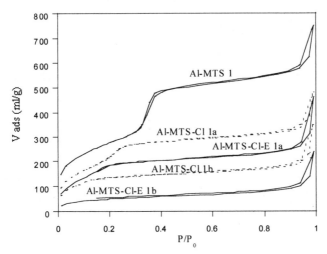

Figure 1 *N_2 adsorption-desorption isotherms of the hybrid materials from the Al-MTS 1 support*

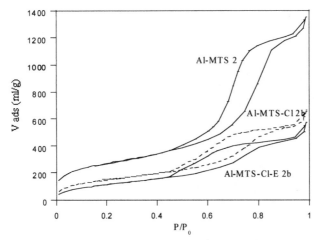

Figure 2 *N_2 adsorption-desorption isotherms of the hybrid materials from the Al-MTS 2 support*

As shown previously,[8-10] the solid prepared according to method **a** shows preservation of the mesoporosity after grafting of the halogens (Al-MTS-Cl **1a**, Figure 1), and substitution by (–)-ephedrine (Al-MTS-Cl-E **1a**, Figure 1). However, a decrease in the mesoporous volume is observed. In the case of the method **b**, the solids Al-MTS-Cl **1b** and Al-MTS-Cl-E **1b** feature microporosity, as shown by the patterns of the nitrogen sorption isotherms (Figure 1). However, increasing the initial pore diameter allows high residual mesoporous volumes to be obtained even after immobilisation of a high density of organic functions by method **b** (Al-MTS-Cl **2b**, Al-MTS-Cl-E **2b**, Figure 2).

3 CATALYTIC ACTIVITY IN THE MODEL REACTION

Addition of diethylzinc to benzaldehyde conducted in the presence of the materials leads to the formation of (R)- or (S)-1-phenyl-propan-1-ol (Scheme 3). All experiments were performed at 0°C using the same weight of solid (0.29g) and 2.3 equivalent of diethylzinc to benzaldehyde.[8-10]

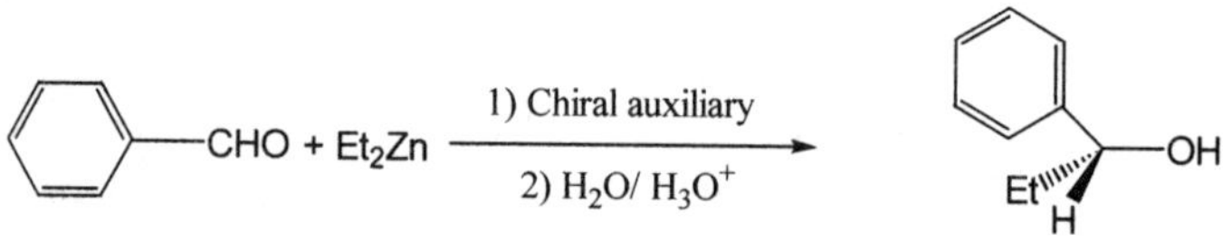

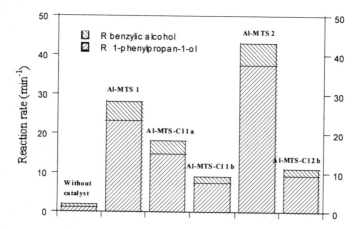

Scheme 3 *Model reaction*

3.1 Mineral Surface Activity

Unfunctionalised and chlorohybrid materials were used in the model reaction in order to determine the activity of the uncovered mineral surface in the formation of racemic 1-phenyl-propan-1-ol. Results are shown in Figure 3 .

Figure 3 *Mineral surface activity*

The inorganic surface of the support catalyses the racemic alkyl transfer. Coverage of the surface by chloropropyl moieties leads to a decrease in the activity of the mineral surface, whatever the support. With the Al-MTS **1** support, mineral surface activity of the hybrid solid synthetised by method **b** (Al-MTS-Cl **1b**), is half the value of that of the hybrid solid synthetised by method **a** (Al-MTS-Cl **1a**). With the Al-MTS **2** support, mineral surface activity is greatly reduced after grafting of CPTMS by method **b** (Al-MTS-Cl **2b**). Thus, the higher the loading of chloropropyl moities is, the higher the coverage of the mineral surface and the lower the activity in the racemic alkyl transfer.

3.2 Catalytic Activity of the Hybrid Chiral Materials

Materials functionalised by (–)-ephedrine were used as chiral auxiliray in the model reaction. Results are shown in Table 2.

All the chiral materials are efficient in the enantioselective transfer of an ethyl group to benzaldehyde. Their efficiency depend on the synthetic method and on the initial pore diameter of the support.

On the Al-MTS **1** support, the activity (kinetic constant of the reaction k_{obs}, is calculated by $k_{obs} = (1/t) \times \ln ([BA]_0 /[BA])$) of the high loading solid (Al-MTS-Cl-E **1b**) is lower than that of the low loading solid (Al-MTS-Cl-E **1a**). Taking into account of the low residual volume of the high loading Al-MTS-Cl-E **1b** (see Figure 1), we believe that accesssibilty to the catalytic sites is limited by diffusionnal constraints and that the reaction mainly proceeds on the external surface and at the pore opening. An increase in the mean pore diameter of the support allows the synthesis of high loading hybrids with conservation of available mesoporous volumes (see Figure 2). Hence, Al-MTS-Cl-E **2b** shows the highest activity obtained up to now. It can be assumed that the catalytic sites located on the inner surface of the pores are active in the reaction.

Enantioselectivities obtained with solids synthetised by method **b** (Al-MTS-Cl-E **1b**, Al-MTS-Cl-E **2b**) are higher than that obtained with solid synthetised by method **a** (Al-MTS-Cl-E **1a**). This result is reliable to the lower reaction rate of the competitive reaction of forming racemic 1-phenyl-propan-1-ol catalysed by the uncovered mineral surface (see Figure 3). The best enantioselctivity is obtained with the Al-MTS-Cl-E **2b** material. This good result may be explained by the the good coverage of the mineral surface concomitant with the good accessibility to the catalytic sites.

Table 2 *Catalytic activity of the hybrid chiral materials*

Catalyst	k_{obs} (h^{-1})	Sel.* (%)	e.e.** (%)	N éphédrine (mmol %)
Al-MTS-Cl-E 1a	0.17	93	47	13.6
Al-MTS-Cl-E 1b	0.11	94	54	15.4
Al-MTS-Cl-E 2b	0.55	98	64	16.8
Homogeneous :				
(–)-ephedrine	0.53	98	62	8.5
(–)N-propyl-ephedrine	0.68	98	76	8.5

* = Selectivity in 1-phenylpropan-1-ol (secondary product : benzylic alcohol)
**= ee in (R) enantiomer

Thus, the best result is obtained with (–)-ephedrine anchored on the Al-MTS **2** support (Al-MTS-Cl-E **2b**) and it is worth noting that ee and activity are close to those obtained in homogenous conditions ((–)-ephedrine , (–)-N-propyl-ephedrine). Moreover, this catalyst can be reused two times without loss of enantioselectivity.

4 CONCLUSION

We performed the synthesis of a new class of hybrid solids by grafting the coupling CPTMS agent by the sol-gel method rather than the silylation method. Supporting (-)-ephedrine on Al-MTS support with large pores by this new grafting method leads to a chiral hybrid material with a high efficiency in the enantioselective alkylation of benzaldehyde by diethylzinc.[15]

In summary, the efficiency of (–)-ephedrine supported on inorganic MTS in the model reaction is determined by two factors: firstly, the activity of the mineral surface towards the formation of racemic 1-phenyl-propan-1-ol wich depends on the coverage of the surface by organics and secondly, the accessibility to the catalytic sites wich depends on the mesopore diameters of the support.

Work is in progress in order to understand the mechanism of the mineral surface activity and to improve the enantioselective efficiency of the hybrid chiral materials.

References

1   N. Ogumi, T. Omi, *Tetrahedron Lett.*, 1984, **25**, 2823.
2   For a review read : K. Soai, S. Niwa, *Chem. Review*, 1992, **92**, 833.
3   S. Itsuno, J.M.J. Frechet, *J. Org. Chem.*, 1987, **52**, 4140.
4   K. Soai, S. Niwa, M. Watanabe, *J. Org. Chem.*,. 1988, **53**, 927.
5   P. Hodges, D. W. L. Sung, P. W. Stratford, *J. Chem. Soc. Perkin Trans. I*, 1999, 2335.
6   S. Itsuno, Y. Sakurai, K. Ito, T. Marumaya, S. Nakahama, J.M.J. Frechet, *J. Org. Chem.*, 1990, **55**, 304.
7   K. Soai, M. Watanabe, A. Yamamoto, *J. Org. Chem.*. 1990, **55**, 4832.
8   M. Laspéras, N. Bellocq, D. Brunel and P. Moreau, *Tetrahedron : Asymmetry*, 1998, **9**, 3053.
9   N. Bellocq, S. Abramson, M. Laspéras, D. Brunel and P. Moreau, *Tetrahedron: Asymmetry*,1999,**10**, 3229.
10   S. Abramson, N. Bellocq, M. Laspéras, *Topics in Catal.*, in the press.
11   F. Di Renzo, B. Chiche, F. Fajula, S. Viale, E. Garrone, *Stud. Surf. Sci. Catal.*, 1996, **101**, 851.
12   F. Di Renzo, A. Galarneau, D. Desplantier-Giscard, L. Mastrantuono, F. Testa, F. Fajula, *Sci. Technol.*, 1999, **81**, 587.
13   W. M. Van Rhijn, D. E. De Vos, B. F. Sels, W. D. Bossaert, P. Jacobs, *Chem. Commun.*, 1998, 317.
14   A. Galarneau, T. Martin, V. Izard, V. Huela, A. Blanc, S. Abramson, F. Di Renzo, F. Fajula and D. Brunel, in preparation.
15   S. Abramson, M. Laspéras, A. Galarneau, D. Desplantier-Giscard, D. Brunel, *Chem. Commun.*, 2000, 1773.

SUPPORTED PERFLUOROALKANEDISULPHONIC ACIDS AS CATALYSTS IN ISOBUTANE ALKYLATION

A. de Angelis *, P. Ingallina, W.O. Parker, Jr., M. G. Clerici and C. Perego

EniTecnologie, via Maritano 26, I-20097 San Donato Milanese, Italy
E-mail: adeangelis@enitecnologie.eni.it

ABSTRACT

Solid acids catalysts based on perfluoroalkanedisulphonic acid supported on silica (PFAS-SiO$_2$) catalyse the alkylation of isobutane with n-butene to yield high-octane gasoline components. PFAS-SiO$_2$ activity is highly dependent on the PFAS employed, on the preparation method and on the reaction conditions adopted. ^{1}H and ^{29}Si MAS NMR spectroscopy were used to evidence the changes in surface silanols caused by impregnating silica with perfluoroethanedisulphonic acid.

1. INTRODUCTION

The alkylation of isobutane with n-butenes to yield saturated-branched octanes is a reaction of great interest in the refining industry. The products have high octane number and are commonly used for gasoline blending. Current production is based on strong mineral acids, i.e. sulphuric and fluoridric acids, used in large amounts. For safety reasons it is highly desirable to substitute these acids with non-corrosive and non-toxic solid catalysts. A number of solid acids have been investigated such as Y zeolite [1], sulphated zirconia [2,3], SbF$_5$ [4], ß zeolite [5], MCM-22 [6] and supported trifluoromethanesulphonic acid [7, 8, 9]. After a high initial activity, all these catalysts deactivate quite rapidly unless they are used in a special reactor [4, 7].

Perfluoroalkanedisulphonic acids (PFAS) are solid and possess strong acid properties both in solid state and in solution. To our knowledge, they have never been used in the alkylation of isobutane. They were obtained as dihydrate and as such were not acidic enough to be active in isobutane alkylation. Furthermore, they possessed low surface areas. The surface area can be increased by supporting PFAS on an amorphous solid, but it is critical that the solid does not attenuate the acid strength. We have found that a method for dehydrating PFAS and for supporting it on silica. The method allows to obtain an new catalyst, PFAS-SiO$_2$, which is active in the alkylation of isobutane.

In this work, the preparation and characterisation of PFAS-SiO$_2$ are described along with the catalytic results for alkylating isobutane.

2. EXPERIMENTAL

2.1 Materials

SO_3, $SOCl_2$, trifluoromethanesulphonic (triflic) anhydride, were used as received. SiO_2 (60 F_{254}), Fluka, was heated at 400°C for 24h under airflow.

2.1.1 Catalyst Preparation. 12 meq of PFAS (chosen between perfluoromethanedisulphonic acid (PFMS), perfluoroethanedisulphonic acid (PFES) and perfluoropropanedisulphonic acids (PFPS)) dihydrate was dissolved in 100 ml of dehydrating agent (chosen between $SOCl_2$, SO_3, $SOCl_2$ + 20% trifluoromethanesulphonic anhydride, $SOCl_2$ + 5%SO_3) under dry nitrogen atmosphere in a glove box. 10 g of silica was added and the suspension was stirred overnight at room temperature. The solvent was then distilled off under vacuum, and the catalyst stored under dry nitrogen.

2.2 Reaction Apparatus

The catalysts were tested in a fixed bed reactor (diameter 0.76-cm, length 26 cm) at the temperature of 25°C and pressure of 20 bars. In each run, 10 ml (ca.11 g) of catalyst was used. The reaction mixture isobutane /1-butene (10/1 ratio) was fed by HPLC pump (mode 1).

In alternated feeding mode, the reagents, pure isobutane and a mixture of isobutane /1-butene (10/1 ratio), were fed alternatively by two HPLC pumps: pure isobutane was fed for twenty seconds and then stopped while the feeding of the mixture of isobutane/1-butene has begun; after ten seconds the feeding of the mixture stopped and the feeding of pure isobutane started again (mode 2).

Under the reaction conditions the feeding was in the liquid phase. WHSV referred to 1-butene was 2.8 h^{-1}.

Reaction samples were taken by cooling the effluent at -80°C and analysed on a poly(dimethylsiloxane) capillary column (SPB-1, Supelchem) and a Hewlett Packard 5890 gas chromatograph.

2.3 Conversion and Selectivity Definition

Conversion and selectivity are defined in the following way:

Conversion: $\dfrac{\text{moles of butene reacted}}{\text{initial moles of butenes}}$

Selectivity: $\dfrac{\text{moles of saturated octanes}}{\text{moles of butene reacted}}$

TMPs selectivity: $\dfrac{\text{moles of trimethylpentanes}}{\text{moles of saturated octanes}}$

2.4 NMR Spectroscopy

The purity (97%) of PFES was determined by solution state ^1H (400 MHz) NMR spectra of its solution in dimethylsulfoxide-d_6 using a Varian VXR-400 spectrometer.

Solid state spectra were obtained using powdered samples (ca. 200 mg), contained in 7mm zirconia rotors undergoing magic angle spinning (MAS) at 5 kHz, and a Bruker ASX-300 spectrometer. The rotors were loaded under nitrogen atmosphere in a glove box. ^1H spectra were collected at 300 MHz, using a 70° rf pulse (4 μs), 5 s relaxation delay, 60 kHz spectral window and 88 scans. These conditions gave quantitative spectra since the longest longitudinal relaxation time (T_1, determined by inversion recovery method) was 3.5 s. ^{29}Si spectra were collected at 59.59 MHz, using high power ^1H decoupling, a 70° rf pulse (5 μs), 5 s relaxation delay, 11 kHz spectral window and 10,000 scans. These conditions did not give quantitative spectra (see results section). Chemical shifts were referenced externally to tetrakis (trimethylsilyl) silane (at -9.8 and -135.2 ppm) for ^{29}Si and to water (at 4.8 ppm) for ^1H.

The amount of silanol groups on the dehydrated silica surface is calculated, from the surface area (350 m^2/g) using the Kiselev-Zhuravlev constant (4.6 OH/nm^2) [10, 11], to be ca. 1.6×10^{21} OH/g.

3. RESULTS AND DISCUSSION

3.1 Catalyst Synthesis and Reactivity

PFAS were obtained with 2 moles of water, for each mole of acid and they could not be dehydrated with physical methods. Hydrated acids, both as such and supported on silica using water as solvent, were not active in isobutane alkylation. Therefore the effect of different dehydrating solvent was studied, in order to remove residual water. The catalysts obtained by supporting perfluoroethanedisulphonic acid on SiO$_2$ (PFES-SiO$_2$) after dissolution in various dehydrating solvents were tested in the reaction and resulted active with high butene conversion (Table 1).

Table 1 *Influence of Dehydrating Solvent*[1]

Dehydrating Solvent	*Conversion*[2]	*Selectivity*[2]	*TMP's selectivity*[2]
	%	%	%
SOCl$_2$	98	69	73
SO$_3$	75	18	66
SOCl$_2$+ SO$_3$	95	8	15
SOCl$_2$+ T. A.	88	43	75

1): Feeding according to mode 1;
2): Samples taken at 2 h t.o.s.

Thionyl chloride was the most adequate solvent yielding a catalyst which gave both the highest butene conversion (98%) and the largest amount of saturated octane in the products (69%). Trimethylpentanes content in octane fraction is fairly elevated (75 %). The catalysts obtained using sulphuric anhydride or mixture of thionyl chloride and sulphuric anhydride as solvent, gave very poor selectivity to saturated octanes. This is due to the high cracking activity of the catalyst: hence the octane formed are cracked to lower molecular weight products. The addition of triflic anhydride to thionyl chloride furthermore decreases the octanes selectivity, likewise increasing cracking activity of the catalyst.

In order to evaluate whether the chain length of perfluoroalkyl group can influence the catalytic properties, perfluoromethanedisulphonic acid (PFMS), perfluoroethanedisulphonic acid (PFES) and perfluoropropanedisulphonic acids (PFPS), previously dissolved in thionyl chloride, were supported on silica (Table 2).

Table 2 *Influence of Chain Length[1]*

Catalyst	Conversion[2]	Selectivity[2]	TMP's selectivity[2]
	%	%	%
PFMS-SiO$_2$	71	58	61
PFES- SiO$_2$	98	69	73
PFPS-SiO$_2$	97	80	69

1): Feeding according to mode 1;
2): Samples taken at 2 h t.o.s.

PFMS-SiO$_2$ was the least active catalyst, while PFES-SiO$_2$ and PFPS-SiO$_2$ gave almost quantitative butene conversion. Both these two acids gave a good content of trimethylpentanes in saturated octane (around 70%), the balance being predominantly dimethylhexanes. No methylheptanes were detected. Cracking activity of the catalyst was not negligible, in fact octane selectivity was between 69% (perfluoroethane) and 80% (perfluoropropane).

In order to reduce cracking activity, reaction temperature was lowered from 25°C to −25°C, using as catalyst PFES-SiO$_2$ obtained by treatment with thionyl chloride (Table 3).

Table 3 *Influence of Reaction Temperature (Perfluoethanedisulphonic Acid)[1]*

Temperature	C5-C7 Selectivity[2]	Selectivity[2]	C9$^+$ Selectivity[2]
°C	%	%	%
25	21	69	10
-25	4	94	2

1): Feeding according to mode 1;
2): Samples taken at 2 h t.o.s.

Lowering reaction temperature, the selectivity to saturate octanes was highly increased reaching the value of 94%, while butene conversion remains complete. Selectivity to trimethylpentanes reaches very high values (always >85%).

The catalytic test using PFES-SiO$_2$ at −25°C was performed until the yield to saturated octanes felt to less than 40% (Table 4).

Table 4 *Catalyst Life (PFES-SiO$_2$)[1]*

Time on Stream	Saturate Octane Yield	TMP's selectivity
h	%	%
1	65	87
2	94	95
3	85	87
4	68	90
5	39	90
6	34	86

1): Feeding according to mode 1.

It is evident that while trimethylpenthanes fraction in saturated octanes is very high for all the catalyst life (always > 85%), saturated octanes yield reaches a maximum at 2 hours of time on stream (94%). As the time on stream increases, catalyst deactivation takes place with a decrease of conversion and selectivity. After 6 hours the yield falls to 34%.

As catalyst life does not exceed 7 hours (9 hours for PFPS-SiO$_2$), catalyst productivity, calculated as grams of saturated octanes per gram of acid, is poor reaching the value of 1.7 (3.6 for PFPS-SiO$_2$). In order to improve catalyst productivity, alternate feed of isobutane and reaction mixture (isobutane/1-butene: 10/1) was performed. The aim of this feeding mode is to facilitate the possible removal of unsaturated compounds which may be responsible of deactivation products formation (Table5).

Table 5 *Influence of Alternated Feed*

Catalyst	Feeding	Catalyst Life h	Productivity g$_{C8}$/g$_{cat.}$
PFES-SiO$_2$	mode 1	7	1.7
PFPS-SiO$_2$	mode 1	9	3.6
PFES-SiO$_2$	mode 2	25	12
PFPS-SiO$_2$	mode 2	126	87

Catalyst productivity is increased greatly using alternate feed: 7 times more for PFES-SiO$_2$ and 24 times more for PFPS-SiO$_2$. The productivity obtained with this method, 87 g$_{octane}$/g$_{catalyst}$ for PFPS-SiO$_2$, is far beyond the influence of isobutane dilution.

In any event, the productivity obtained with PFPS-SiO$_2$ is comparable to those obtained by other supported catalysts for isobutane alkylation (18.6 g$_{octane}$/g$_{catalyst}$ in the case of CF$_3$SO$_3$H-SiO$_2$ [8], 102.4 g$_{octane}$/g$_{catalyst}$ in the case of BF$_3$-Al$_2$O$_3$ [12]).

3.2 Characterisation

The interaction between the PFES and silica was investigated by NMR spectroscopy. ^{1}H and ^{29}Si MAS NMR spectra of SiO$_2$ and PFES-SiO$_2$ were made under the same

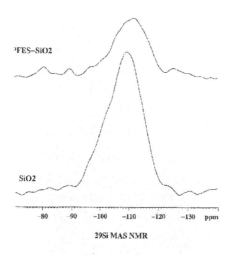

Figure 1 *^{29}Si NMR spectra of SiO$_2$ and PFES-SiO$_2$*

experimental conditions to allow a quantitative comparison between the two samples.

[29]Si MAS spectra are plotted with the same vertical scale in Figure 1. The spectrum of silica shows a broad composite signal centered near -110 ppm with a poor signal to noise ratio. Spectra collected with much longer recycle times (e.g. 60 s) indicated a very lengthy longitudinal relaxation time for the [29]Si nuclei. The only efficient source of relaxation is via the silanol protons, since molecular oxygen was excluded during the sample preparation. Thus, the signals in spectra Figure 1 arise from nuclei that are incompletely relaxed (i.e. the recycle time was not long enough to allow complete relaxation between scans). Silicon nuclei bearing one hydroxyl (silanol) give a signal near -100 ppm, which is an apparent shoulder to the main signal. Geminal silanols (-90 ppm) do not survive preparation treatment (heating at 400°C). After impregnation the signal area decreases and this is attributed to the loss of silanol protons which reduces the relaxation rate of all Si nuclei even further.

^{1}H MAS NMR gives more detailed information on the surface silanols. As seen in Figure 2, a composite signal near 2 ppm arises from the silanol protons. No water molecules are present due to the drying treatment and exclusion of humidity during manipulations. After impregnation with PFES (10% by weight) the total ^{1}H signal area

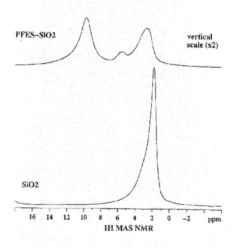

Figure 2 ^{1}H *NMR spectra of SiO$_2$ and PFES-SiO$_2$*

decreases to half its original value. This loss of signal area is not due to insufficient relaxation, as for [29]Si, but rather to a real loss of protons during de-hydroxylation reactions driven by SOCl$_2$. The PFES-SiO$_2$ sample gives 3 signals at 9.6, 5.5 and 2.7 ppm. The last signal is attributed to isolated silanols, which did not undergo the condensation reaction, as for vicinal silanols, caused by SOCl$_2$. The small signal at 5.5 ppm could be due to moderately acidic (or H-bonded) protons. The signal at 9.6 ppm is due to strongly acidic protons. This signal area corresponds to 0.2 equivalents of the silanol protons originally present on the silica. The amount of silanol groups on the dehydrated silica surface is calculated, from the surface area ($350 \ m^2/g$) using the Kiselev-Zhuravlev constant ($4.6 \ OH/nm^2$) [10], to be ca. $1.6 \ x \ 10^{21} \ OH/g$. So the amount of acidic protons of PFES-SiO_2 is approximately $3.2 \ x \ 10^{20} (0.2 \cdot 1.6 \ x \ 10^{21})$. The original proton content of PFES was $1.2 \ meq/g = 7.2 x 10^{20} \ H^+/g$, which is about twice the amount of H^+, measured through quantitative NMR. Therefore each supported PFES molecule has only one strong acid H^+, instead of the original two. This is consistent with a covalent interaction between PFES and SiO_2 since an ionic, H-bonded or physical interaction would leave each PFES with at least 2 H.

4. CONCLUSIONS

PFAS can be supported on silica using a dehydrating solvent, obtaining an active catalyst. Spectroscopic studies suggest that the interaction between PFAS and the support is covalent. PFAS-SiO_2 catalyses the alkylation of isobutane with n-butenes to yield a mixture of products, of which saturated octanes are the major fraction. Trimethylpentanes are the main constituent of this fraction. The best results were obtained using thionyl chloride as the dehydrating solvent.

All PFAS-SiO_2 are active; PFES-SiO_2 and PFPS-SiO_2 gave complete butene conversion with good (69-80%) selectivity to saturated octane. This selectivity is greatly increased (above 90%) by lowering the reaction temperature from 25°C to –25°C.

The selectivity to octane and trimethylpentanes is satisfactory, but the lifetime of this catalyst is too short for industrial usage. However, with alternated isobutane/reaction mixture feeding the catalyst lifetime and productivity are increased several times (7 times for PFES and 35 times for PFPS).

References

1. A.Corma, A. Martinez, and C. Martinez *J.Catal.* 1994, **146**, 185-192.
2. A.Corma, A. Martinez, and C. Martinez *J.Catal.* 1994, **149**, 52-60.
3. Liang, Chin-Huang, *Alkylation of isobutane with n-butenes over solid catalysts*, (U.M.I., Dissertation N 9411297, 1994).
4. C.S. Crossland, A.Johnson, E.G.Pitt, *US Patent 5,157,196,* 1992.
5. A.Corma, V.Gomez, A. Martinez, *Appl. Catal.,* 1994, **119**, 83-96.
6. A. Corma, A. Martinez and C. Martinez, *Catal. Lett.,* 1994, **28**, 187-201.
7. S.I. Hommeloft, H. F.A. Topsoe, *US Patent 5,245,100,* 1993.
8. M.G. Clerici, C. Perego, A.de Angelis, L.Montanari *US Patent 5,571,762,* 1993.
9. M.G.Clerici, A.de Angelis, P.Ingallina *IT Patent 1,272,926* , 1995.
10. H.E. Bergna in "*The Colloidal Chemistry of Silica*", Washington, DC, ACS, 1994, 28.;
11. L.T. Zhuravlev, React. *Kinet. Catal. Lett.,* 1993, **50**, 15-25.
12. M. Cooper, D. King, W. Sanderson, *WO9204977,* 1992.

POLYMER IMMOBILISED TEMPO (PIPO): AN EFFICIENT CATALYTIC SYSTEM FOR ENVIRONMENTALLY BENIGN OXIDATION OF ALCOHOLS

A. Dijksman, I. W. C. E. Arends and R. A. Sheldon

Laboratory for Organic Chemistry & Catalysis, Department of Biotechnology
Delft University of Technology
Julianalaan 136, 2628 BL Delft, The Netherlands
E-mail: secretariat-ock@stm.tudelft.nl

The mild and selective conversion of primary alcohols into aldehydes and secondary alcohols into ketones forms a key step in organic synthesis.[1] The use of stable nitroxyl radicals, such as TEMPO, as catalysts for the oxidation of alcohols to aldehydes, ketones and carboxylic acids is well documented.[2] Typically, these transformations employ 1 mol% of the nitroxyl radical and a stoichiometric amount of a terminal oxidant, e.g. sodium hypochlorite,[3] MCPBA (m-chloroperbenzoicacid),[4] sodium bromite,[5] sodium chlorite,[6] trichloroisocyanuric acid,[7] oxone[8] and oxygen in combination with CuCl[9] or RuCl$_2$(PPh$_3$)$_3$.[10] In particular, the TEMPO-bleach protocol using bromide as co-catalyst introduced by Anelli et al.[3] is finding wide application in organic synthesis (figure 1).

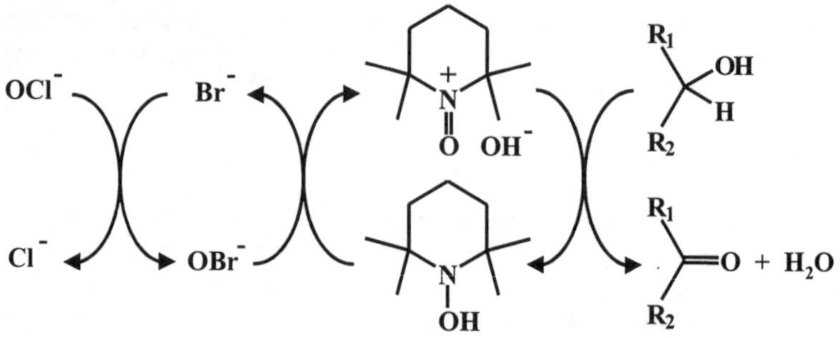

Figure 1 *TEMPO-catalysed bleach-oxidation of alcohols using bromide as co-catalyst*

Although only a small amount of catalyst is used, recyclability is an issue and several heterogeneous TEMPO systems have been reported.[11] For example, MCM-41[12] and silica-supported TEMPO[13,14] were applied in oxidation reactions using hypochlorite as the oxidant. The preparation of these catalysts involves initial functionalisation of the support followed by covalent attachment of a 4-substituted TEMPO via an amide, amine or ether linker (figure 2).

Here we report the use of a readily prepared polymer immobilised TEMPO as a catalyst for alcohol oxidations.[15] It was derived from a commercially available oligomeric, sterically hindered amine, poly[[6-[(1,1,3,3-tetramethylbutyl)amino]-1,3,5-triazine-2,4-diyl][2,2,6,6-teramethyl-4-piperidinyl)-imino]-1,6-hexane-diyl[(2,2,6,6-tetramethyl-4-piperidinylimino]], better known as Chimassorb 944 (MW ~ 3000; see figure 3 for structure). This compound is used as an antioxidant and a light stabiliser for plastics. It contributes significantly to the long-term heat stability of polyolefins and has broad approval for use in polyolefin food packaging.[16]

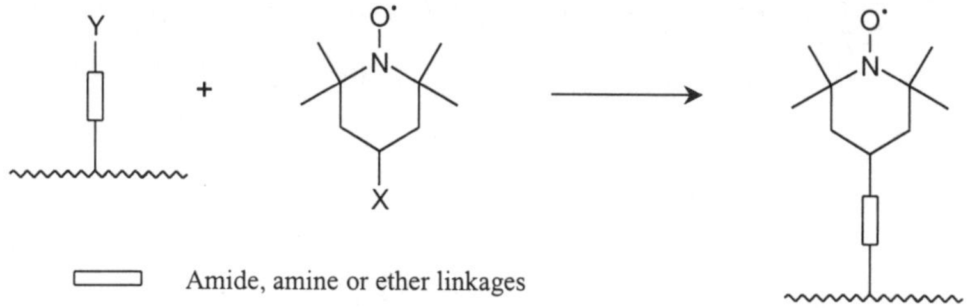

Figure 2 *Immobilisation of TEMPO on SiO₂ or MCM-41*

Nitroxyl radicals are normally prepared by treating the analogous secondary amine with hydrogen peroxide and a catalytic amount of $Na_2WO_4 \cdot 2H_2O$.[17] In the case of Chimassorb 944, the same procedure was applied resulting in the formation of an oligomeric TEMPO (figure 3). Probe-MS data revealed that the mass of each segment increased with 30 owing to transformation of two secondary amine moieties into the corresponding nitroxyl radicals. This new polymer immobilised TEMPO, further referred to as PIPO (Polyamine Immobilised Piperidinyl Oxyl), proved to be an effective catalyst for oxidations of alcohols with hypochlorite using the Anelli protocol.[3,18]

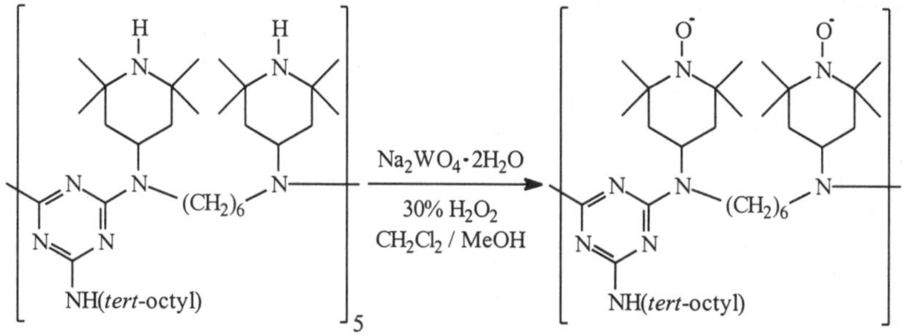

Figure 3 *Synthesis of PIPO*

Primary and secondary aliphatic and benzylic alcohols were smoothly converted to the corresponding aldehydes and ketones in CH_2Cl_2 (table 1). Under these conditions the system was homogeneous, as PIPO is soluble in dichloromethane. In contrast, in the absence of solvent (entry 3) PIPO was an active heterogeneous catalyst. The heterogeneous nature of the catalyst was confirmed in a filtration experiment,[19] in which the reaction mixture was filtered after 10 minutes The filtrate showed no activity at all during 1 hour after filtration. The residue, however, could be reused at least 2 times as catalyst (entries 4 and 5). The minor loss of activity (<5%) observed is probably due to mechanical losses occurring during filtration of the small amount of catalyst.

Further investigation revealed that the use of bromide was not necessary. Thus, in contrast to the conventional TEMPO-bleach oxidations, which use dichloromethane as solvent and bromide as a co-catalyst,[3] PIPO catalyses the oxidation of a variety of alcohols in the absence of organic solvent and using only a hypochlorite solution (0.35M, pH 9.1) as the oxidant (table 2). However, under these conditions primary aliphatic alcohols such as octan-1-ol, gave low selectivities to aldehydes owing to over-oxidation of octanal to octanoic acid (entry 1). This problem was circumvented by using MTBE as the organic solvent, in which PIPO is not soluble, affording an increase in selectivity to 94% (entry 2). Here again filtration experiments[19] confirmed that this system was heterogeneous analogous to the solvent-free conditions.

Table 1   *PIPO-catalysed oxidation of alcohols with bromide/hypochlorite* [a]

Entry	substrate	product	time (min)	conv. (%) [b]	sel. (%) [b]
1	octan-1-ol	octanal	20	>99	>99
2	octan-2-ol	octan-2-one	20	>99	>99
3 [c]			45	95	>99
4 [d]			45	92	>99
5 [e]			45	90	>99
6	benzylalcohol	benzaldehyde	20	>99	>99
7	1-phenylethanol	acetophenone	20	>99	>99

[a] 0.8 mmol substrate, 2.5 mg PIPO (1 mol% nitroxyl), 2 ml CH_2Cl_2, 0.16 ml 0.5M KBr-sol. (10 mol%), 0.14 g $KHCO_3$ (for pH 9.1), 2.86 ml 0.35M hypochlorite-sol. (1.25 equiv.), 0°C; [b] conversion and selectivity determined by GC using *n*-hexadecane as internal standard; [c] no CH_2Cl_2 (solvent-free); [d] recycling of catalyst in entry 3; [e] recycling of catalyst in entry 4.

In addition to primary and secondary aliphatic alcohols (entries 2-7), benzylic alcohols were also efficiently oxidised (entries 9 and 10), complete conversion being observed within 30 minutes. In competition experiments, the catalyst showed a marked preference for primary alcohols (entries 8 and 11). This is analogous to the already reported homogeneous[3] and heterogeneous[13] TEMPO systems. A stereogenic centre at the α-position is not affected during oxidation as shown by the selective oxidation of (S)-2-methylbutan-1-ol to (S)-2-methylbutanal (entry 12).[20]

For comparison, the previously described silica and MCM-41 supported TEMPO catalysts were also employed in the bleach-oxidation of octan-2-ol under the chlorinated hydrocarbon solvent- and bromide-free conditions. As shown in figure 4, PIPO is the most active catalyst under these conditions. All silica and MCM-41 supported TEMPO systems gave comparable conversions and are more active than homogeneous TEMPO. On the other hand, the activities obtained under these environmentally benign conditions were lower than in the case of dichloromethane/bromide.

Table 2   *PIPO-catalysed oxidation of alcohols with hypochlorite* [a]

Entry	Substrate	Product	Time (min)	Conv. (%) [b]	Sel. (%) [b]
1	octan-1-ol	octanal	45	90	50 [d]
2 [c]			45	80	94
3 [c]	hexan-1-ol	hexanal	45	89	95
4	octan-2-ol	octan-2-one	45	99	>99
5	hexan-2-ol	hexan-2-one	45	99	>99
6	octan-3-ol	octan-3-one	45	70	>99
7	cyclooctanol	cyclooctanone	45	100	>99
8 [c]	octan-1-ol/ octan-2-ol	octanal	45	86/<1	96
9	benzylalcohol	benzaldehyde	30	100	>99
10	1-phenylethanol	acetophenone	30	100	>99
11	benzylalcohol / 1-phenylethanol	benzaldehyde/ acetophenone	30	95/4	>99
12	(S)-2-methyl-butan-1-ol	(S)-2-methyl-butanal	45	90	>99

[a] 0.8 mmol substrate, 2.5 mg PIPO (1 mol% nitroxyl), 0.14 g $KHCO_3$ (for pH 9.1), 2.86 ml 0.35M hypochlorite-sol. (1.25 equiv.), 0°C; [b] conversion and selectivity determined by GC using *n*-hexadecane as internal standard; [c] 2 ml MTBE as solvent; [d] octanoic acid and octyloctanoate as side products.

For PIPO, the time for complete conversion increased from 20 to 45 minutes. The silica-supported TEMPO system reported by Bolm et al.[13] gave 74% conversion in 2 hours, whereas using the Anelli protocol (dichloromethane/bromide) this activity was already reached within 30 minutes. With homogeneous TEMPO, the differences were even more dramatic, *i.e.* complete conversion was reached within 10 minutes using the Anelli protocol,[3] whereas only 45% conversion was observed in 2 hours under the chlorinated hydrocarbon solvent- and bromide-free conditions.

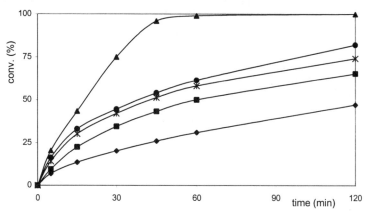

Figure 4 *Bleach oxidation of octan-2-ol using 1 mol% of nitroxyl catalyst:* (▲) *PIPO (3.19 mmol/g; amine linker)[15]*; (●) *MCM-41 TEMPO (0.60 mmol/g; ether linker)[12]*; (✱) *SiO₂ TEMPO (0.87 mmol/g, amine linker)[13]*; (■) *SiO₂ TEMPO (0.40 mmol/g, amide linker)[14]*; (♦) *TEMPO*

The activities of the various catalysts were also compared using *in situ* IR under diffusion limiting conditions,[21] by following the intensity of the carbonyl stretch-vibration of octan-2-one (1718 cm^{-1}) in time (figure 5). Under the diffusion limiting conditions, the same trends as above were observed. PIPO was the most active catalyst and all the other heterogeneous systems had comparable activities (table 3).

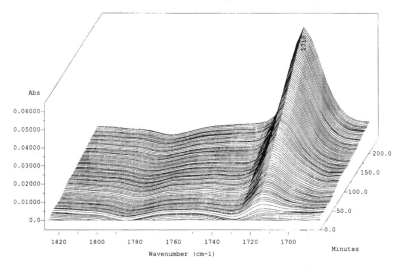

Figure 5 *Nitroxyl-catalysed bleach-oxidation of octan-2-ol followed by in situ IR*

Table 3   *Nitroxyl-catalysed bleach oxidation of octan-2-ol followed by in situ IR* [a]

Entry	nitroxyl	time (h) [b]	TOF$_o$ [c]
1	PIPO	5	4.0
2	TEMPO	7.5	3.4
3	MCM-41 TEMPO [12]	7	3.5
4	SiO$_2$-TEMPO [13]	6	3.7
5	SiO$_2$-TEMPO [14]	6.5	3.7

[a] 12.6 mmol octan-2-ol, 0.03 mol% nitroxyl, 11.4 ml 0.35M hypochlorite-sol. (4 mmol), 25 °C, 500 rpm; [b] for 30% conversion of substrate (100% conversion of hypochlorite); [c] TOF$_o$ = initial turn over frequency (mmol of product per mmol of catalyst per minute).

Carbohydrates are also oxidised by the PIPO/NaOCl system analogous to TEMPO/NaOCl.[22] For example, methyl α-D-glucopyranoside afforded methyl α-D-glucopyranosiduronate in 70% yield (figure 6).[23] As in the case of the oxidation of simple alcohols, filtration experiments confirmed that the catalyst is truly heterogeneous. Further investigation in the field of carbohydrate oxidations using PIPO/NaOCl is in progress and will be reported on in due course.

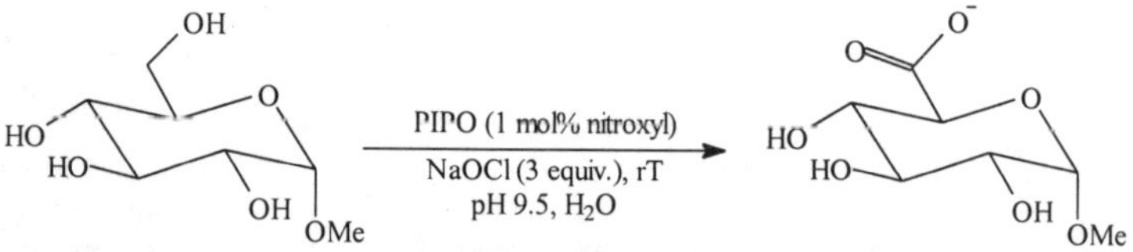

Figure 6   *Oxidation of methyl α-D-glucopyranoside using PIPO/NaOCl*

Besides hypochlorite, oxygen can also be used as the oxidant.[9,10] Unfortunately, in contrast to homogeneous TEMPO the combination of PIPO and RuCl$_2$(PPh$_3$)$_3$ in chlorobenzene[10] is not able to catalyse the aerobic oxidation of octan-2-ol, probably owing to coordination of ruthenium to the polyamine. On the other hand, in combination with CuCl in DMF,[9] it catalyses the aerobic oxidation of benzylalcohol to benzaldehyde within 1.5 hours (entry 2; table 4).[24] The activity of PIPO is comparable to that of TEMPO (entries 4 and 5) and is superior to that of the previously described heterogeneous TEMPO systems (entries 6,7 and 8). CuCl/PIPO also catalyses the aerobic oxidation of benzylalcohol under solvent-free conditions (entry 3).

Table 4   *CuCl/nitroxyl-catalysed aerobic oxidation of benzylalcohol to benzaldehyde* [a]

Entry	nitroxyl	S/C ratio	time (h)	conv. (%) [b]	TOF [c]
1	PIPO	100	20	45	2.3
2		10	1.5	96	6.4
3 [d]		300	20	12	1.8
4	TEMPO	100	24	97	4
5		10	1	96	9.6
6	MCM-41 TEMPO [12]	100	24	20	0.9
7	SiO$_2$-TEMPO [13]	100	20	20	1.0
8	SiO$_2$-TEMPO [14]	100	20	4	0.2

[a] 10 mmol benzyl alcohol, CuCl/nitroxyl = 1/1 (mol/mol), 25 ml DMF, 25°C, oxygen atmosphere; [b] conversion determined by GC using *n*-hexadecane as internal standard (selectivity in all cases: >99%); [c] TOF = turn over frequency (mmol of product per mmol of catalyst per hour); [d] 30ml benzyl alcohol (solvent-free).

The differences displayed above are probably caused by coordination/bonding of copper to the free silanol groups on the surface of MCM-41 and silica. Besides these free silanol groups, the silica supported TEMPO system reported by Brunel et al.[14] also contains unreacted amine linkers, which inactivate the catalyst almost completely (entry 8). Therefore, the activity of the MCM-41 and silica supported TEMPO systems may be improved by blocking the free silanol groups and amine linkers on the surface.

In summary, we have developed a recyclable heterogeneous catalyst for the bleach oxidation of alcohols and polyols. In contrast to previously reported systems, neither a chlorinated hydrocarbon solvent nor a bromide cocatalyst is necessary to achieve good activity. Besides bleach-oxidation, PIPO is also effective in the CuCl/nitroxyl catalysed aerobic oxidation of benzyl alcohol. A further advantage of our system is that PIPO is readily prepared from inexpensive and commercially available raw materials. We believe that it will find wide application in organic synthesis.

We gratefully acknowledge IOP (Innovation-Oriented Research Program) for financial support and C. Bolm, D. Brunel and H. van Bekkum for the kind donation of their supported TEMPO systems. We thank M. Verhoef for valuable discussions.

Notes and references

1. R.A. Sheldon and J.K. Kochi, "Metal Catalysed Oxidations of Organic Compounds", Academic Press, New York, 1981; S.V. Ley, J. Norman, W.P. Griffith and S.P. Marsden, *Synthesis*, 1994, 639; M. Hudlicky, "Oxidations in Organic Chemistry", ACS, Washington, DC, 1990 and references cited therein.
2. A. E. J. de Nooy, A. C. Besemer and H. van Bekkum, *Synthesis*, 1996, 1153 and references cited therein; J. M. Bobbitt and M. C. L. Flores, *Heterocycles*, 1988, **106**, 509.
3. P. L. Anelli, C. Biffi, F. Montanari and S. Quici, *J. Org. Chem.*, 1987, **52**, 2559.
4. J. A. Cella, J. A. Kelley and E. F. Kenehan, *J. Org. Chem.*, 1975, **40**, 1860; S. D. Rychovsky and R. Vaidyanathan, *J. Org. Chem.*, 1999, **64**, 310.
5. T. Inokuchi, S. Matsumoto, T. Nishiyama and S. Torii, *J. Org. Chem.*, 1990, **55**, 462.
6. M. Zhao, J. Li, E. Mano, Z. Song, D. M. Tschaen, E. J. J. Grabowski and P. J. Reider, *J. Org. Chem.*, 1999, **64**, 2564.
7. C. -J. Jenny, B. Lohri and M. Schlageter, *Eur. Pat.*, 1997, 0775684A1.
8. C. Bolm, A. S. Magnus and J. P. Hildebrand, *Org. Lett.*, 2000, **2**, 1173.
9. M. F. Semmelhack, C. R. Schmid, D. A. Cortés and S. Chou, *J. Am. Chem. Soc.*, 1984, **106**, 3374.
10. A. Dijksman, I. W. C. E. Arends, R. A. Sheldon, *Chem. Comm.*, 1999, 1591.
11. (a) T. Osa, U. Akaba, I. Segawa and J. M. Bobbitt, *Chem. Lett.*, 1988, 1423; (b) Y. Kashiwagi, H. Ono and T. Osa, *Chem. Lett.*, 1993, 257; (c) T. Osa, Y. Kashiwagi and Y. Yanagisawa, *Chem. Lett.*, 1994, 367; (d) F. MacCorquodale, J. A. Crayston, J. C. Walton and J. Worsfold, *Tetrahedron Lett.*, 1990, **31**, 771; (e) T. Osa, Y. Kashiwagi, J. M. Bobbitt and Z. Ma, in *Electroorganic Synthesis*, Marcel Dekker Inc., New York, 1991, p.343; (f) A. Heeres, H. A. van Doren, K. F. Gotlieb and I. P. Bleeker, *Carbohydr. Res.*, 1997, **299**, 221.
12. M. J. Verhoef, J. A. Peters and H. van Bekkum, *Stud. Surf. Sci. Catal.*, 1999, **125**, 465.
13. C. Bolm and T. Fey, *Chem. Commun.*, 1999, 1795.
14. D. Brunel, P.Lentz, P. Sutra, B. Deroide, F. Fajula and J. B. Nagy, *Stud. Surf. Sci. Catal.*, 1999, **125**, 237.
15. A. Dijksman, I. W. C. E. Arends and R. A. Sheldon, *Chem. Comm.*, 2000, 271.
16. http://www.pidc.org.tw/enst/e-11.htm
17. E. G. Rozantsev and V. D. Sholle, *Synthesis*, 1971, 190.
18. General procedure for the bleach-oxidation of alcohols with PIPO as catalyst: In a glass reaction vessel was placed 2.5 mg of PIPO (8 μmol based on complete functionalisation; degree of functionalisation = 3.2 mmol g^{-1}). Then a CH$_2$Cl$_2$

solution (2ml) of the alcohol (0.4M) and *n*-hexadecane (0.12 M; as internal standard) was added followed by an aqueous solution (0.16 ml) of KBr (0.5M). After the cooling of the reaction mixture to 0 °C, 2.86 ml of aqueous NaOCl (0.35 M and buffered by the addition of 0.14 g KHCO$_3$ to pH 9.1) was added. Then, the reaction mixture was vigorously shaken for 45 minutes. After destroying the excess of hypochlorite with Na$_2$SO$_3$, the reaction mixture was extracted with ether, dried over Na$_2$SO$_4$ and analysed on GC [Chrompack CP-WAX 52 CB column; 50 m x 0.53 mm].

19. In a filtration experiment, the reaction mixture was filtered after 10 minutes. The filtrate was reacted further and the residue was washed with water and reused as catalyst.

20. For bleach oxidation of (*S*)-2-methylbutan-1-ol with homogeneous TEMPO, see: P. L. Anelli, F. Montanari and S. Quici, *Org. Synth.*, 1990, **69**, 212.

21. Like ref. [18], but with: 11.4 ml (4 mmol) hypochlorite solution, 2 ml (12.6 mmol) octan-2-ol and 4 μmol nitroxyl catalyst at 25 °C and 500 rpm. The IR probe is put in the octan-2-ol layer.

22. A. E. J. de Nooy, A. C. Besemer and H. van Bekkum, *Tetrahedron*, 1995, **51**, 8023; A. E. J. de Nooy, A. C. Besemer and H. van Bekkum, *Carbohydr. Res.*, 1995, **269**, 89; N. J. Davis and S. L. Flitsch, *Tetrahedron Lett.*, 1993, **34**, 1181.

23. Procedure for the oxidation of methyl α-D-glucopyranoside with PIPO as catalyst: In a glass reaction vessel was placed 15.7 mg of PIPO (50 μmol based on complete functionalisation; degree of functionalisation = 3.2 mmol g^{-1}) and 200 mg methyl α-D-glucopyranoside (1.03mmol). Then, 8 ml of aqueous NaOCl (0.56 M and brought to pH 9.5 with a 1M aqueous hydrochloride solution) was added. During the reaction the pH was kept constant at 9.5 by automatic titration with a 0.1 M potassium hydroxide solution. When the hydroxide consumption stopped (1.3 equivalents of hydroxide were consumed), Na$_2$SO$_3$ was added to destroy the excess of hypochlorite. The crude mixture was analysed using HPLC.

24. General procedure for the aerobic oxidation of benzylalcohol with CuCl/PIPO as catalyst: In a glass reaction vessel was placed benzyl alcohol (1.08 g; 10 mmol), *n*-hexadecane (2 mmol), CuCl (10.0 mg; 0.1 mmol), PIPO (31.8 mg; 0.1 mmol based on nitroxyl) and 25 ml DMF. The resulting reaction mixture was stirred at 25 °C under an oxygen atmosphere. The samples were analysed on GC [Chrompack CP-WAX 52 CB column; 50 m x 0.53 mm].

THE PREPARATION AND FUNCTIONALISATION OF (VINYL)POLYSTYRENE POLYHIPE. SHORT ROUTES TO BINDING FUNCTIONAL GROUPS THROUGH A DIMETHYLENE SPACER

A.Mercier[a], H.Deleuze[a*], B.Maillard[a], O.Mondain-Monval[b]

[a]Laboratoire de Chimie Organique et Organométallique
UMR 5802-CNRS Université Bordeaux I, 33405 Talence, France
[b]Centre de Recherche Paul Pascal UPR 8641-CNRS, 33600 Pessac, France

1 INTRODUCTION

Functionalised polystyrenes, initially used as ion-exchange resins and supports for solid phase peptide synthesis, have progressively expanded in importance as insoluble supports for catalysts and organic reagents[1]. Currently, most of these reactive polymers are prepared from polychloromethylstyrene, the so-called Merrifield resin, by a nucleophilic attack onto the chlorine atom. However, these supports present two main drawbacks:
- They possess a gel-type structure due to their styrenic macromolecular chains lightly crosslinked by 1 to 2 % of divinylbenzene. Therefore, the Merrifield resin and its derivatives are able to swell significantly only on good solvents of the polystyrene chain. Their efficiency is therefore strongly lowered in more polar solvents such as alcohol or water.
- They can suffer of facile nucleophilic displacements of the heteroatom present on the labile benzylic position[2].
The simple way to cancel these two limitations appears to synthesise a material with a permanent porosity, allowing a good access of the grafted species in any kind of solvent. In addition, the presence of two or more methylene units between the heteroatom of the functional group and the phenyl ring of the polymer backbone would avoid the problem of the benzylic position. There are different ways to prepare a polymer supports having permanent porosity, but only some of them could be used for our purpose:
- Foams having a higher level of porosity would be obtained by gas blowing[3]. However, in that case the produced structure has closed cells, unsuitable for use in supported chemistry.
- Macroporous resins can be obtained by a chemical phase separation process[4]. In that case, the level of crosslinking agent has to be rather high, 10 to 80%, and a solvent called porogen is added to the organic phase to allow the precipitation of the growing macromolecular chain, thus generating the porosity (large proportion of mesopores from 20 to 500 μm). Nevertheless, the diffusion of the species present in the solution toward the grafted functions can therefore be restricted, thus reducing the efficiency of the functional resin.

- Polymeric foams, called polyHIPE®, has been developed by Unilever researchers[5]. The production of these porous materials was based on the polymerisation of high internal phase emulsion (HIPE)[6]. The system is composed of two phases: an organic phase -called the continuous phase- containing the monomers and a suitable amount of emulsifier and an aqueous phase -called the dispersed phase- containing the radical initiator (scheme 1).

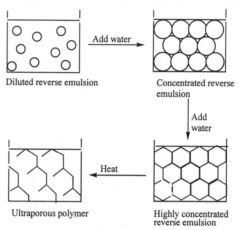

Scheme 1 Preparation of polyHIPE

By introducing droplets of aqueous phase in the organic mixture, under constant agitation, a dilute reverse (water-in-oil) emulsion is produced. If the amount of water is increased a highly concentrated emulsion is obtained. The structure is now analogous to soap bubbles, with thin films surrounding and separating the drops. During the polymerisation step, holes are formed in the thin films separating the droplets and an open structure is formed. The water is removed to produce a foam of the corresponding structure.

The present work reports such an approach with the synthesis of (vinyl)polystyrene polyHIPEs and their functionalisation by free radical addition of thiols to the remaining unsaturations. The interest of such materials is proven by the presentation of an application of a thiol functional support in free radical reduction of alkyl halides.

2 EXPERIMENTAL SECTION

In a typical experiment, a volume V_{org} of organic phase constituted of commercial grade solution of DVB (80:20 DVB:EVB) and sorbitan monooleate (Span®80) as emulsifying agent (20% wt of the organic phase) was placed in a reactor. The mixture was stirred with a rod fitted with a D-shaped paddle, connected to an overhead stirrer motor, at approximately 300 rpm. A V_{aq} volume of aqueous phase was prepared separately by dissolving the initiator, potassium persulfate $K_2S_2O_8$, and sodium chloride NaCl (1.5% wt of the aqueous phase) in distilled water. This solution was added dropwise, under constant mechanical stirring, to the organic solution in order to obtain a thick white homogeneous emulsion without apparent free water. Once all the aqueous phase had been added, stirring was continued for a further 15 min to produce an emulsion as uniform as possible. The high internal phase emulsion obtained was then placed in a polyethylene bottle. The polymerisation occurred by immersing the plastic bottle in a water bath, heated to 60°C for 10 hours. The container was then cut away to collect the resulting polymeric monolith. This one was extracted with acetone in a Soxhlet apparatus for 48 hours, then was dried under vacuum at 60°C for 48 hours. The resulting monolith was cut in cubes (approx. 5 mm per side). The polyHIPE thus synthesised is characterised by a volumic fraction of pore precursor (water) $\phi=V_{aq}/(V_{aq}+V_{org})$. The materials obtained are rigid opaque monolith with good mechanical properties

The functionalisation by thiols was performed as follows: Small cubes of (vinyl)polystyrene polyHIPE [3.00 mmol C=C/g (1 g, 3 mmol C=C)] were impregnated with toluene by freeze-drying under vacuum and suspended in toluene (20 mL). The thiol (8 mmol), AIBN (5 mg, 1%mol. / C=C) were then added. The suspension was heated at 70°C for 48 h. The polymer was filtered off, extracted with acetone overnight on a Soxlet apparatus and dried in vaccum at 60°C, overnight.

The reduction of the 6-bromohex-1-ene was performed by heating at 80°C a degassed solution of dilauroyl peroxide (0.004 M), triethylsilane (1.5 M), unsaturated bromide (0.0386M), dodecanethiol (0.003 M) or thiol supported polymer(0.002 M) in heptane.

3 RESULTS AND DISCUSSION

3.1 Synthesis of PS-SH

The free radical polymerisation of commercial divinylbenzene led to a crosslinked macroporous monolith. The overall interconnected open-cellular macrostructure of the material can be clearly seen by scanning electron microscopy (fig.1).
It is worthwhile to note the relatively large cells (up to 10-20 μm in diameter) and the large number of about 1 μm 'windows' between adjacent cells. The large pore volume results directly from the large internal volume of water incorporated in the HIPEs (more than 95%). The large average cell size corresponds roughly to the size of the water droplets in the HIPEs.
Free radical copolymerisation of divinylbenzene gives crosslinked resins that have been shown to often still bear many unreacted pendant vinyl groups[7]. These remaining pendant vinyl bonds as well as the crosslinking level can be quantified by FTIR. The value obtained for the produced support is about 3.0 mmol/g of pendant vinyl bonds.

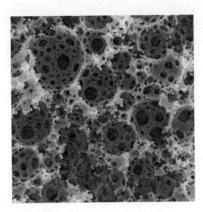

Fig.1. Scanning electron micrograph of (vinyl)polystyrene polyHIPE.

The functionalisation of these residual pendant vinyl group by its extremity would be a convenient way to incorporate a dimethylene spacer for more stable attachments of the grafted species through a non-benzylic position. Thiols adding to olefins under an anti-Markovnikov way by free radical mechanism[8], this approach was used to diversely functionalise our polyHIPEs (scheme 2).

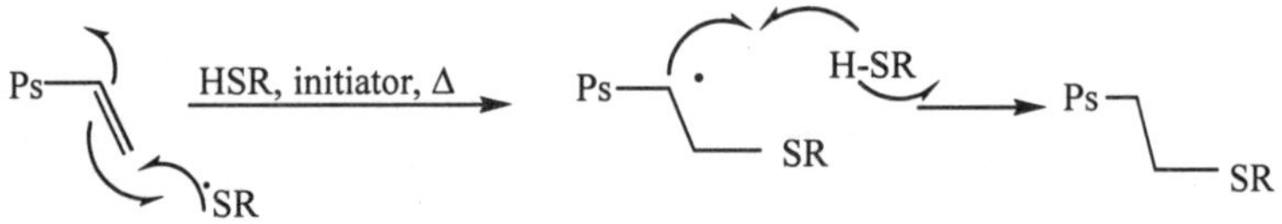

Scheme 2 : Free-radical addition of thiols to (vinyl)polystyrene polyHIPE

Several functionalities were introduced onto the resin such as amine, alcohol, ester, acid and thioacetic acid (scheme 3).

Degrees of functionalisation of the modified resins were in the range 10-60% according to the thiol and the method employed the grafted function contain was between 0.25 and 1.9 mmol/g of polymer.

Different applications of these functional materials can be designed like their use as scavenger –in particular for supported amine and acids- and as supported catalysts. An example of such a last application could be found in the reduction of alkyl halides by triethylsilane.

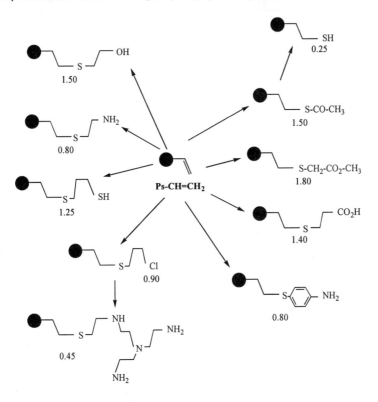

Scheme 3 : Functionalisation of (vinyl)polystyrene polyHIPE by thiols (numerical data are the functionalisation levels in mmol/g).

The removal of a functional group G from an organic compound R-G and its replacement by hydrogen to give R-H is a basic transformation of considerable importance in synthetic chemistry. Tributyltin hydride is preeminent amongst reagents for bringing about homolytic reductive removal of functional groups. However, organotin compounds are toxic and often difficult to remove completely from the desired reaction products[9]. The system triethylsilane / thiol, as described by Roberts[10] is a very interesting alternative (scheme 4).

This reaction looks attractive because triethylsilane is non toxic and its by-products are low boiling point compounds easily removed by distillation. However, the alkanethiols have a strong bad smell and the remaining traces could be difficult to eliminate from the solution in certain cases, which make them undesirable in the synthesis of fine chemicals. Therefore, it appeared of interest to test the use of thiols supported on polyHIPE in these radical chain reductions.

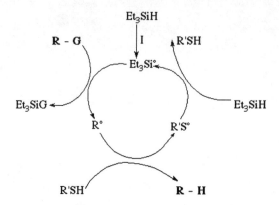

Scheme 4 : Reduction of alkyl halides by the thiol/triethylsilane system.

The efficiency of this approach was first studied through the radical reduction of 1-bromoadamantane by Et_3SiH in presence of the polyHIPE-supported thiol catalyst (table 1).The conversion of 1-bromoadamantane is complete after 1 hour at 80°C with a catalytic amount of dodecanethiol and two equivalents of triethylsilane. In the same conditions, 70% of the bromide were reduced in the presence of the supported thiol, showing a lower reactivity of this last one. The absence of reduction without thiol proved the role of the supported thiol.

Entry	T (°C)	Catalyst	Reducing Agent (eq)	Cat./RBr	Reaction Time (h)	Conversion (%)
1	80	none	Et_3SiH (2)	0	1	0
2	80	$C_{12}H_{25}SH$	Et_3SiH (2)	0.02	1	100
3	80	PS-SH	Et_3SiH (2)	0.02	1	70

Table 1 : Free-radical reduction of bromoadamantane

In a second time, the reduction of an unsaturated halide was an interesting goal because such a reaction could be an easy access to cyclic molecules via a radical cyclisation in competition with the hydrogen transfer. 6-bromohex-1-ene is one of the compound whose free radical reduction was the most extensively studied, according to the radical clock status of the 6-hex-1-enyl radical[11]. Therefore, it appeared of interest to study it with the above system. The results obtained are reported in table 2.
A low conversion of 6-bromohex-1-ene, without formation of the expected products, was observed using dodecanethiol under low concentration of triethylsilane (table 2, entry 1). Roberts and his coworkers[10], pointed out the weak efficiency of a similar

system for the reduction of this unsaturated halide, attributing it to the possible addition of the thiol to the unsaturations present in the medium.

Entry	T (°C)	Catalyst	Reducing Agent (eq)	Reaction Time (h)	Methyl-cyclopentane	Hexene
1	80	$C_{12}H_{25}SH$ (0.05)	Et_3SiH (2)	10	0	0
2	80	$C_{12}H_{25}SH$ (0.05)	Et_3SiH (40)	1	100	0
3	80	PS-SH (0.05)	Et_3SiH (2)	12	0	0
4	80	PS-SH (0.05)	Et_3SiH (40)	10	100	0

Table 2 : Free-radical cyclisation of 6-bromohex-1-ene

Taking into account the existence of a competition for the thiyl radical between the abstraction of an hydrogen to the triethylsilane and the addition to the double bond of the unsaturated halide, one can see that the first process would be favoured by the increase of the relative ratio triethylsilane / 6-bromohex-1-ene. The use of a fortyfold excess of the silane allowed to obtain the total conversion of the alkenyl bromide to yield methylcyclopentane within 1 hour (table 2, entry 2). The selective formation of methylcyclopentane showed the existence of a relatively low rate of hydrogen transfer with this system ([dodecanethiol] = 0.003M). The replacement of dodecanethiol by the supported thiol led to the expected compound but with a longer reaction time (table 2, entry 4), proving the interest of this catalyst and leading to envisage its use for synthetic purpose. The high excess of triethylsilane necessary could be considered as a drawback, even if it could be easily recovered by distillation in the case of the reduction of unsaturated halides having a high boiling point. Trials are currently undertaken in our laboratory to decrease the relative amount of the silane and to propose other silanes.

4 CONCLUSION

This work describes a new type of monolithic support offering a highly interconnected permanent porosity possessing pendant double bond. This support may provide a better accessibility to active sites and allows the use of a wider range of solvents than classical gel type beads prepared by suspension polymerisation. The free radical addition of functional thiols led to the production of functional supports (acid, ester, alcohol, amine, thiol…).
The polyHIPE supported thiol is a good catalyst for radical reduction of alkyl halides and reductive cyclisations of ε-bromoalkenes by triethylsilane.

5 REFERENCES

1. D.C. Sherrington, P. Hodge, *Syntheses and Separation Using Functional Polymers*, John Wiley, New York, (1988).
2. B.R Stranix, J. P. Gao, R. Barghi, J. Salha, G.D. Darling, J. Org. Chem., 1997, **62**, 8987.
3. W. R. Evans, D. P. Gregory, MRS Bulletin, 1994, **4**, 29.
4. D. C. Sherrington, Chem. Commun., 1998, **21**, 2275.
5. D. Barby, Z. Haq, European Patent 0,060,138 (1982) (to Unilever).
6. N. R. Cameron, D. C. Sherrington, Adv. Polym. Sci., 1996, **126**, 163.
7. K. L. Hubbard, J. A. Finch, G. D. Darling, React. Funct. Polym., 1998, **36**, 1.
8. a) K. Griesbaum, Angew. Chem. Int. Ed. Engl., 1970, **9**, 373, b) F. W. Stacey, J. F. Harris, Org. React., 1963, **13**, 150.
9. a) D. P. Curran, C. T. Chang, J. Org. Chem., 1989, **54**, 3140; b) D. Milstein, J. K. Stille, J. Am. Chem. Soc., 1978, **100**, 3636; c) D. Crich, S. Sun, J. Org. Chem., 1996, **61**, 7200.
10. S. J. Cole, J. N. Kirwan, B. P. Roberts, C. R. Willis, J. Chem. Soc. Perkin Trans. 1, 1991, 103.
11. D. Griller, K. V. Ingold, Acc. Chem. Res. 1980, **13**, 317.

POLYNITROGEN STRONG BASES AS IMMOBILIZED CATALYSTS

G. Gelbard, * F. Vielfaure-Joly.

Institut de Recherches sur la Catalyse - CNRS, 2 Avenue Albert Einstein,
69626 Villeurbanne Cédex (France)
fax: +334 72 44 53 99 ; e-mail : gelbard@catalyse.univ-lyon1.fr

1 INTRODUCTION:

Guanidines are strong bases whose use as catalysts is convenient because of their solubility in organic media. Since their basicity lies in the range of common inorganic bases, such as alkaline hydroxides and carbonates, they are suitable for use in a wide range of base-directed reactions such as Michael additions, esterifications and transesterifications.

Biguanides, which are even stronger bases, open new perspectives but they require special preparations. Novel syntheses of poly-N-alkylated biguanides have been devised from carbodiimides, and recent examples of catalyses with soluble and immobilised guanidines and biguanides are presented with emphasis on the transesterification of triglycerides from vegetable oils.

1.1 Guanidines

As organic strong bases are soluble in most common solvents, they are often selected preferentially to metallic oxides, hydroxides and carbonates ; the most common ones are bicylic amidines such as 1,8-diazabicyclo[5.4.0]undec-7-ene **1** and 1,4-diazabicyclo[4.3.0]non-5-ene (DBU and DBN respectively)[1] and the bicyclic guanidines 1,5,7-triazabicyclo[4.4.0]dec-5-ene **2** and its 7-methyl analogue (TBD and MTBD respectively).[2] However, simpler and less expensive guanidines such as tetramethylguanidine **3** (TMG) are also used (Figure 1).

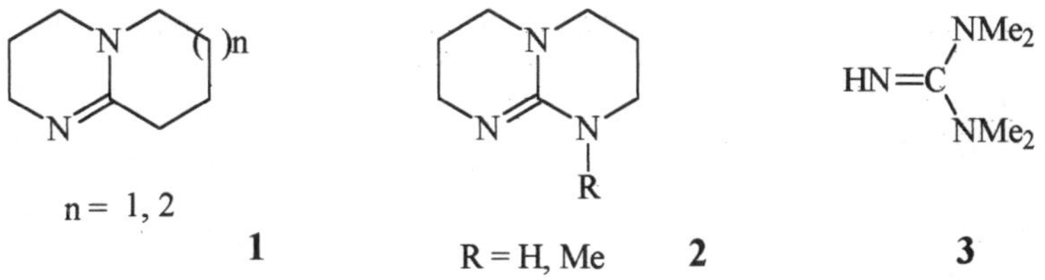

$$n = 1, 2 \qquad \mathbf{1} \qquad R = H, Me \quad \mathbf{2} \qquad \mathbf{3}$$

Figure 1 *Amidines and guanidines*

In order to increase the efficiency of the catalysts, stronger bases are required such as the highly N-substituted biguanides of type **4** or **5** (Figure.2).

A short review of the uses of polymeric and immobilised guanidines and biguanides is given along with recent results on the preparation and properties of biguanides.

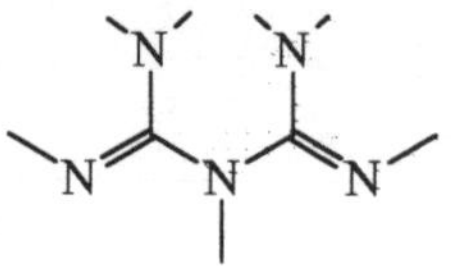

Figure 2 *Cross-conjugated 4 and fully-conjugated biguanides 5.*

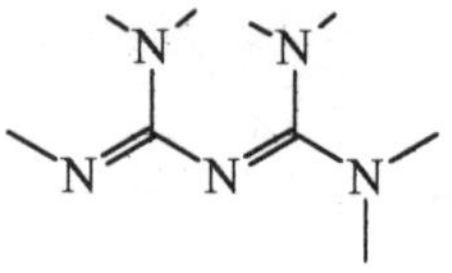

2 SOLUBLE GUANIDINES AND POLYGUANIDINES

Henry and Michael-Henry addition of nitroalkanes to aldehydes, alicyclic ketones and α,ß-unsaturated ketones were performed at r.t. with TMG **3** as catalyst ; moderate then good [3a] yields were obtained in direct and conjugated addition products .

Under similar conditions, the same authors performed the addition of diethylphosphite to aldehydes and ketones (a Pudovik-like reaction) and to Michael acceptors such as imines, acrylonitrile and α,ß-unsaturated ketones[3b] to give functionalised phosphonates (Figure 3).

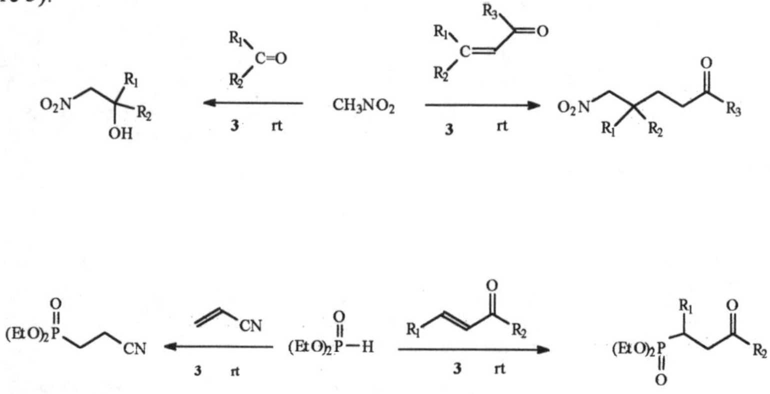

Figure 3 *Michael additions of carbon and phosphorous acids to ketones and enones.*

Asymmetric esterification of benzoic acid with racemic (1-bromoethyl)benzene was performed with chiral cyclic guanidines as stoechiometric bases[4] ; the ester was obtained with ee below 15%. The chiral guanidines **6** were obtained from a chiral diamine of C_2 symmetry which was tranformed first to urea and then to several guanidines through an intermediate chloroformamidinium chloride (Figure 4).

Similar chiral catalysts **7** were used in the Michael addition of glycine derivatives to acrylic esters to obtain functionalised α-aminoacids[5] ; here, the guanidines were prepared from chiral amines or diamines in one step but using cyanogen bromide, which is a noxious reagent (Figure 4).

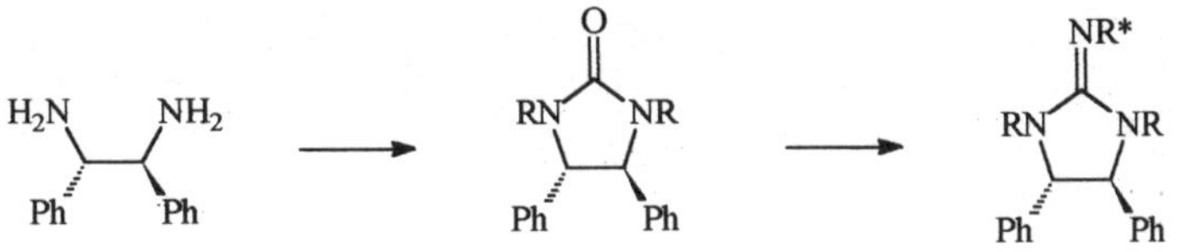

Figure 4 *Chiral guanidines from C$_2$ symmetry diamines*

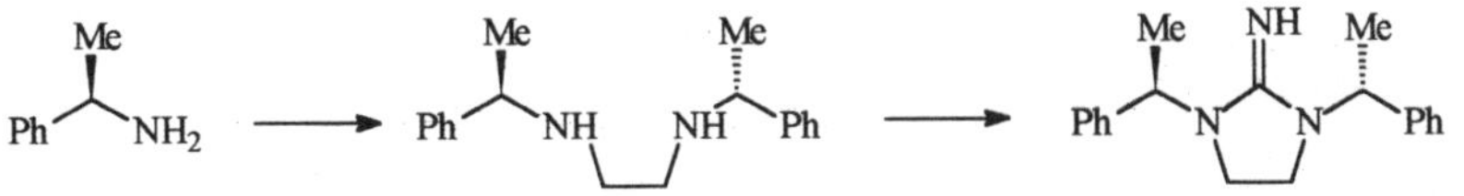

The aminoacids were obtained with ee below 28% according to the reaction :
$$Ph_2C=N-CH_2-CO_2R^1 + CH_2=CH-CO_2R^2 \longrightarrow Ph_2C=N-CH^*(CO_2R^1)-CH_2CH_2-CO_2R^2$$
For the base-catalysed curing of epoxy resins, several multi-N-alkylated biguanide were used, but the preparation of 1,2,2,4,4,5,5-heptamethylbiguanide, the more signicant catalyst, was not clearly given.[6]

When carbodiimides are subjected to anionic polymerisation, the oligomers are formed with repeating type **4** biguanide units (Figure 5). The polymer chain has restrictions in its conformational degreees of freedom and adopts an helical conformation.[7a] If the N-substituents of the carbodiimide are chiral, there is intensive induction to yield one of the two diasteroisomeric chiral helices, as shown by optical studies.

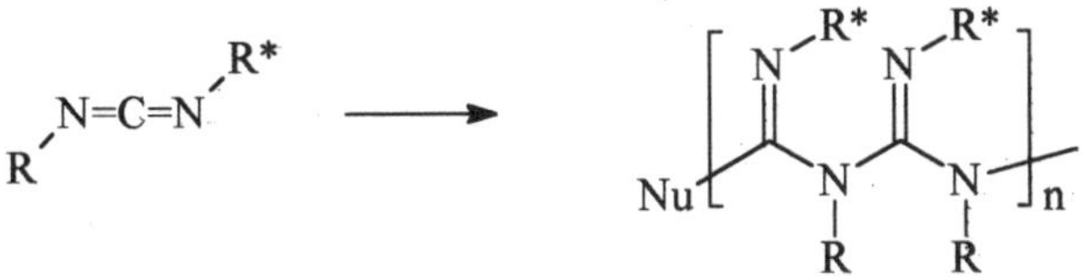

6/1 helix

Figure 5 *Polycondensation of carbodiimides.*

To date, however, there is no exampleof uss of these polymers as a potentially chiral catalysts.

3 SILICA- AND POLYSTYRENE-GRAFTED GUANIDINES

When used as soluble catalysts, guanidines must be removed from the reaction mixture at the same time that the products are isolated. Thus, tethering the base molecule to an insoluble and porous carrier offers advantages in work-up. It is also expected that the catalyst might be reused several times. To date hardly any recycling experiments have been performed.

The first example of grafting a guanidine to silica was described as earlyas 1986[8] with 2-[3-(trimethoxysilyl)-propyl]-1,1,3,3-tetramethylguanidine **8** (Figure 6). The alkylation of TMG **3** with 3-chloropropyl(trimethoxy)silane **7** affords the suitable soluble guanidine[9] which is then attached on heating to the surface of a silica.

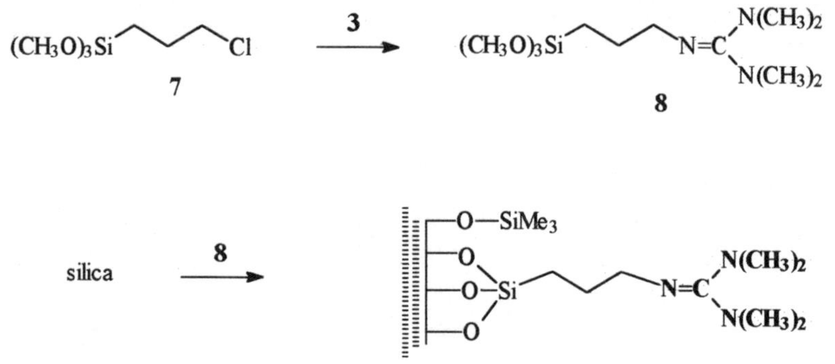

Figure 6 *Grafting of tetramethylguanidine to silica.*

The catalyst was checked in the transesterification of methyl acetate to ethyl acetate :
$$CH_3CO_2CH_3 + C_2H_5OH \longrightarrow CH_3CO_2C_2H_5 + CH_3OH$$
Similarly, the aminated silane **9**, obtained by alkylation of *n*-butylamine with the chlorosilane **7**, can be transformed into a guanidine or a guanidinium salt.[10] Here the corresponding grafted guanidinium salt was used as a Lewis-acid catalyst in the phosgenation of carboxylic acids (Figure 7).
$$RCO_2H + COCl_2 \longrightarrow RCOCl + HCl + CO_2$$

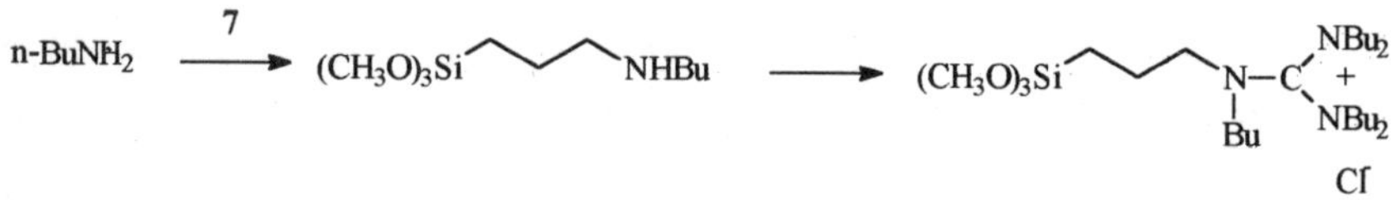

Figure 7 *Grafting of a guanidinium cation to silica.*

An other approach was followed with a micelle-templated silica of the MCM-41 type. These mesoporous carriers were first grafted with the chlorosilane **7** and then aminated with a guanidine such as TBD **2** [11] (Figure 8).

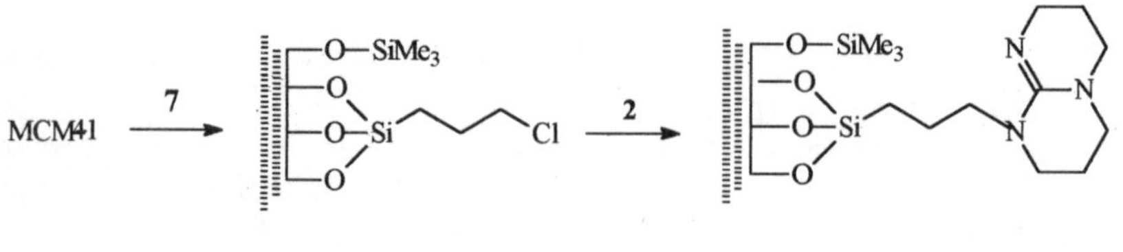

Figure 8 *Grafting of TBD to silica.*

The catalytic properties were checked in the transesterification of ethyl propionate with *n*-butanol and in the Knoevenagel condensation of benzaldehyde with ethyl cyanoacetate :

$$C_2H_5CO_2C_2H_5 + n\text{-}C_4H_9OH \longrightarrow C_2H_5CO_2C_4H_9 + C_2H_5OH$$
$$C_6H_5\text{-}CHO + CNCH_2CO_2C_2H_5 \longrightarrow C_6H_5\text{-}CH=C(CN)\,CO_2C_2H_5$$

The same TBD **2** was also grafted to silica but by means of a different electrophilic linker : [3-(glycidyloxy)propyl]trimethoxysilane was first grafted and then TBD reacted with the oxirane group to give the hydroxy guanidine (Figure 9).

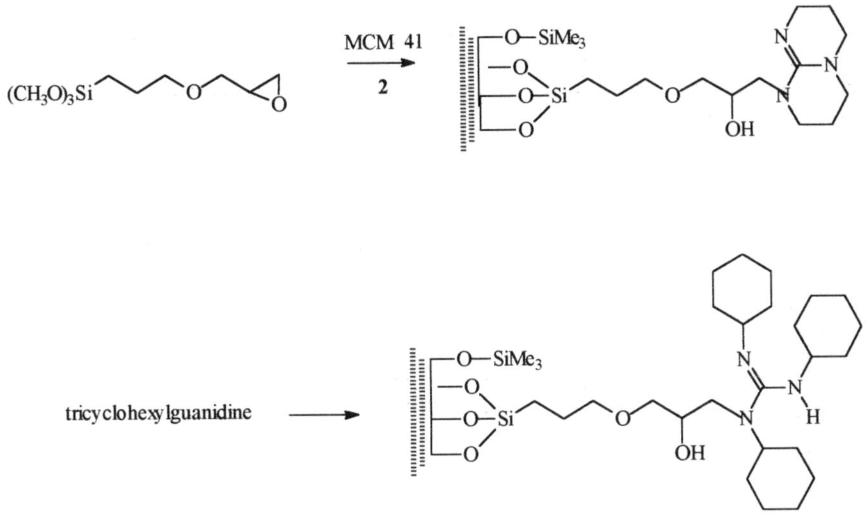

Figure 9 *Grafting of guanidines to silica by a glycidol function.*

In Michael additions of ethyl cyanoacetate, diethylmalonate and 1,3-diones to acrolein, methylvinylketone and other enones, the catalytic activity on the system was demonstrated: [12]

$$CH_2=CHCHO + CNCH_2CO_2C_2H_5 \longrightarrow CHOCH_2CH_2CH(CN)CO_2C_2H_5$$

The same [3-(glycidyloxy)propyl]trimethoxysilane was also used as a linker but was reacted with tricyclohexyl guanidine to give a similarly grafted catalyst (Figure 9). The presence of an hydroxyl group in the spacer may buffer the basicity of the biguanide (BiG) according to the equilibria : R-OH + BiG ↔ R-O⁻ + BiGH⁺. This aspect is usually ruled out with the other spacers by end-capping the residual SiOH functions with a silane.

The aforementioned tricyclohexyl guanidine was alternatively constructed inside the pores of a zeolite. Dicyclohexycarbodiimide was first impregnated in the porous material and reacted with cyclohexylamine. This gave a trapped base. A seive effect keeps the molecules inside the material.

Both compounds were checked in the aldolisation of acetone with benzaldehyde ; the reaction gives mainly the crotonisation product :[13]

$$C_6H_5\text{-}CHO + CH_3COCH_3 \longrightarrow C_6H_5\text{-}CH=CHCOCH_3.$$

It should be stressed however that, from an historical point of view, such organic bases were immobilised first on organic polymers, not on silica. The well known bicyclic amidine DBU **1** (an amidine, not a guanidine), was lithiated and grafted to ω-chloroalkyl-polystyrenes (Figure 10).

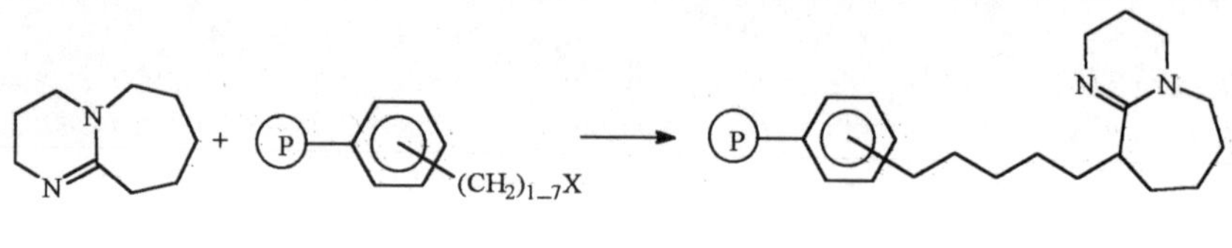

Figure 10 *Polystyrene-supported DBU.*

Several reaction were examined : the dehydrobromination of bromoalkanes, the esterification of benzoic acid with 1-bromobutane and the macrolactonisation of ω-bromocarboxylic acids.[14]

In the area of hydrometallurgy, Cu^{2+} and $Au(CN)_2^-$ salts could be selectively sorbed from aqueous solutions with polyacrylamide-based copolymers containg aminoguanidine units. Porous copolymers of acrylonitrile, divinylbenzene and vinylacetate were derivated with aminoguanidine.[15] Alternatively, methacrylate-based copolymers were transamidated with the same base[16] to give polyacrylamide-type resins (Figure 11).

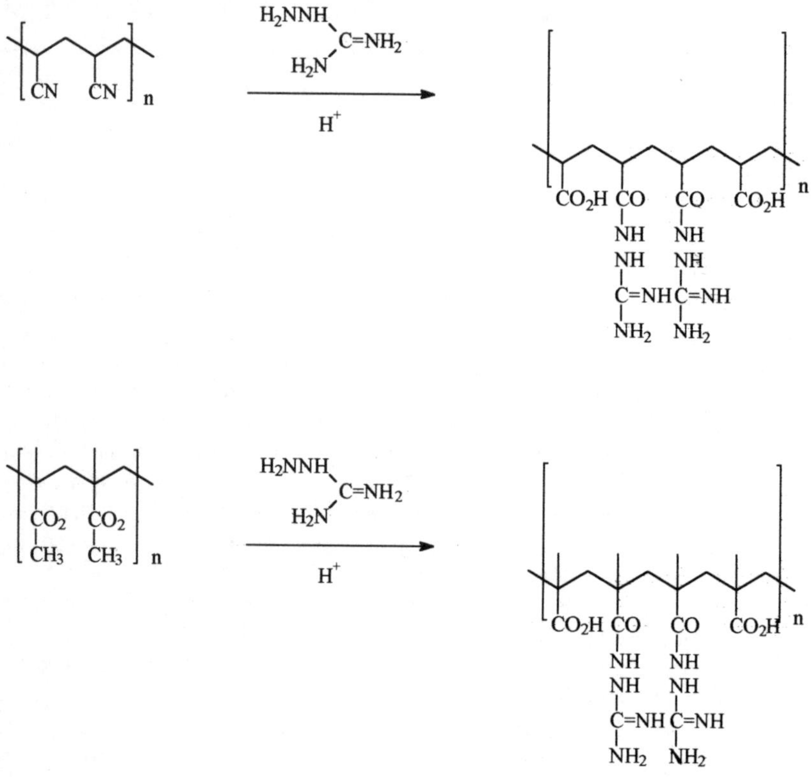

Figure 11 *Polyamides of aminoguanidines for metal complexation.*

Unfortunately, all these supported base catalysts were considered only from the point of view of the ease of separation from reaction mixture and not at all in terms of possible repeated use of these expensive materials. Furthermore, the known long-term instability of silica and silica-based materials in basic media is likely to restrict technological use of these species and this aspect seems to have been completely neglected.

More suitable materials would be more chosen : more base stable inorganic compounds or selected organic copolymers.

More recently, guanidines were sucessfuly introduced onto polystyrene matrices by different approaches : N-alkylation of TMG or TBD by ω-chloroalkylpolystyrenes or by the addition of the amine function of a ω-aminolkylpolystyrenes to a dicyclo-hexylcarbodiimides (Figure 12);

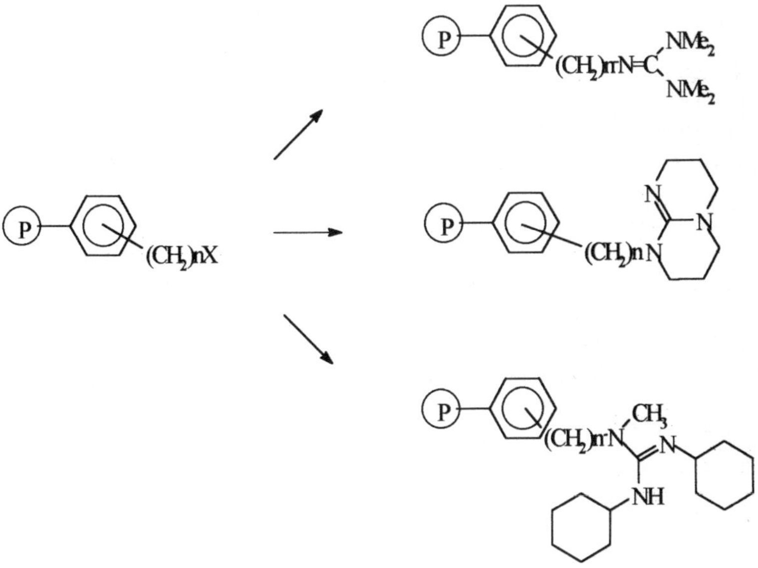

Figure 12 *Polystyrene-supported guanidines.*

These catalysts were found to be efficient catalysts in the transesterification of triglycerides of rapeseed and soyabean oils with methanol, as shown below,[17] and recycling experiments were performed (Figure 13).

Figure 13 *Transesterification of triglycerides with methanol.*

4 POLYSTYRENE-GRAFTED BIGUANIDES

For the transesterification of vegetable oils, even stronger bases are required, such as the biguanides **4** and **5**. These are the higher analogues of guanidines and their increased basicity resultsd of the higher conjugation of the protonated form, i.e. an extended delocalisation of the positive charge. Protonated type **4** biguanides are cross-conjugated and protonated type **5** are fully conjugated.

So as to get the strongest possible basicity, these biguanides were prepared with the highest possible degree of N-alkylation. However, no satisfactory precedure was described in the literature. So original synthetic methods were devised for this purpose. These rely essentially on the addition of guanidines to carbodiimides or of amines to cationic dimeric carbodiimides.[18]

For the purpose of recycling, some of these strong bases where immobilized on gel-type and macroporous polystyrene. The influence of the morphology on the reactivity was checked : a six carbon chain-spacer was introduced between the main chain and the biguanide group.

The synthesis of the polystyrene-supported biguanides was performed in the same way as for the soluble ones. The starting polymers were gel-type polystyrenes with chloromethyl or 6-bromohexyl groups prepared according to Merrifield or Tundo respectively.[19] These new catalysts were obtained either by sequential syntheses on aminated polystyrene or by covalent grafting of ready made biguanides.

4.1 Building biguanides on polymers.

4.1.1 From carbodiimides : The aminated polymers (from haloalkylpoystyrenes and potassium phtalimide) were converted into urea with isocyanate and dehydrated to carbodiimides ;[20] addition of tetramethylguanidine gave the type **5** N-hexalkylbiguanides **9** (Figure 14).

4.1.2 From cyanoguanidines : Aminated polystyrenes were reacted to cyanoguanidines (from secondary amines and dicyanamides) to give type **4** N-trisubstituted biguanides **10** (Figure 14).

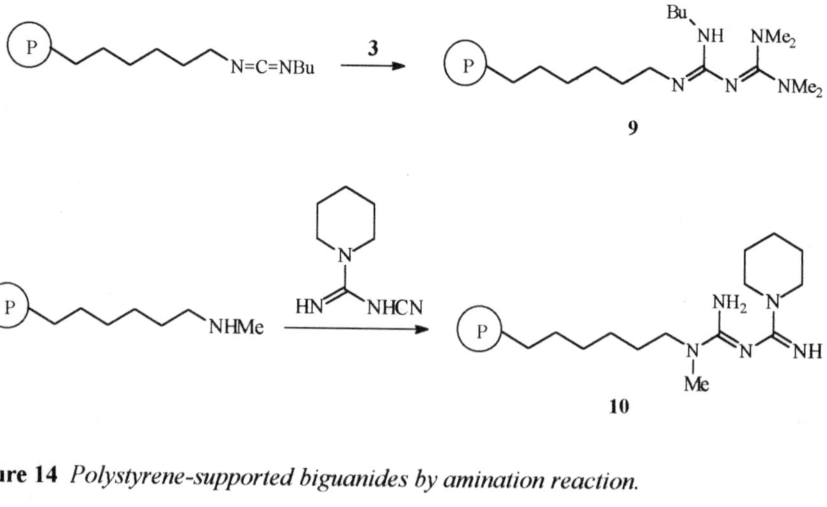

Figure 14 *Polystyrene-supported biguanides by amination reaction.*

4.1.3 From diazetidinium salts : The same aminated polystyrenes as above were reacted with and excess of diazetidinium triflate (from methyltriflate and diisopropylcarbodiimide) to get type **5** N-heptaalkylated biguanides **11** (figure 15).

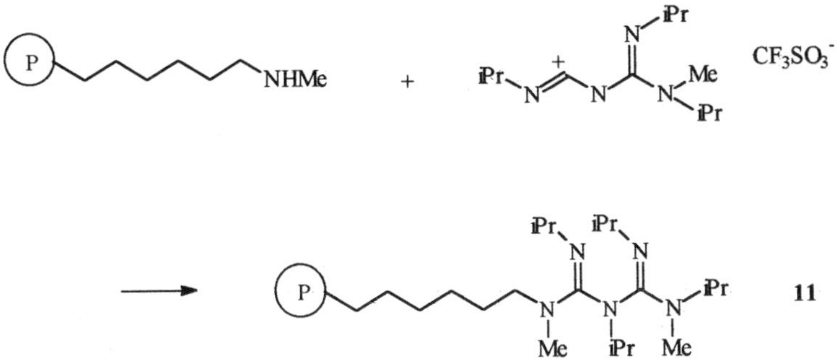

Figure 15 *Polystyrene-supported biguanides by alkylation reaction.*

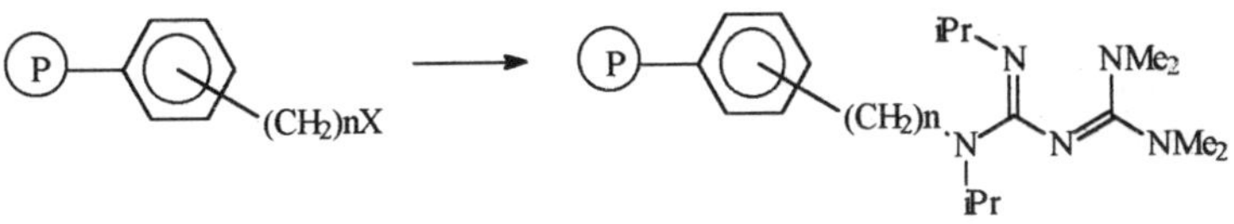

4.2 Grafting ready-made biguanides

Halogenoalkyl polystyrenes, gel-type, macroporous chloromethylpolystyrenes and Tundo-type, were used as N-alkylating agents for some biguanides. The grafting reaction increases the degree of N-alkylation to give type **4** N-heptalkyl biguanides **12** (Figure 15) with good yields with the exception of the macroporous-type resins.

5 TRANSESTERIFICATION REACTIONS

These grafted strong bases were checked in their both soluble and immobilised form as catalysts for the transesterification of several vegetable oils such as linseed, rapeseed, sunflower and palm oil. When methanol is used, the reaction affords a mixture of the methyl esters, used as fuel for diesel engines[21, 22] (Figure 13).

The transesterification reactions were performed at 70°C with molar ratio of methanol / ester function / catalyst = 2.3 / 1 / 0.02 and at atmospheric pressure. The rates were monitored by ^{1}H NMR.[23] The results are given in Table 1.

Table 1 *Comparative efficiencies of soluble and immobilised guanidines and biguanides in the transesterification of rapeseed oil with methanol* [a].

Catalysts	Reaction *time* (mn) and **yield** (%)					
	5	*15*	*30*	*60*	*180*	*300*
Soluble biguanide **4, 5**	**81**	**85**	**94**			
Polystyrene supported biguanide **11, 12**	**62**	**81**	**90**	**94**		
Pentamethylguanidine [20a]			**30**	**46**	**60**	
Polystyrene-supported guanidine [20b]				**63**	**87**	**92**

a) see text for conditions

With the soluble catalysts, yields above 80% are obtained after 10 mn and 90% after 30 mn reaction time. These biguanides are about thirty times more reactive than soluble guanidines : 60% yield after 5 mn instead of the 180 mn.

The polystryrene-bound biguanides were checked under exactly the same conditions and exhibited excellent catalytic properties (Table 1). The immobilization of the biguanide units induced only a very limited decrease in the reactivity. Yields above 90% were obtained in less than 15 mn. The yields of methyl esters are above 94% even before 15 mn reaction time.

These biguanide-supported catalysts are far more reactive and more stable than the previously described guanidines.[17b] This confirms the higher basicity of biguanides vs. guanidines and that their immobilisation induces only a very limited decrease in reactivity : 90% instead of 94% yield after 30 mn reaction time.

Even more interestingly, the recycling of the same sample of polymer-bound catalysts **11** or **12** could be performed seventeen times (Figure 16). The efficiency remains unaffected far more than ten cycles, after which a slow fall in activity begins to appear. This remakable behaviour occurs only with hexa- and heptaalkylated biguanides. The less N-alkylated species degrade after only a few cycles.

Elemental analysis shows the decrease of N% corresponding to a partial cleavage of the biguanide units. ^{1}H NMR experiments performed with soluble biguanide in d_4-methanol proved the methanolysis of the molecule at the C=N bond.[24]

As far as we know, this is the first report of a polymer-supported strong base catalyst with which recycling has been performed so successfully. These properties are very promising for a lot of other base-catalysed reactions ; some are currently under investigation in our Laboratory.

AKNOWLEDGMENT

We thank the CNRS and ADEME (contract 4/01/0044) for financial support and ONIDOL for a fellowship to one of us (F.V.J.).

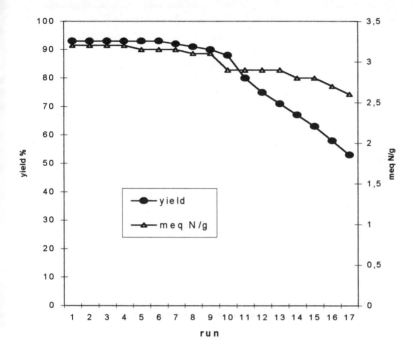

Figure 16 Sucessful recycling of biguanide-based catalysts

REFERENCES AND NOTES

1. H. Oediger, F. Möller, K. Eiter, *Synthesis* , 1972, 591.
2- R. Schwesinger, *Chimia* , 1985, **39**, 269 ; *Angew. Chem. Int. Ed. Engl.* 1987, **28**, 1164 *Nachr. Chem. Tech. Lab.*, 1990, **38**, 1224.
3- a) D. Simoni, F.P. Invidiata, M. Manfredi, I. Lampronti, R. Rondanon, M. Roberti, G.P. Pollini, *Tetrahedron Lett.* 38, 1997, 2749 ; b) 1998, **39**, 7615.
4- T. Isobe, K. Fukuda, T. Ishikawa, *Tetrahedron : Asym.*, 1998, 9, 1729.
5- D. Ma, K. Cheng, *Tetrahedron Asym.*, 1999, **10**, 713.
6- K.G. Flynn, D.R. Nemortas, *J. Org. Chem.*, 1963, **28**, 3227.
7 a)- A. Goodwin, B. M. Novak, *Macromolecules* , 1995, **27**, 5520 ; b)- D.S. Schlitzer, B. M. Novak, *J. Am. Chem. Soc.* , 1998, **120**, 2196.
8- M.J. Green, BP Chemicals, European Patent 168167 (15/01/86).
9- T. Tagako, Shin-Etsu Chemical Co, US Patent 4248992 (20/09/79).
10 - P. Gros, P. Le Perchec, J.P. Senet, *Reactive Funct. Polym.*, 1995, **26**, 25.
11 a)- D. Brunel, *Micropor. Mesopor. Mater.* , 1999, **27**, 329 ; b)- A. Derrien, G. Renard, D. Brunel, "Mesoporous Molecular Sieves", L. Bonneviot, F. Béland, C. Danumath, S.Giasson and S. Kaliaguine, edit., in Stud. Surf. Sc. Catal., Elsevier B.V., 1998, vol 117, p 445.

12 - Y.V. Subba-Rao, D.E. De Vos, P.A. Jacobs, *Angew. Chem. Int. Ed. Engl.* , 1997, **36**, 2661 ; 3rd International Symposium, "Supported Reagents and Catalysts in Chemisty", Limerick, Ireland, 1997, OC12.

13- R. Sercheli, R.M. Vargas, R.A. Sheldon, U. Schuchardt, *J. Mol. Cat.* 1999, **148,** 173.

14- a) M. Tomoi, Y. Kato, H. Kakuichi, *Makromol. Chem.* , 1984, **185**, 2117 ; b) M. Tomoi, T. Watanabe, T. Suzuki, H. Kakuichi, *Makromol. Chem.* , 1985, **186**, 2473.

15- B.N. Kolarz, J. Jezierska, D. Bartkowiak, A.Gontarczyk, *Reactive Polym.*, 1994, **23**, 53.

16- B.N. Kolarz, D. Bartkowiak, A.W.Trochimczuk, W. Apostoluk, B. Pawlow, *Reactive Funct. Polym.*, 1998, **36**, 185.

17- a) U.F. Schuchardt, R.M. Vargas, G. Gelbard , *J. Mol. Cat.* ,1995, **99**, 65 ; b) 1996, **109**, 37.

18- a) G. Gelbard, F. Vielfaure-Joly, *Tetrahedron Lett.* , 1998, **39**, 2743 ; b) *C.R. Acad. Sc. Ser. Chim.*, 2000, in press.

19- P. Tundo, *Synthesis* , 1978, 315.

20- a) N.M. Weinschenker, C.M. Shen, *Tetrahedron Lett.* , 1972 , 3281 ; b) D.H. Drewry, S.W. Gerritz, J.A. Linn, *Tetrahedron Lett.* , 1997, **38**, 3377.

21- Known as "Biodiesel" or Diester® ; essentially from rapeseed oil in Europe.

22- P. Gateau, J.C. Guibet, G. Hillion, R. Stern, *Rev. Instit. Fr. Petrole* , 1985, **40**, 509.

23- G. Gelbard, O. Brés, F. Vielfaure, U.F. Schuchardt, R.M. Vargas, *J.Amer. Oil Chem. Soc.* , 1995, **72**, 1239.

24- G. Gelbard, F. Vielfaure, unpublished results.

SELECTIVE SYNTHESIS OF 2-ACETYL-6-METHOXYNAPHTHALENE OVER HBEA ZEOLITE

E.Fromentin, J-M Coustard and M. Guisnet

Laboratoire de Catalyse en Chimie Organique, UMR CNRS 6503
Université de Poitiers, Faculté des Sciences
40 avenue du Recteur Pineau
86022 Poitiers Cedex, France

1 INTRODUCTION

Aromatic ketones are important intermediates in the production of fine chemicals and pharmaceuticals[1,2]. Thus, the anti-rheumatic Naproxen is produced by the Friedel-Crafts acetylation of 2-methoxynaphthalene into 2-acetyl-6-methoxynaphthalene and subsequent Willgerodt-Kindler reaction. Commercial acylation processes involve over-stoechiometric amounts of metal chlorides (e.g. $AlCl_3$) as catalysts and acid chlorides as acylating agents, which results in a substantial formation of by-products and in corrosion problems. This is why the substitution of these corrosive catalysts by solid acid catalysts and of acid chlorides by anhydrides or acids is particularly desirable.

This paper deals with the selective synthesis of 2-acetyl-6-methoxynaphthalene, precursor of Naproxen, over zeolite catalysts and especially over HBEA zeolites. As has been previously observed[3-8], acetylation of 2-methoxynaphthalene occurs preferentially at the kinetically controlled 1-position with formation of 1-acetyl-2-methoxynaphthalene (I). The desired isomer, 2-acetyl-6-methoxynaphthalene (II) and the minor isomer, 1-acetyl-7-methoxynaphthalene (III), are the other primary products. However, it will be shown that in presence of 2MN, isomerization of I can occur allowing a selective production of II, the desired product; the effect of the operating conditions (solvent, temperature) and of the acidity and porosity of the zeolite catalyst will be presented.

2 ACETYLATION OF 2-METHOXYNAPHTHALENE: REACTION SCHEME

Acetylation of 2-methoxynaphthalene (2MN) with acetic anhydride (AA) was carried out in a batch reactor over a HBEA zeolite with a framework Si/Al ratio of 15 (HBEA 15). The standard operating conditions were the following: temperature of 120°C, 500mg of catalyst previously activated at 500°C overnight under dry air flow, solution containing 35 mmol of 2MN ($C_{2MN} = 3,43 mol.l^{-1}$), 7 mmol of AA ($C_{AA} = 0,68 mol.l^{-1}$) and 4 cm^3 of nitrobenzene.

Acetic anhydride is completely consumed after 45 minutes reaction: total yield in acetylmethoxynaphthalene (AMN) equal to 100% (Figure 1). As expected, isomer I is formed in larger amount (60%) than isomer II and particularly than isomer III. After 50 hours reaction the yield in isomer I is equal to 21% and those in isomers II and III to 62 and 7% respectively, which demonstrates that I can isomerize into II and III. A decrease in the total yield in AMN is also detected : 10% after 50 hours reaction (Figure 1) indicating a deacylation process probably of isomer I. Indeed, deacylation is known to occur easily with ketones hindered in the ortho position[9].

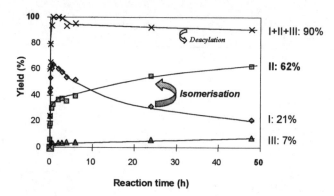

Figure 1   *Acetylation of 2-methoxynaphthalene with acetic anhydride over a HBEA zeolite (Si/Al = 15). Total yield in acetylmethoxynaphthalene (I + II + III) and yields in isomers I, II and III versus reaction time.*

Therefore the transformation of the 2-methoxynaphthalene acetic anhydride mixture over HBEA 15 can be proposed to occur through two main steps : fast acetylation of 2-methoxynaphthalene preferentially into isomer I then slow isomerization and deacylation of this product (Figure 2).

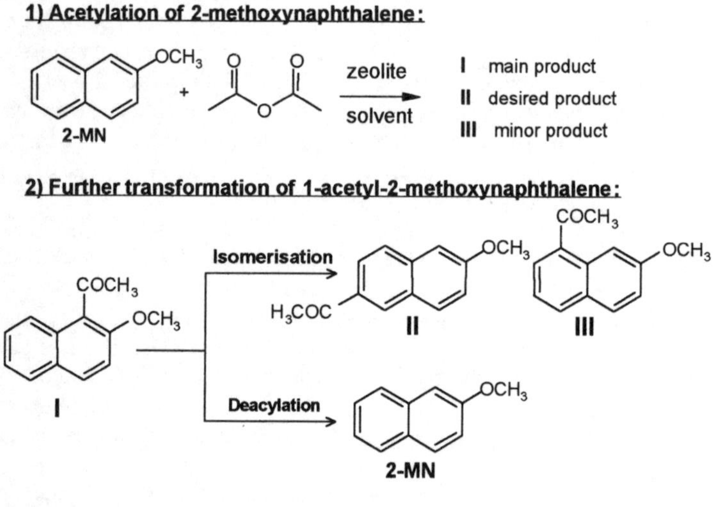

1) Acetylation of 2-methoxynaphthalene:

2) Further transformation of 1-acetyl-2-methoxynaphthalene:

Figure 2   *Transformation of 2-methoxynaphthalene with acetic anhydride. Reaction scheme.*

3   ISOMERIZATION MECHANISM

Isomerization of 1-acetyl-2-methoxynaphthalene (I) is a key step in the formation of the desired product (II). In order to optimize the production of II it is therefore important to understand how occurs this latter reaction.

The transformation of pure I was firstly investigated under the following conditions: 500 mg of HBEA, 20 mmol of I and 9,7 mmol of nitrobenzene, which corresponds to a total volume of 5 cm^3 and a concentration of isomer I 14 times higher than the maximum concentration obtained during 2MN acetylation with acetic anhydride. Under these conditions, the isomerization of I was found to be very slow : only 4% of II and III can be obtained after 40h, whereas during acetylation with acetic anhydride, 29% of these isomers could be formed during the same time through isomerization (Figure 1). This large difference in isomerization rates can be due to the absence of 2-methoxynaphthalene in the reaction mixture contrary to what is found in the acetylation mixture. To check this hypothesis, the transformations of I in the presence and in the absence of 2MN were compared under identical conditions of concentrations of I and of nitrobenzene (this polar solvent can indeed compete with reactants for adsorption on the acidic sites). The following conditions were chosen : 120°C, 4 mmol of isomer I, 9,7 mmol of nitrobenzene, and either 20 mmol of 2MN (A) or of 1-methylnaphthalene (B), (1-methylnaphthalene is little polar and undergoes no acetylation reactions). Figure 3 shows that isomerization is much faster in the presence of 2MN than in its absence : 10 times for the formation of isomer II and 5 times for the formation of III. Therefore, it can be concluded that both formation of isomers II and III occur essentially through transacylation of 2MN by isomer I :

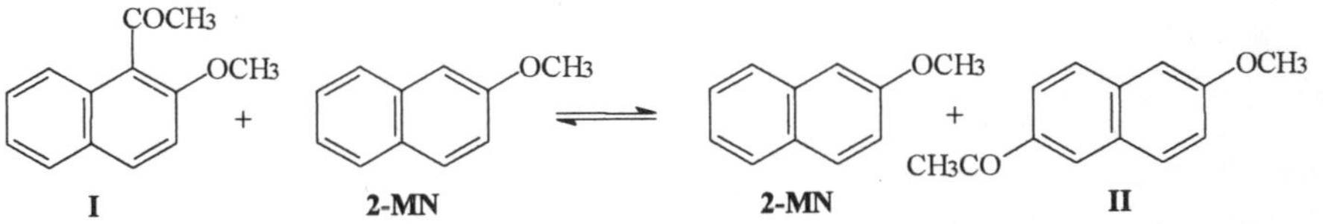

Furhtermore, the less pronounced effect of 2MN on the formation of III suggests that a small part of isomerization of I into III may occur through an intramolecular mechanism.

The predominantly intermolecular nature of acetylmethoxynaphthalene isomerization was confirmed by using a mixture of isomer I with a deuterated methoxy group (OCD_3) and of light 2MN as a reactant[10].

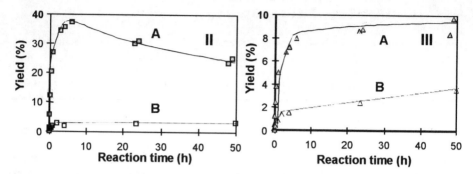

Figure 3   *Isomerization of 1-acetyl-2-methoxynaphthalene (I). Yields in isomers II and III versus reaction time in the presence (A) and in the absence of 2-methoxynaphthalene (B).*

4   INFLUENCE OF OPERATING CONDITIONS

4.1   Effect of Solvent

The effect of solvent polarity on the activity and selectivity of the HBEA zeolite was examined on the acetylation of 2MN with acetic anhydride as well as on the transacylation of 2MN with isomer I. The reactions were carried out in solvents of various polarities under the following conditions: temperature of 120°C, 500mg of HBEA and 35mmol of 2MN (3,43 mol.l^{-1}), 7mmol of AA (0,68mol.l^{-1}) and 4 cm^3 of solvent (sulfolane, nitrobenzene, 1,2-dichlorobenzene or 1-methylnaphthalene) for acylation and 20 mmol of 2MN (4 mol.l^{-1}), 4 mmol of I (1 mol.l^{-1}) in 3,3 cm^3 of solvent for transacylation. The E$_T$ parameter defined by Dimroth et al[11] was used for characterizing the solvent polarity.

Acetylation was found to be faster in 1,2-dichlorobenzene (E$_T$ = 0.225) than in the other solvents with higher polarity[12]: nitrobenzene (E$_T$ = 0.324) and sulfolane (E$_T$ = 0.41) and with lower polarity : 1-methylnaphthalene (E$_T$ = 0.142). Transacylation was also maximum for a solvent of intermediate polarity (nitrobenzene), a very low rate being obtained with the more polar solvent. Similar observations were made by Espeel et al.[12] for alkylation of phenol with cyclohexene and by Jayat et al.[14] for acylation of phenol by phenylacetate. According to these authors, the molecules of very polar solvents compete with the reactant molecules for diffusion inside the pores and for adsorption on the acid sites, reducing therefore significantly the rate of the main reaction. On the other hand, in the less polar solvents, the primary reaction products remain for a long time in the zeolite pores with consequently significant secondary reactions and an apparently slower main reaction. In agreement with this latter proposal, a significant deacylation is observed during transacylation in the 1-methylnaphthalene solvent and practically no deacylation in sulfolane.

4.2   Effect of Temperature

The effect of temperature was determined on acetylation only, the reaction being carried out at 120 and 170°C under the standard conditions. Temperature has a significant effect not only on the rate of 2MN transformation but also on the selectivity (Figure 4). A significant increase in selectivity to the desired isomer (II) is observed, due at least partly to a greater activation energy for isomerization than for acetylation. Unfortunately temperature increases also the rate of deacylation. However, Figure 4 shows that a high yield in isomer II ($\approx$ 83 %) can be obtained by operating at 170°C and at short reaction time (0.5 h).

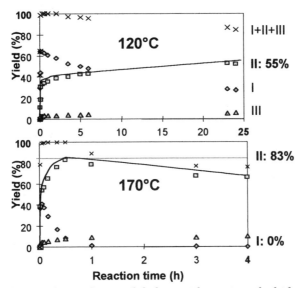

Figure 4   *Acetylation of 2-methoxynaphthalene with acetic anhydride over a HBEA zeolite (Si/Al = 15) at 120°C and at 170°C. Yield in acetylmethoxynaphtha- lenes (I + II + III) and yields in isomers I, II and III versus reaction time.*

5   INFLUENCE OF THE CATALYST

5.1   Influence of Pore Structure

Acetylation of 2MN with acetic anhydride was compared over two large pore zeolites: HFAU and HBEA and over an average pore size zeolite: HMFI. The framework Si/Al ratios of these zeolites were equal to 19, 15 and 40 respectively. Table 1 shows that the activity per protonic site (able to retain pyridine adsorbed at 150°C) i.e. the turnover frequency TOF, for the formation of all the acetylmethoxynaphthalenes decreases with the pore size. Thus, TOF for formation of isomer I on HFAU 19 is 7 times greater than on HBEA 15 and 21 times than on HMFI 40. For isomer II which is less bulky, the difference between TOF for HFAU and HBEA is smaller whereas the difference between HFAU 19 and HMFI 40 is similar than for isomer I.

However while HFAU is the most active for acetylation, the subsequent isomerization of I into II is slower and deacylation faster than with HBEA. Thus after 25 hours reaction, there is 22% of deacylation with HFAU 19 compared to 9% with HBEA 15; the yield in isomer II is only equal to 30% with HFAU 19 against 55% with HBEA.

Table 1   *Acetylation of 2-methoxynaphthalene over various zeolites. Turnover frequency (TOF) for formation of acetylmethoxynaphthalene isomers I, II and III.*

Si/Al fram	TOF (I+II+III) (h^{-1})	TOF (I) (h^{-1})	TOF(II) (h^{-1})	TOF (III) (h^{-1})
HFAU 19	2300	2100	170	30
HBEA 15	420	310	100	10
HMFI 40	105	95	10	0

5.2 Influence of the Si/Al ratio of the HBEA zeolite

The acetylation of 2MN was carried out under the standard conditions over a series of HBEA zeolites resulting from dealumination of a commercial sample with a framework Si/Al ratio of 15, by treatment with chlorhydric acid[15] or with ammonium hexafluorosilicate[16]. The initial reaction rate passes through a maximum for a number of Al atoms per unit cell between 2 and 3 (i.e. for a framework Si/Al ratio between 20 and 40) and this whatever the mode of dealumination. An important remark is that on these dealuminated zeolites, the number of acidic Lewis sites is very low compared to the non dealuminated sample. This confirms that acidic Lewis sites do not participate in acetylation of 2MN[16].

Table 2 shows that high yields of II and III (> 70%) can be obtained with HBEA 20 and 40 than with HBEA15 and this for shorter reaction times.

Table 2   *Acetylation of 2-methoxynaphthalene with acetic anhydride. Distribution of acetyl-methoxynaphthalene isomers (I, II and III), total yield, yields in I and II+III as a function of reaction time.*

Catalyst	Reaction Time (h)	Distribution (%)			Yield (%)		
		I	II	III	Total	I	(II+III)
	0	75.5	22.5	2.0	0	0	0
HBEA 15	0.75	63.6	33.2	3.2	100	63.6	36.4
	25	34.3	59.4	6.3	91.9	31.5	60.4
	50	**23.4**	**68.4**	**8.2**	**90.9**	**21.3**	**69.6**
	0	79	19	2	0	0	0
HBEA 20	0.5	68.4	28.4	3.2	100	68.4	31.6
	25	**16.7**	**73.2**	**10**	**88.7**	**14.8**	**73.9**
	49	6.7	81.2	12.1	63.7	4.3	59.4
	0	80.5	17.5	2	0	0	0
HBEA 40	0.5	67.4	29.2	3.4	100	67.4	32.6
	25	**12.9**	**77.3**	**9.8**	**82.3**	**10.6**	**71.7**
	49.5	5.4	82.8	11.8	75.8	4.1	71.7

6 CONCLUSION

Over HBEA zeolites, acetylation of 2-methoxynaphthalene with acetic anhydride leads mainly to 1-acetyl-2-methoxynaphthalene. However, the desired product, i.e. 2-acetyl-6-methoxynaphthalene, precursor of Naproxen is obtained at long reaction time by an intermolecular irreversible isomerization process. A very selective production of II (83%) can be obtained by acetylation of 2-methoxynaphthalene over a commercial HBEA zeolite (Si/Al = 15) at 170°C, with nitrobenzene as a solvent. With dealuminated HBEA samples (framework Si/Al ratio between 20 and 40), better results could be expected. Furthermore, preliminary experiments showed that this selective synthesis of 2-methoxynaphthalene can be carried out in a flow reactor system.

References

1. H. G. Franck, J. W. Stadelhofer, *Industrial Aromatic Chemistry*, Springer-Verlag, Berlin/Heidelberg, 1988.
2. H. W. Kouwenhoven and H. van Bekkum in *Handbook of Heterogeneous Catalysis*, G. Ertl, H. Knözinger and J. Weitkamp, Eds., Wiley, 5, p. 2358, 1997.
3. G. Harvey and G. Mäder, *Collect. Czech. Chem. Commun.*, 1992, **57**, 862.
4. G. D. Yadav, M., M.S. Krishnan, *Chemical Engineering Science*, 1999, **54**, 4189.
5. G. Harvey, G. Binder and R. Prins, *Stud. Surf. Sci. Catal.*, 1995, **94**, 397.
6. H. K. Heinichen and W.F. Hölderich, *J. Catal.*, 1999, **185**, 408.
7. E. A. Gunnewegh, S. S. Gopie, H. van Bekkum, *J. Mol. Catal.*, 1996, **106**, 151.
8. D. Das, S. Cheng, *Appl. Catal. A : General*, 2000, **201**, 159.
9. Al-Ka'bi, J., Farooqi, J. A., Gore, P. H., Nassar, A. M. G., Saad, E. F., Short, E. L., Waters, D. N., *J. Chem. Soc. Perkin Trans. II*, 1988, 943.
10. E. Fromentin, J-M Coustard and M. Guisnet, *J. Catal.*, 2000, **190**, 433.
11. K. Dimroth, C. Reichardt, T. Siepmann and F. Bohlmann, *Liebigs Ann. Chem.*, 1963, **661**, 1.
12. E. Fromentin, J-M Coustard and M. Guisnet, *J. Mol. Catal.*, accepted for publication.
13. P. H. J. Espeel, K. A. Vercruysse, M. Debaerdemaker, P. A. Jacobs, *Stud. Surf. Sci. Catal.*, 1994, **84**, 1457.
14. F. Jayat, M.J. Sabater Picot, D. Rohan and M. Guisnet, *Stud. Surf. Sci. Catal.*, 1997, **108**, 91.
15. M. Guisnet, P. Ayrault, C. Coutanceau, M.F. Alvarez, J. Datka, *J. Chem. Soc., Faraday Trans.*, 1997, **93**, 1661.
16. A. Berreghis, P. Ayrault, E. Fromentin and M. Guisnet, *J. Catal.*, accepted for publication.

THE INFLUENCE OF "SUPERACIDIC" MODIFICATION ON ZrO_2 AND Fe_2O_3 CATALYSTS FOR METHANE COMBUSTION

A. S. C. Brown[a,1], J. S. J. Hargreaves[a,2,*], M-L. Palacios[b], S. H. Taylor[b].

a Catalysis Research Laboratory, Department of Chemistry and Physics, Nottingham Trent University, Clifton Lane, Nottingham NG11 8NS, UK.

b Department of Chemistry, Cardiff University, PO Box 912, Cardiff CF10 3TB, UK.

[1]present address : School of Chemistry, Queen's University of Belfast, Belfast BT41 1PB, UK.
[2]no longer at this address.

[*]to whom correspondence should be addressed, e-mail justin-hargreaves@supanet.com

Abstract.
A range of metal oxides have been compared as methane combustion catalysts. The effect of modification to generate "superacidic" behaviour on ZrO_2 and Fe_2O_3 systems has been studied. It has been shown that whilst sulfation lowers the activity of Fe_2O_3, sulfation and, particularly, molybdation enhance the performance of ZrO_2. Despite enhancing the activity of the unmodified base oxides, the addition of low levels of platinum has been demonstrated to poison the activity of "superacidic" zirconias. Potential reasons for these observations are discussed.

Introduction.
In recent years, the ability of some metal oxides to catalyse reactions characteristic of very strong acids when sulfated, molybdated or tungstated has attracted much interest [1]. Initially, primarily based upon Hammett acidity indicator and reaction measurements it was proposed that such systems were "superacidic", although the consensus is now emerging that this is not the case [2]. Whatever the origin of activity, it is clear that these systems have high efficacy for hydrocarbon activation, with n-butane isomerisation at room temperature being reported in some cases [3]. With this in mind, we have applied a series of such catalysts to methane oxidation, a reaction where methane activation is generally rate determining. We have investigated both selective oxidation to produce methanol, a reaction known to be limited by the contribution of gas-phase radical chemistry [4], and catalytic combustion which is of interest for the NO_x free production of energy [5]. Details have been reported elsewhere, with sulfation promoting methanol production in the case of iron oxide catalysts [6,7] and molybdation of zirconia at close to monolayer level enhancing combustion performance [8,9].

In the current contribution, we give an overview of the performance of "superacidic" materials and compare their activity with simple and mixed oxides reported to be of potential interest for catalytic combustion of methane. The effect of platinum doping has also been investigated, since it is known that supported platinum is an effective methane combustion catalyst in its own right [5], and it has been reported that Pt doping can increase the isomerisation activity of "superacidic" metal oxides [10]. Here we limit our discussion to modified Fe_2O_3 and ZrO_2 catalysts, since the overall data set has been discussed elsewhere [9].

Experimental.
Catalyst Preparation.
Catalyst preparation has been described elsewhere [9]. For the sake of brevity only the preparation of materials specifically discussed in this manuscript is given.
ZrO_2 systems. Zirconium hydroxide was precipitated from a 0.5 M solution of zirconium basic carbonate (MEL Chemicals) in 1 L of a 1:1 HCl:distilled H_2O mixture by the addition of a 35% ammonia solution until pH 10 was achieved. The resultant precipitate was then filtered, washed with 4 x 1L of distilled water and dried at 110 C overnight. Molybdenum oxide doping was performed using appropriate amounts of ammonium heptamolybdate tetrahydrate (Aldrich, 98%) in distilled water (0.445 ml g^{-1} $Zr(OH)_4$) to obtain loadings corresponding to 2, 5 and 10 wt% MoO_3/ZrO_2. The samples were dried at 110°C overnight prior to calcination at 800°C for 12 h in static air. ZrO_2 was prepared by calcination of zirconium hydroxide at 800 C for 12h in static air. Sulfated zirconia was prepared by calcining sulfated zirconium hydroxide (MEL Chemicals) at 800°C for 12h in static air.
Fe_2O_3 systems. Fe_2O_3 (Aldrich, 99%) was calcined at 800C for 12h in static air.
Sulfated iron oxide was prepared via goethite as follows: 100 ml of a 1 M solution of iron (III) nitrate nonahydrate (Avocado, 98+%), prepared using distilled water, was added to a polythene screw top bottle. To this, 180 ml of a 5 M sodium hydroxide solution was added with stirring. This solution was then diluted to 500 ml with distilled water and aged in an oven at 70°C for 60 h. The resultant yellow precipitate was then filtered, washed with distilled water and dried in a vacuum oven for two days. X-ray diffraction analysis showed the resultant material to be goethite. Two grams of this material was then sulfated by immersion in 0.5 M sulfuric acid for 30 minutes, following which the suspension was filtered and the filtrate was dried in an oven at 110°C overnight. The sample was then calcined at 550°C for 3 h and at 800°C for a further 12h in static air.
Pt doping. Platinum doping of samples was achieved using the impregnation of hexachloroplatinic acid (Aldrich, 8 wt% aqueous solution of Pt) corresponding to a 0.5 wt% loading of Pt. The samples were then dried at 110°C overnight and calcined in a 100 ml min^{-1} flow of helium (Air Products, 99.999%) at 300°C for 3 h prior to activity testing.
All materials tested were pelleted and sieved to yield 0.6 – 1.0 mm particles prior to activity testing.

Table 1 *Catalyst Surface Areas Prior to Reaction*

Material	Surface Area $(m^2\ g^{-1})$
ZrO_2	17
SO_4^{2-}/ZrO_2	17
2 wt.% MoO_3/ZrO_2	26
5 wt.% MoO_3/ZrO_2	33
10 wt.% MoO_3/ZrO_2	48
MoO_3	9
$Zr_xCe_{1-x}O_2$	27
10 wt.% CeO_2/ZrO_2	16
Fe_2O_3	5
CuO	2
0.5 wt.% Pt/ZrO_2	17
0.5 wt.% $Pt/SO_4^{2-}/ZrO_2$	17
0.5 wt.% $Pt/5$ wt.% MoO_3/ZrO_2	33
0.5 wt.% Pt/Fe_2O_3	5

Catalyst Testing.

Catalyst performance was evaluated in a fixed-bed microreactor. A stainless steel jacketed quartz reactor tube was used in which 0.75 ml of catalyst was held centrally within the heated zone of a furnace by quartz wool plugs. Methane (Air Products, 99%), oxygen (Air products, 99.6%) and helium (Air Products, 99.999%) were flowed over the catalysts in the ratio 10:40:250 ml min^{-1} to give a GHSV of ~24000 h^{-1} using Brooks 5850 TR mass – flow controllers. All lines downstream of the reactor were trace heated to a temperature in excess of 150°C to prevent condensation of the products. Analysis was performed on-line using a Varian Saturn GCMS equipped with a thermal conductivity detector. Megabore Poraplot GS-Q and Megabore Molsieve columns were used to perform the separation.

In all cases the reactor was allowed to stabilise for 1h under the conditions reported and the results are the mean of three analyses made at steady state. Unless otherwise stated, the carbon balances of data reported are 100 ±3%. Methane conversion has been calculated on the basis of the difference of inlet and outlet concentration.

Catalyst Characterisation.

Catalyst surface areas prior to reaction were determined by application of the BET method to nitrogen physisorption isotherms determined at 77K.

Temperature programmed reduction studies were conducted using a flow rate of 40 ml min^{-1} of a 5% hydrogen in argon mixture (Air Products) with a ramp rate of 5°C min^{-1} up to a temperature of 1000°C. The data reported correspond to a sample mass of 0.26g for 2 and 5 wt% MoO_3/ZrO_2 and 0.13g of 10wt% MoO_3/ZrO_2.

Laser Raman spectroscopy was performed with powder specimens using a Renishaw System 1000 laser Raman microscope. An argon ion laser (514.5 nm) was used as an excitation source, typically the laser was operated at relatively low power of 20 mW, minimising sample damage. Spectra were collected using a back scattering geometry with a 180°C angle between the illuminating and the collected light. The type of detector used was a CCD (charge coupled device) camera and the laser was focused onto the sample by means of an Olympus BH2-UMA optical microscope. *In situ* studies were carried out using a Linkam TS 1500 environmental cell. The sample was

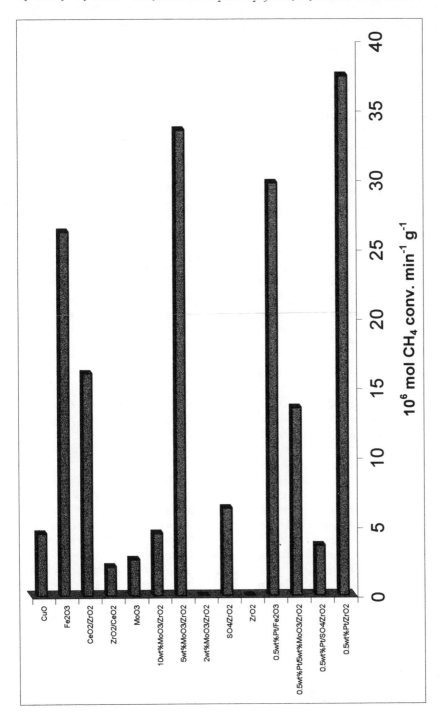

Figure 1 *Mass normalised activity*

contained in a ceramic furnace with a thermocouple placed in intimate contact. A ramp rate of 1°C min $^{-1}$ to the appropriate temperature was used, and the temperature was maintained constant whilst Raman spectra were acquired with 10 accumulations each of 20 seconds each. A feed ratio of $CH_4:O_2:He$ of 1:4:25 was used, being the total gas flow rate 45 ml min^{-1}. Gas flow over the sample was regulated by electronic thermal mass flow controllers and the laser focused on the sample through a water-cooled silica window in the cover of the cell.

Powder x-ray diffraction, used for phase composition analysis, was performed using a Hiltonbrooks modified Phillips powder diffractometer using Cu Kα radiation operating at 42.5 kV and 18.0 mA. Samples were prepared by compaction into a glass-backed aluminium sample holder and were scanned in the rage 5-80 degrees two theta using a step size of 0.02 degrees and a counting rate of 0.5 s/step. The contribution of Cu Kα$_2$ broadening was removed by application of the Rachinger correction.

Results and Discussion.

The catalyst surface areas prior to reaction and the mass normalised and surface area normalised combustion activities are reported in Table 1 and Figures 1 and 2 respectively. In all cases catalysts displayed 100% selectivity towards carbon dioxide, with the exceptions of 10wt% MoO_3/ZrO_2 (where there was a 30% selectivity to carbon monoxide), 2wt% MoO_3/ZrO_2 (for which there was a poor carbon balance and evidence for the formation of a carbonaceous deposit during reaction) and ZrO_2 (for which no activity was observed). As mentioned previously, preliminary discussion of the overall activity patterns reported in Figures 1 and 2 has been presented elsewhere [9]. Here we present a discussion of the various classes of "superacid" studied.

(i) ZrO_2 based systems.

"Superacidity" generated on ZrO_2 by both sulfation and molybdation procedures markedly enhances its activity for combustion. The effect of molybdation is particularly pronounced, with the intermediate loading producing the highest activity. This has previously been ascribed by us to the formation of an active monolayer of MoO_3 species supported on ZrO_2 during reaction [8].

Figure 3 shows the phase composition of ZrO_2 (as determined by application of Toraya's method [11] to powder x-ray diffraction patterns) as a function of MoO_3 content for the catalysts prior to reaction. In accordance with previous literature, it can be seen that both surface area and the tetragonal phase content of ZrO_2 are stabilised by the inclusion of increasing amounts of MoO_3. It is possible to explain the observed relationship between these two parameters on the basis of the work of Garvie at al. [12], which proposes that, due to surface free energy considerations, the phase composition of ZrO_2 is crystallite size dependent, with smaller sizes favouring the tetragonal polymorph. Since powder X-ray diffraction patterns did not evidence discrete molybdenum containing phases, in further characterisation studies the MoO_3/ZrO_2 catalysts have been studied by Raman spectroscopy. The spectra of catalysts prior to reaction are given in Figure 4, where it is apparent that the form of molybdenum oxo species present is dependent upon loading. On the basis of previous literature, features at ca 915 and ca. 870 cm^{-1}, which are observed in all three catalysts, have been assigned to co-ordinated surface molybdates [13]. As loading increases, additional bands at ca. 988 and ca. 816 cm^{-1} become apparent. These can be assigned

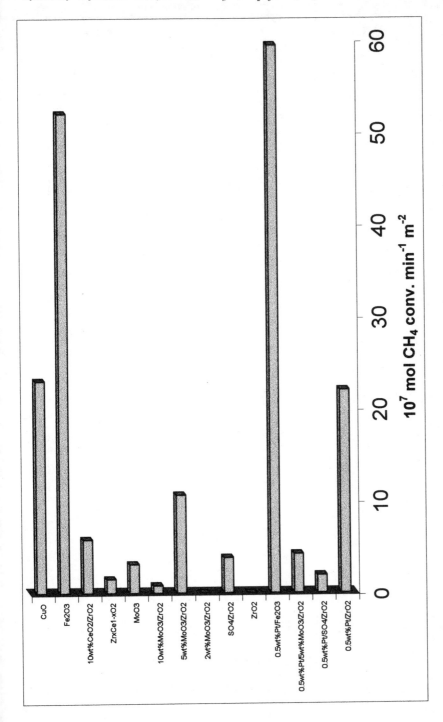

Figure 2 *Surface area normalised activity*

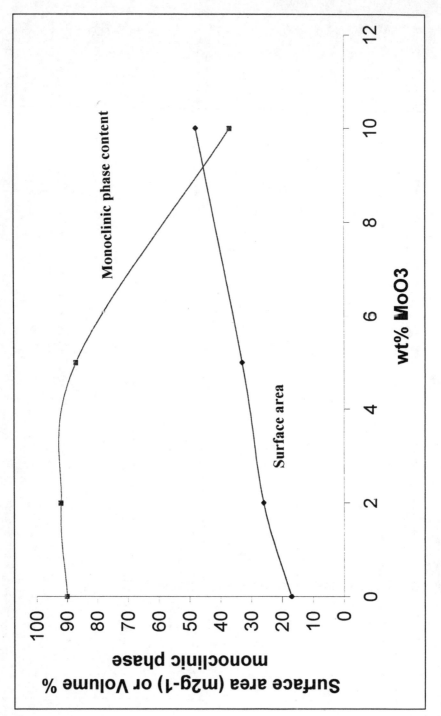

Figure 3 *BET and phase composition of Mo₃/ZrO₂ catalysts as a function of MoO₃ content*

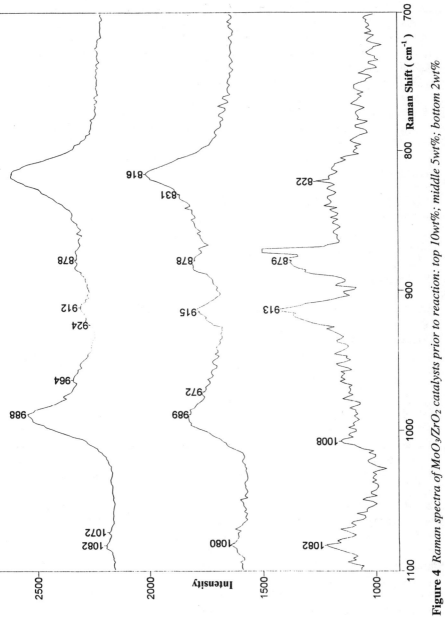

Figure 4 *Raman spectra of MoO$_3$/ZrO$_2$ catalysts prior to reaction: top 10wt%; middle 5wt%; bottom 2wt%*

to Mo=O and Mo-O-Mo stretches of bulk crystallites respectively [14] and it can therefore be concluded that non-XRD detectable crystallites are present in both the 5 and 10 w% MoO_3/ZrO_2 catalysts prior to reaction. Furthermore, on the basis of the relative intensities of these two bands, it can be concluded that differences in the dispersion between the 5 and 10wt % materials are small [8]. Although the form of molybdenum oxide is similar in both systems, its reducibilty is different. Figure 5 presents temperature programmed reduction profiles for the three MoO_3/ZrO_2 catalysts. Blank experiments have demonstrated that there is no contribution of ZrO_2 reduction to these profiles. Consistent with literature observations [15], the reduction is a two stage process and occurs at lower temperatures than for bulk MoO_3. In the case of the 10wt% sample, an asymmetry is present on the lower temperature maximum and there is an additional high temperature peak. These observations can be ascribed to the mixture of the polymorphological forms of the ZrO_2 support present (63% tetragonal and 37% monoclinic). We therefore conclude that the reducibilty of MoO_3/ZrO_2 is dependent upon the phase of the ZrO_2 support present and is greater for the monoclinic form.

Given that the Tamman temperature of MoO_3 is only 534K [16], under our reaction conditions it is probable that the MoO_3 crystallites evidenced in the Raman studies are highly mobile during reaction. In this context, and in view of the pronounced dependence of combustion activity upon MoO_3 loading, we propose that it is significant that we have calculated the content of MoO_3 to be 48, 95 and 131 % of a monolayer for the 2, 5 and 10 wt% MoO_3/ZrO_2 samples respectively [8]. It is our view that the active form of the catalyst comprises a fluid monolayer under reaction conditions. On this basis the dependence of activity as a function of MoO_3 content can be explained, with the lower activity of the 10wt% sample being a combination of site blocking and the intrinsically lower reducibility of MoO_3 on the tetragonal polymorph of ZrO_2. In the case of 2wt% MoO_3/ZrO_2 which possesses a similar phase composition to the optimum catalyst, the initial form of the molybdenum species is different as evidenced in Figure 4. In-situ Raman spectra of the catalysts taken during methane oxidation and presented in Figure 6 indirectly support our view of the working catalyst as it is evident that all the molybdenum structural features have been lost.

In the case of sulfate addition where enhancement of the activity of ZrO_2 is also observed, it is tempting to ascribe this effect to enhanced acidity facilitating methane activation. However, the consensus is now emerging that sulfation does not in fact increase the acid strength of ZrO_2 markedly [2]. With this in mind, one possible explanation may be that the enhancement of activity we observe is due to the oxidation activity of residual sulfate groups themselves, although it is known that most of the sulfate groups will be lost at the temperature employed for the calcination in this study. A fuller investigation examining the activity dependence on sulfate content , morphological effects of sulfation and also the nature of sulfate binding to the ZrO_2 surface would be required to establish our suggestion.

Despite being reported to enhance low temperature isomerisation activity, it is apparent from Figures 1 and 2 that the inclusion of Pt in the SO_4^{2-}/ZrO_2 and 5wt% MoO_3/ ZrO_2 catalysts has reduced their activity for methane combustion. This may be somewhat surprising because platinum is known to be an effective catalyst for the reaction in its own right and also Pt/ZrO_2 can be seen to be one of the most effective catalysts in the

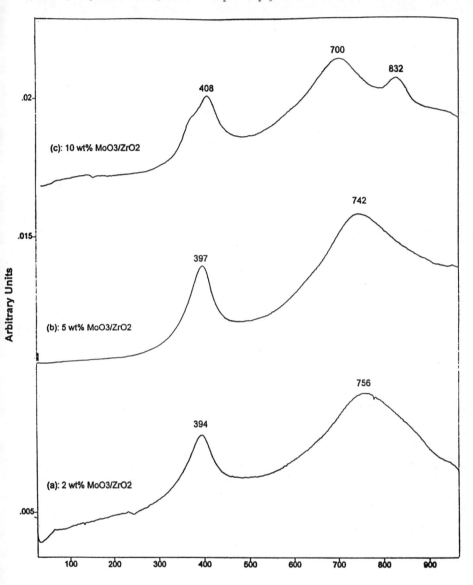

Figure 5 *Temperature programmed reduction studies of MoO$_3$/ZrO$_2$ catalysts*

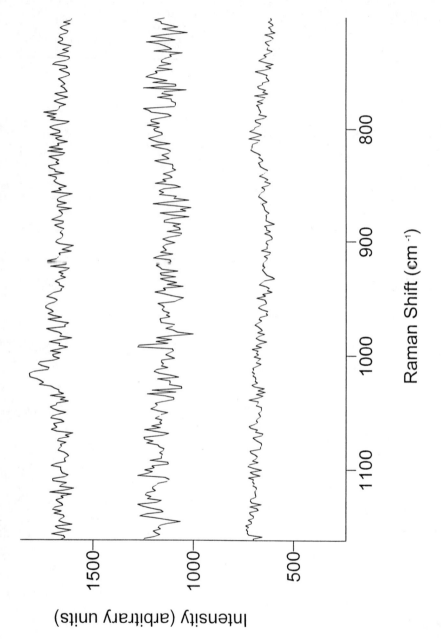

Figure 6 *In situ Raman spectra of MoO₃/ZrO₂ catalysts at 800 °C: top 2wt%; middle 5wt%; bottom 10wt%*

present study. Our data indicates that the inclusion of Pt in "superacidic" forms of ZrO$_2$ actually removes sites active for combustion. We have previously explained this on the basis of stabilisation of higher oxidation states of platinum and/or the formation of inactive compounds with the active phase of the catalyst.   In this context, it is particularly interesting to note that a recent study by Yori and Parera [17] of platinum doping on the related WO$_3$/ZrO$_2$ "superacidic" system has concluded that the form of Pt (i.e. metallic versus non-metallic) is dependent upon the polymorphological form of the ZrO$_2$ support. It is argued that the form of Pt associated with tetragonal ZrO$_2$ catalyst systems is non-metallic. In our case it is clear that the inclusion of platinum is exerting a negative effect on the overall conversion and that, therefore, it is somehow either destroying or reducing the activity of sites. In our studies, Pt inclusion is observed to actually alter the phase composition of the support in the 5wt% MoO$_3$/ZrO$_2$ catalysts, enhancing the presence of the tetragonal form (5wt % MoO$_3$/ZrO$_2$ being 87% monoclinic and 13% tetragonal whereas 0.5wt% Pt 5wt% MoO$_3$/ ZrO$_2$ is 69% monoclinic and 31% tetragonal from Toraya's method). This observation may explain, at least in part, the deleterious effect of Pt since, as we have discussed above, the reducibility of MoO$_3$ is lower on tetragonal ZrO$_2$. Despite the similarity in phase composition (69 and 63% tetragonal respectively) it is evident from Figure 2 that the specific activity of 0.5wt% Pt 5wt% MoO$_3$/ZrO$_2$ is still higher than 10wt% MoO$_3$/ZrO$_2$ by a factor of approximately 4, which indicates that monolayer capacity may still be an important consideration.   The lack of activity associated with the Pt component is possibly associated with the formation of compounds containing stabilised higher oxidation states.

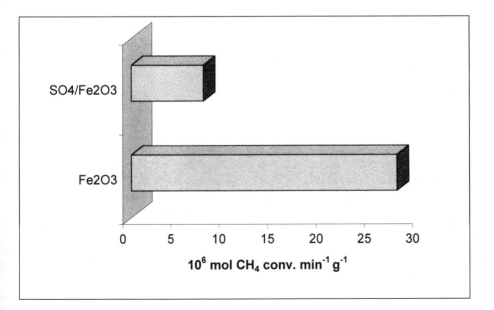

Figure 7 *Mass normalised activity*

(ii) Fe₂O₃ based systems.

In previous studies of selective methane oxidation conducted at lower temperatures and elevated pressures, we have observed that sulfation enhances the oxidation activity of iron oxide at the upper range of temperature applied (500 and 550°C) [16,17]. It was therefore of interest to investigate the effect of sulfation on the activity of Fe_2O_3 for methane combustion and the results are presented in Figure 7. Unlike ZrO_2, the base Fe_2O_3 was observed to possess activity for reaction and, on a surface area normalised basis, was the most active non-platinum doped oxide in our study. Again unlike ZrO_2, sulfation is observed to dramatically lower activity. It is possible to rationalise this observation in terms of the bare Fe_2O_3 possessing sites which are highly active for the reaction at the temperature studied and which are blocked by residual sulfate groups. Powder X-ray diffraction has demonstrated that sulfation does not effect phase composition of Fe_2O_3, which is haematite (α-Fe_2O_3) in both cases. Even though we have proposed that such sulfate groups are active in the case of ZrO_2, if they possess a lower intrinsic activity than the sites they are blocking in Fe_2O_3, it is clear that their presence will reduce the overall activity of the catalysts. Sulfation therefore promotes the activity of an inactive or low activity catalyst whilst lowering the performance of a highly active catalyst. Again, as for the case of ZrO_2, the addition of platinum to the base oxide is observed to promote combustion efficacy which is consistent with its known activity.

Although the inclusion of sulfate in the Fe_2O_3 catalyst has a negative effect on catalytic activity in this study, we have shown elsewhere that sulfation of goethite (α-FeOOH) precursors may be a potentially simple route to the preparation of functionalised nanoporous haematite [18,19].

Conclusion.

The effect of chemical modification to enhance the "superacidity" of various metal oxides for methane combustion activity has been investigated. It has been observed that such modification produces different relative effects, depending upon the original activity of the unmodified oxide. ZrO_2, which was observed to be inactive, is promoted by sulfation and molybdation whereas Fe2O3, which is amongst the most active single oxides tested, is poisoned by sulfation. This observation has been interpreted on the basis of the comparative oxidation activities of the additional components – in the case of ZrO_2 it is argued that SO_4^{2-} and MoO_3 entities have a higher activity than bare ZrO_2, whereas in Fe_2O_3, SO_4^{2-} groups block sites of higher intrinsic activity. Furthermore, the MoO_3/ZrO_2 catalysts are considered to consist of mobile molybdenum oxide phases under reaction conditions, with the most active catalyst comprising a fluid monolayer in contact with a predominantly monoclinic phase support. Poisoning of this catalyst, and also SO_4^{2-}/ZrO_2, by the inclusion of low levels of Pt which is generally observed to enhance the activity of other systems has been explained on the basis of inactive/lower activity phase formation involving the MoO_3 and SO_4^{2-} derived sites respectively. In the case of MoO_3/ZrO_2 poisoning, it is proposed that the enhanced tetragonal ZrO_2 phase content observed on the addition of Pt may, at least, partly explain its negative effect.

Acknowledgement.

We gratefully acknowledge MEL Chemicals for kindly providing some of the materials used in this study. In particular, we are grateful to Peter Moles, Gary Monks and Colin Norman.

References.
[1] K. Arata, Appl. Catal. A 1996 146 3.
[2] A. S. C. Brown and J. S. J. Hargreaves, Green Chem. 1999 1 17.
[3] C-Y. Hsu, C. R. Heimbuch, C. T. Armes and B. C. Gates, J. Chem. Soc.,
Chem. Commun. 1992 1645.
[4] T. J. Hall, J. S. J. Hargreaves, G. J. Hutchings, R. W. Joyner and S. H.
Taylor, Fuel. Proc. Technol. 1995 42 151.
[5] J. W. Geus (Ed.), Catal. Today Vol. 47 No. 1.
[6] A. S. C. Brown, J. S. J. Hargreaves and B. Rijniersce, Catal. Lett. 1998 53 7.
[7] A. S. C. Brown, J. S. J. Hargreaves and B. Rijniersce, Topics in Catal. 2000
11/12 181.
[8] A. S. C. Brown, J. S. J. Hargreaves and S. H. Taylor, Catal. Lett. 1999 57
109.
[9] A. S. C. Brown, J. S. J. Hargreaves and S. H. Taylor, Catal. Today 2000 59
403.
[10] M. Hino and K. Arata, J. Chem. Soc., Chem. Commun. 1995 789.
[11] H. Toraya, M. Yoshimura and S. Somiya, J. Am. Ceram. Soc. 1984 67 C-
119.
[12] R. Garvie, J. Phys. Chem. 1965 69 1238.
[13] M. R. Smith, L. Zhang, S. A. Driscoll and U. S. Ozkan, Catal. Lett. 1993 19
1.
[14] E. M. Gaigneaux, D. Herla, P. Tsiakaras, U. Roland, P. Ruiz and B. Delmon
in Heterogeneous Hydrocarbon Oxidation, Am. Chem. Soc. Symposium Series, Vol.
638, eds B. K. Warren and S. T. Oyama 1996.
[15] A. Parmaliana, F. arena, F. Frusteri, G. Martra, S. Coluccia and V. D.
Sokolovskii, Stud. Surf. Sci. Catal. 1997 110 347.
[16] Y. Chenand L. Zhang, Catal. Lett. 1992 12 51.
[17] J. C. Yori and J. M. Parera, Catal. Lett. 2000 65 205.
[18] M. A. Edwards, A. S. J. Baker, A. S. C. Brown, J. S. J. Hargreaves and C.
J. Kiely, Inst. of Phys. Conf. Ser. 1999 161 549.
[19] A. S. J. Baker, A. S. C. Brown, M. A. Edwards, J. S. J. Hargreaves, C. J.
Kiely, A. Meagher and Q. A. Pankhurst, J. Mater. Chem. 2000 10 761.

STRUCTURE AND REACTIVITY OF POLYMER-SUPPORTED CARBONYLATION CATALYSTS

Anthony Haynes, Peter M. Maitlis, Ruhksana Quyoum, Harry Adams and Richard W Strange

Department of Chemistry
University of Sheffield
Brook Hill, Sheffield S3 7HF

1 INTRODUCTION

The use of transition metal catalysts dissolved in a liquid reaction medium is a feature of many commercial processes. One of the best known and most important examples is the manufacture of acetic acid from methanol and carbon monoxide. The "Monsanto process" developed in the late 1960's, uses a rhodium catalyst with an iodide promoter to produce acetic acid with very high selectivity (> 99% based on MeOH).[1-3] This rhodium-based technology became the predominant method for manufacture of acetic acid, and was eventually acquired by BP Chemicals in 1986. Despite the great success of the Monsanto process, efforts to find an improved catalyst for methanol carbonylation have continued. The most significant breakthrough in recent times has been the introduction by BP Chemicals of the *Cativa*™ process in 1995, which utilises a homogeneous promoted iridium catalyst that gives even higher activity and selectivity than rhodium.[4] An important additional requirement for all homogeneous processes, however, is that the dissolved catalyst must be separated from the liquid product and recycled to the reactor without significant catalyst loss. A major goal of catalytic chemists is to immobilise (or "heterogenise") the homogeneous catalyst on a solid support in order to confine the catalyst to the reactor and overcome the need for catalyst recycle.

1.1 Immobilised carbonylation catalysts

A number of types of solid support have been employed to heterogenise rhodium catalysts for methanol carbonylation. These were reviewed[2] by Howard *et al* in 1993 and include activated carbon, inorganic oxides, zeolites and a range of polymeric materials. One frequent approach to catalyst immobilisation is covalent attachment, in which the support (*e.g.* carbon or polystyrene) is modified to contain a pendant group (usually a phosphine) capable of acting as a ligand for the metal complex.[5] However, the susceptibility of covalently bound systems to metal-ligand bond cleavage makes irreversible leaching of the catalyst from the solid support a serious problem. This is particularly the case for methanol carbonylation, as pendant phosphines are prone to degradation by the aqueous acidic medium and high methyl iodide levels.

This paper describes some of the results of a collaborative project which has explored an alternative strategy for catalyst immobilisation, through ion-pair interactions between ionic catalyst complexes and polymeric ion exchange resins. Mechanistic studies of the rhodium carbonylation catalyst, in particular by Forster and co-workers at Monsanto,[1] and more recently by the Sheffield group,[3,6] have given a detailed understanding of the catalytic cycle, depicted in Figure 1. All four rhodium complexes in this cycle are anionic, which makes this system an attractive candidate for ionic attachment. This was first realised in 1980 by Drago *et al*, who published results suggesting the effective immobilisation of the rhodium catalyst on polymeric supports based on methylated polyvinylpyridines.[7] The activity was reported to be equal to the homogeneous system at 120 °C with minimal leaching of the supported catalyst, which was identified as $[Rh(CO)_2I_2]^-$ by infrared spectroscopy.

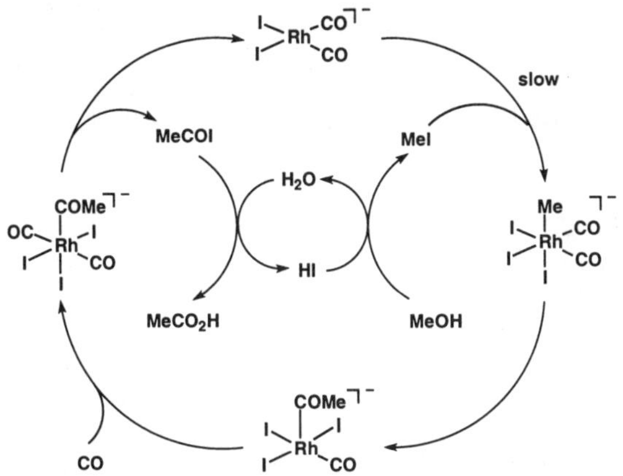

Figure 1 *Cycle for rhodium/iodide catalysed methanol carbonylation*

There has recently been a resurgence of interest in the ionic attachment strategy, both from academic groups[8] and industry.[9] Most significantly, in 1998 Chiyoda and UOP announced their *Acetica*™ process, which uses a polyvinylpyridine resin tolerant of elevated temperatures and pressures.[10] The process, as reported, claims increased catalyst loading in the reactor and reduced by-product formation as a consequence of lower water concentration.

The research we describe in this paper addresses the fundamental nature of the active catalytic species in ionically supported systems. Structure and reactivity are probed using a variety of techniques, in order to make a detailed comparison of the supported and homogeneous catalysts. An important aspect of the study was to develop an *in situ* technique to follow the key organometallic reactions of the polymer-supported catalyst, and to obtain quantitative kinetic data for these reactions.

2 RESULTS AND DISCUSSION

2.1 Preparation and characterisation of polymer-supported $[M(CO)_2I_2]^-$ (M = Rh, Ir)

The solid supports used in this study were macroporous co-polymers of vinylpyridine and styrene crosslinked with divinylbenzene. Polymers of this type in the form of beads are available commercially (e.g. Reillex™ 425) and were also prepared for this study by Purolite. For spectroscopic studies, a more convenient sample morphology was required and thin-film polymers of similar stoichiometry were synthesised by the group of Sherrington at the University of Strathclyde. Full details of the methods used to prepare thin film polymers are reported elsewhere.[11] To generate the ion exchange resin, the pyridyl functionalities of the polymer were quaternised with methyl iodide (Eq 1).

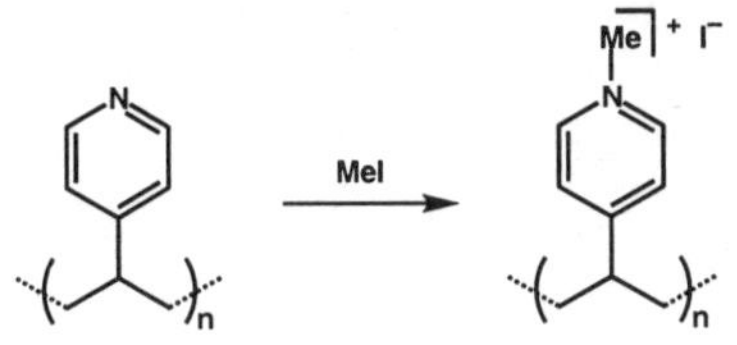

(1)

Rhodium complex was loaded onto the quaternised polymer support by the reaction with $[Rh(CO)_2I]_2$ in hexane (Eq 2). The resulting polymer beads or films showed the characteristic yellow colour of $[Rh(CO)_2I_2]^-$. An infrared spectrum of the powdered beads (KBr disk) showed two weak $\nu(CO)$ absorptions of similar intensity at 2056 and 1984 cm^{-1}, consistent with the presence of the *cis*-dicarbonyl complex, $[Rh(CO)_2I_2]^-$ (2059, 1988 cm^{-1} in CH_2Cl_2). Spectra of a much higher quality and intensity were obtained from polymer films loaded with rhodium complex. These observations of polymer supported $[Rh(CO)_2I_2]^-$ match those reported in the original study of Drago *et al.*[7]

$$(2)$$

A slightly different approach was required to load the corresponding iridium complex onto the polymer since the iridium analogue of $[Rh(CO)_2I]_2$ is not available. Therefore, instead of reacting a *neutral* precursor complex with the polymer-bound I$^-$ to generate the desired anion, we employed the pre-formed *anionic* complex, $[IrI_2(CO)_2]^-$ in an ion-exchange reaction (Eq 3).

$$(3)$$

The infrared spectrum ($\nu(CO)$ 2046 and 1969 cm^{-1}) was again consistent with the presence of the desired polymer-bound complex, $[Ir(CO)_2I_2]^-$. ICP mass spectroscopic analysis of the metal loaded resins showed the metal contents to be 0.56-0.66 % (Rh) and 0.79-0.86 % (Ir) by weight. These values are in line with the observation that virtually all the metal complex is taken up from solution under the conditions of these loading experiments, and that 15-20% of the pyridinium sites are loaded with metal in the products.

2.2 Structural characterisation of polymer supported complexes using EXAFS

The structures of the supported complexes described above were probed using rhodium and iridium extended X-ray absorption fine structure (EXAFS) measurements. The results of the curve-fitting analysis of the EXAFS traces for supported $[M(CO)_2I_2]^-$ are illustrated in Figure 2. The Rh K-edge EXAFS shows very strong oscillations extending more than 1000 eV above the Rh absorption edge energy. The first (weaker) Fourier transform peak at ca 1.8 Å corresponds to the carbons of the CO ligands. The peak at *ca* 2.7 Å is associated with the heavy I atoms. Multiple scattering from the CO ligands also contributes to this peak, at about 3 Å. The basic simulation (dashed line Figure 2a) using two iodide and two carbonyl ligands almost completely accounts for the EXAFS data. The geometric data and fit parameters are given in Table 1.

The EXAFS analysis for polymer-supported $[Ir(CO)_2I_2]^-$ (Figure 2b) shows similar characteristics to the Rh system, with an intense peak at *ca.* 2.7 Å. This peak and the corresponding EXAFS amplitude is larger than for the Rh sample, causing the peak in the due to the C shell to appear smaller. The data are again well-simulated using two iodide and two carbonyl ligands with the geometric parameters given in Table 1. The scattering amplitudes of the more distant shells of the atoms are weak making it very difficult to assign atoms to these peaks with confidence. However, for both metals there are features

Table 1 *Coordination parameters of polymer-supported $[M(CO)_2I_2]^-$ derived from EXAFS measurements*

Atoms	Distance (Å) $M = Rh^\dagger$	Debye-Waller (Å^2)	Distance (Å) $M = Ir^\dagger$	Debye-Waller (Å^2)	Distance (Å) sol^n $M = Rh^{12}$
2 C	1.84	0.006	1.86	0.025	1.845(4)
2 I	2.67	0.005	2.68	0.003	2.645(1)
2 O	3.00	0.005	3.02	0.009	2.961(2)

† fit index 1.29 (Rh) and 1.07 (Ir); R-factor 38.1% (Rh) and 43.6% (Ir)

between the first and second peaks in the Fourier transforms which suggest the presence of additional scattering atom(s). Only light atoms such as N or O are possible candidates, due to the short distance to Rh. A single additional N atom at a distance of 2.11 Å (Rh) or 2.10 Å (Ir) gave small improvements in the fits. However, the short distance to Rh is not compatible with an N atom from a pyridinium unit of the polymeric counter-ion; attempts to fit this N to longer distances did not improve the simulation. An alternative explanation could be the presence of some residual water molecules which interact with the rhodium centre. However, there is no other evidence that $[M(CO)_2I_2]^-$ binds solvent ligands in the vacant axial coordination sites and these features in the EXAFS data may be an artefact. It is difficult to rationalise the identity of a small atom at such close distance to the metal in realistic chemical terms. EXAFS measurements have been reported previously by Cruise and Evans for $Ph_4P[Rh(CO)_2I_2]$, dissolved in methanol.[12] Their geometric data and fit parameters, given in the final column of Table 1, are very similar to those for the polymer supported complex.

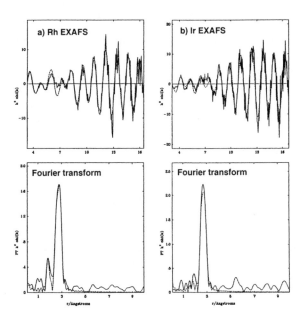

Figure 2 *EXAFS and Fourier transforms of polymer-supported $[M(CO)_2I_2]^-$ (M = (a) Rh, (b) Ir)*

2.3 X-ray crystal structures for $[4\text{-}R\text{-}C_5H_4NMe][RhI_2(CO)_2]$ (R = H, Et)

Rather surprisingly, there is no literature report of a crystal structure for the carbonylation catalyst, $[Rh(CO)_2I_2]^-$. Previous attempts in our laboratory to characterise salts of the anion crystallographically have been frustrated by disorder in the solid state.

During this study we synthesised the model compounds, [4-R-C$_5$H$_4$NMe][Rh(CO)$_2$I$_2$] (R = H (**1**) and Et (**2**)) as structural mimics of the polymer-supported system. These salts crystallised from CH$_2$Cl$_2$/Et$_2$O as deep yellow blocks suitable for X-ray diffraction studies. The stacking pattern of anions and cations for **1** is illustrated in Figure 3 and single ion pairs are shown in Figure 4 for both **1** and **2**. Geometrical data are compared with the EXAFS data in Table 2. The coordination geometry around rhodium is square planar with the two carbonyls *cis*, consistent with infrared spectroscopic data. The average Rh-C bond length and Rh-I bond length are in excellent agreement with those obtained by EXAFS for [Rh(CO)$_2$I$_2$]$^-$ in solution and supported on a polymer. The packing of ions in the lattice structures of these salts is also of interest. For **1** the [Rh(CO)$_2$I$_2$]$^-$ anions and N-methylpyridinium cations are stacked in an alternating fashion, with their planes essentially parallel to each other (Rh-N distance 3.66 Å). In **2** the 4-ethyl substituent on the N-methylpyridinium cation appears to force the cation to twist away from a parallel ion-pair interaction with the rhodium complex and this is reflected by a somewhat longer Rh-N distance (3.85 Å). One could speculate that a similar arrangement of cation and anion might also exist for polymer supported [Rh(CO)$_2$I$_2$]$^-$. In both X-ray structures the Rh-N distances are much longer than the "fifth ligand" suggested by the EXAFS analysis for polymer supported complex.

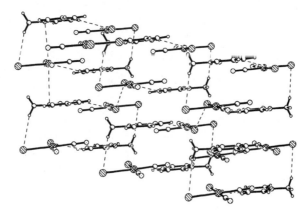

Figure 3 *Stacking pattern of cations and ions in [C$_5$H$_5$NMe][Rh(CO)$_2$I$_2$] (**1**)*

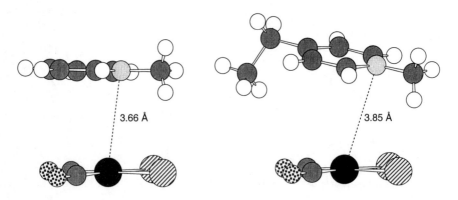

Figure 4 *X-ray crystal structures of [4-R-C$_5$H$_4$NMe][Rh(CO)$_2$I$_2$] (R = H (**1**) and Et (**2**))*

Table 2 *Geometrical parameters of $[Rh(CO)_2I_2]^-$ from X-ray crystal structures[‡] of **1** and **2** and comparison with EXAFS data*

	X-ray (1)	X-ray (2)	EXAFS (polymer)	EXFAS (solution)[12]
		Bond length (Å)		
Rh-I (av)	2.658	2.665	2.67	2.645
Rh-C (av)	1.855	1.848	1.84	1.845
C-O (av)	1.129	1.119		
Rh---O	2.984	2.967	3.00	2.961
Rh---N (py)	3.66	3.85		
		Bond angle (°)		
I-Rh-I	92.6	94.9		
C-Rh-C	93.4	93.3		
I-Rh-C (av)	87.0	85.9		

‡ Full X-ray crystallographic data will be reported elsewhere. Summary for **1** (150 K): yellow; triclinic; P-1, a = 7.137(2) Å, b = 7.503(3) Å, c =12.697(3) Å, α = 78.392(18)°, β = 87.860(17)°, γ = 75.72(3)°; Z = 2, R_1 = 0.0309; wR_2 = 0.0848; GOF 1.186. For **2** (150 K): yellow; monoclinic; $P2_1/n$ a = 9.127(2) Å, b = 13.553(3) Å, c = 12.251(3) Å, β = 92.465(5)°; Z = 4, R_1 = 0.0471; wR_2 = 0.1140; GOF 1.042.

2.4 Reactions of polymer-bound $[M(CO)_2I_2]^-$

Although Drago *et al*[7] used IR spectroscopy to show that the supported complex in their study was, $[Rh(CO)_2I_2]^-$, no further investigations of the stoichiometric reactions of this supported complex were undertaken. We have carried out both qualitative and quantitative measurements on the reactivity of polymer supported $[M(CO)_2I_2]^-$. It is known that in solution, for M = Rh, the stoichiometric reaction with MeI leads to the acetyl complex $[Rh(CO)(COMe)I_3]^-$ *via* oxidative addition followed by rapid migratory CO insertion. The acetyl product exists as an iodide bridged dimer in the solid state[13] and as a solvated monomer in coordinating solvents.[14] In contrast, for M = Ir, oxidative addition step is much faster and gives a stable methyl complex, $[Ir(CO)_2I_3Me]^-$.

When a polymer film loaded with $[Rh(CO)_2I_2]^-$ was treated with excess methyl iodide (neat or diluted in CH_2Cl_2) the film changed from yellow to red-brown, similar to the colour change observed for the analogous reaction in solution. An IR spectrum of the resulting film showed a terminal $\nu(CO)$ band at 2053 and a broad absorption around 1700 cm^{-1} for the acetyl carbonyl, similar to the solution spectrum of $[Rh(CO)(COMe)I_3]_n^{n-}$. The spectroscopic data do not indicate whether the polymer-bound species is monomeric or dimeric. When a dry polymer film containing the Rh acetyl complex was treated with gaseous carbon monoxide, very slow conversion into a product with infrared bands at 2074 and 1700 cm^{-1} was observed, consistent with the *trans*-dicarbonyl complex $[Rh(CO)_2(COMe)I_3]^-$ found in solution studies. This carbonylation step was much faster if the CO was bubbled through a solvent (e.g. CH_2Cl_2) in which the polymer film was immersed, suggesting that molecules of CO enter the pores of the polymer much more efficiently when dissolved in a solvent. If left over several hours, the dicarbonyl acetyl species slowly decomposed to give $[Rh(CO)_2I_2]^-$, presumably via reductive elimination of acetyl iodide. The reaction of polymer-supported $[Ir(CO)_2I_2]^-$ with methyl iodide gave the expected iridium-methyl product, $[Ir(CO)_2I_3Me]^-$ with $\nu(CO)$ bands at 2098 and 2046 cm^{-1}. No metal species were detected by infrared or ICP-MS analysis of the liquid phases recovered from these experiments, indicating that the complexes remained bound to the polymer, with no measurable leaching into solution under these conditions. These results show that the reactivity of $[M(CO)_2I_2]^-$ supported on an ion exchange resin is qualitatively similar to the well-established solution chemistry. The same organometallic steps of the homogeneous catalytic carbonylation cycle illustrated in Figure 1 have been shown to occur under mild conditions in the supported system.

2.5 Kinetic studies on oxidative addition of MeI to polymer-supported [Rh(CO)$_2$I$_2$]$^-$

To extend the comparison between homogeneous and supported systems, we have carried out kinetic measurements on the reactions of polymer-supported [Rh(CO)$_2$I$_2$]$^-$ complexes with MeI. In order to monitor the reactions *in situ*, polymer films were inserted between the windows of a conventional infrared liquid cell, fitted with a thermostatted jacket. The cell was then filled with neat MeI or a solution of MeI in CH$_2$Cl$_2$, and transmission spectra recorded directly through the polymer film immersed in the MeI solution. An example of a series of spectra recorded in this way is shown in Figure 5. The absorption bands of [Rh(CO)$_2$I$_2$]$^-$ are replaced cleanly by those of the product acetyl complex, [Rh(CO)(COMe)I$_3$]$^-$, with the high frequency band of the reactant almost coinciding with the terminal v(CO) band of the product.

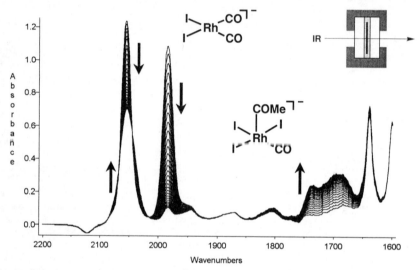

Figure 5 *Series of IR spectra from a kinetic experiment, monitoring the reaction of neat MeI with [Rh(CO)$_2$I$_2$]$^-$ supported on a polymer film (25 °C)*

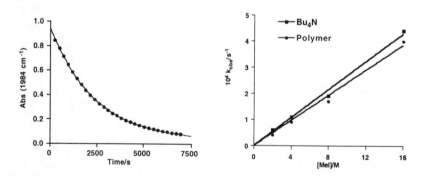

Figure 6 (a) *Decay of IR absorption at1984 cm^{-1} with exponential curve fit*
(b) plots of k$_{obs}$ vs. [MeI] for oxidative addition reactions of [Polymer][Rh(CO)$_2$I$_2$] and Bu$_4$N[Rh(CO)$_2$I$_2$] (25 °C)

Table 3 *Kinetic parameters for oxidative addition of MeI to [Rh(CO)$_2$I$_2$]$^-$*

Cation	k_{obs} / s^{-1} (neat MeI)	$\Delta H^{\ddagger}$ / kJ mol^{-1}	$\Delta S^{\ddagger}$ / J mol^{-1} K^{-1}
Polymer$^+$	4.0 x 10^{-4}	61 ± 4	-126 ± 15
Bu$_4$N$^+$	4.4 x 10^{-4}	54 ± 2	-151 ± 5
C$_5$H$_5$NMe$^+$	7.1 x 10^{-4}	58 ± 4	-133 ± 10
4-Et-C$_5$H$_4$NMe$^+$	9.2 x 10^{-4}	53 ± 3	-149 ± 8
4-Bz-C$_5$H$_4$NMe$^+$	7.9 x 10^{-4}	55 ± 3	-142 ± 8

A plot of absorbance *vs.* time for the low frequency reactant band at 1984 cm^{-1} (Figure 6a) is well fitted by an exponential decay curve, showing that the reaction is first order in the concentration of [Rh(CO)$_2$I$_2$]$^-$. Pseudo first order rate constants, k_{obs}, were obtained in this manner from a series of reactions at different methyl iodide concentrations. Figure 6b shows that k_{obs} has a first order dependence on [MeI], and hence the reaction is second order overall. A second order rate constant, k_2, can be obtained from the slope of the plot of k_{obs} *vs.* [MeI]. The apparent k_2 obtained in this way (2.57 x 10^{-5} M^{-1} s^{-1}, 25 °C)) is very similar to the value measured for Bu$_4$N[Rh(CO)$_2$I$_2$] in solution (2.71 x 10^{-5} M^{-1} s^{-1} in CH$_2$Cl$_2$). The k_{obs} values for the homogeneous reaction are also plotted in Figure 6b for comparison. Care, however, must be taken in comparing second order rate constants in homogeneous and heterogeneous systems, since the effective concentration of a liquid reactant within the pores of a polymer may differ from that in the bulk solution.[15] So in this case, [MeI]$_{polymer}$ is not necessarily the same as [MeI]$_{sol}$. Depending on whether MeI is preferentially concentrated or depleted in the polymer relative to the liquid phase, the real value of k_2 for the polymer-supported system may be higher or lower than that given above. Nevertheless, it is clear that k_{obs} for a given [MeI]$_{sol}$ is not markedly different for the polyvinylpyridine-supported complex and the Bu$_4$N$^+$ salt.

Previous studies on the solution reactivity of [Rh(CO)$_2$I$_2$]$^-$ salts indicated that cation effects were small.[16] Of the small number of cations tested, none contained a pyridyl group. In order to provide a closer mimic of the environment within the polymer, we therefore measured reaction kinetics for salts of [Rh(CO)$_2$I$_2$]$^-$ with cations of the type [4-R-C$_5$H$_4$NMe]$^+$ (where R = H, Et, Bz). Interestingly, the pseudo first order rate constants obtained in neat MeI for these reactions (listed in Table 3) are approximately double those obtained for either the Bu$_4$N$^+$ salt or the polymer supported complex. Variable temperature studies were also carried out and Eyring plots of the resulting data gave the activation parameters listed in Table 3. Values of $\Delta H^{\ddagger}$ and $\Delta S^{\ddagger}$ are similar for all the cations tested and are typical for the oxidative addition reaction. The large negative entropy of activation found in each case is consistent with an initial S$_N$2 attack by the complex on MeI.[17] In view of the significant error bars, the faster reactions of the pyridyl salts cannot be ascribed with confidence to an enthalpic or entropic effect. A chemical explanation for the very subtle differences between cations is not clear.

3 CONCLUSIONS

Spectroscopic and EXAFS studies show that the structure of [Rh(CO)$_2$I$_2$]$^-$ supported on polyvinylpyridine-based resins is essentially identical to that found in solution and in the solid state. Two crystal structures of salts of [Rh(CO)$_2$I$_2$]$^-$ containing N-methyl pyridinium show a close and almost parallel arrangement of the anion and cation, which may reflect the ion-pair structure in the polymer supported system. Reactivity studies how that [M(CO)$_2$I$_2$]$^-$ undergoes oxidative addition with MeI in a qualitatively similar fashion to the homogeneous reaction, yielding an acetyl complex (after rapid migratory insertion) for M = Rh and a stable methyl complex for M = Ir. Kinetic studies on the Rh system show that oxidative addition is first order in [MeI]$_{sol}$ and [Rh(CO)$_2$I$_2$$^-$], as in solution. Observed rate constants are very similar to those found for Bu$_4$N[Rh(CO)$_2$I$_2$] but smaller than for salts containing monomeric N-methylpyridinium cations. The comparable oxidative addition rates found for homogeneous and heterogeneous reactions suggest that the permeability of the polymer to MeI is high, and that the reaction rate is not mass-transfer limited.

4 EXPERIMENTAL

4.1 Materials and synthetic methods

Metal complexes [Rh(CO)$_2$I]$_2$, Bu$_4$N[Rh(CO)$_2$I$_2$] and Ph$_4$As[Ir(CO)$_2$I$_2$] were prepared by established methods.[16,18] The N-methylpyridinium salts [4-R-C$_5$H$_4$NMe][Rh(CO)$_2$I$_2$] were prepared in a similar procedure to the Bu$_4$N$^+$ salt, by reaction of the appropriate N-methylpyridinium iodide with [Rh(CO)$_2$I]$_2$. Recrystallisation from CH$_2$Cl$_2$/diethyl ether mixtures afforded [4-R-C$_5$H$_4$NMe][Rh(CO)$_2$I$_2$] as yellow solids which were dried *in vacuo* and stored under an atmosphere of CO at -10 °C. Crystals of [C$_5$H$_5$NMe]-[Rh(CO)$_2$I$_2$] (1) and [4-Et-C$_5$H$_4$NMe][Rh(CO)$_2$I$_2$] (2) suitable for X-diffraction were obtained by slow diffusion of diethyl ether into a CH$_2$Cl$_2$ solution of the compound at 0 °C. Crystals of 2 were found to melt on warming to room temperature, whereas those of 1 remained solid. Elemental analysis for 1 (%) C 19.19, H 1.58, N 2.67; expected for C$_8$H$_8$NI$_2$O$_2$Rh: C 18.93, H 1.58, N 2.76.

Poly(4-vinylpyridine-co-styrene-co- divinylbenzene) were supplied in the form of beads (Reillex 425 and Purolite) or prepared as thin films (dimensions *ca.* 76 x 14 mm x 40-120 μm) at the University of Strathclyde.[11] Quaternisation of the pyridine groups was carried out by slow addition of the polymer (typically 1 g) to ethanol (40 cm^3) To this was added methyl iodide (2 cm^3) and the mixture was heated to 50 °C for 2 hours and then cooled. The quaternised polymer was washed with ethanol followed by acetone and then vacuum dried.

Polymer was loaded with rhodium complex as follows: The quaternised polymer (1 g) was added to a solution of [Rh(CO)$_2$I]$_2$ (15 mg, 26 μmol) in hexane (20 ml) and stirred for 4 hours (until the solution in contact with the polymer beads or films became colourless). The polymer beads/films were filtered and washed with hexane, dried *in vacuo* and stored under an atmosphere of CO at -10 °C. Iridium complex was loaded in a similar manner using Ph$_4$As[Ir(CO)$_2$I$_2$] (23 mg, 26 μmol) in CH$_2$Cl$_2$ (20 cm^3).

4.2 Kinetic measurements

The reactions of polymer supported [Rh(CO)$_2$I$_2$]$^-$ with methyl iodide were monitored using FTIR spectroscopy in a solution cell (CaF$_2$ windows, 0.5 mm path length) fitted with a thermostatted jacket. A strip of the metal loaded polymer film was placed between the CaF$_2$ plates and MeI (neat or diluted to known concentration with CH$_2$Cl$_2$) was added. IR spectra were recorded on a Mattson Genesis Fourier Transform Spectrometer or a Magna 560 Nicolet IR spectrometer and stored electronically. A series of spectra were collected with a constant time interval; data processing involved subtraction of the solvent spectrum to give a data set of absorbance against time for each kinetic run. Kinetic measurements were made by following the decay of the low frequency ν(CO) absorption of [Rh(CO)$_2$I$_2$]$^-$ (1984 cm^{-1} in neat MeI). The *pseudo*-first-order-rate constants were found by fitting exponential decay curves to the experimental data.

4.3 EXAFS measurements

EXAFS data (Rh K-edge ((23220 eV) or Ir L$_{III}$-edge (13419 eV)) were collected in transmission mode on station 9.2 of the Daresbury Synchrotron Radiation Source, operating at 2 GeV with an average current of 150 mA. A water-cooled Si(220) double crystal monochromator was used, with its angle calibrated by running an edge scan of a 5 μm Rh or Ir foil. For each sample 2-10 scans were recorded at room temperature in the range from *ca.* 200 eV below the edge to *ca.* 600 eV above; data range up to 13 Å^{-1}; count time used k^3 weighting; spectra analysed using *EXCURV97*.

Acknowledgements

The research described in this paper was carried out as part of a collaborative LINK project, funded by the EPSRC and the DTI. We thank our co-workers on the project, Prof. David Sherrington, Dr. Paul Findlay and Dr. Salla-M. Leinonen (University of Strathclyde) for providing polymeric materials, Mr. Philip Howard, Dr Mike Simpson and Dr Mike Jones (BP Chemicals, Hull), Mr Simon Collard and Dr. Andrew Chiffey (Johnson Matthey, Royston) for performing the ICP-MS measurements and Dr. Jim Dale (Purolite UK, Pontyclun) for providing polymeric materials. We acknowledge the provision of time on DARTS, the UK national synchrotron radiation service at the CLRC Daresbury Laboratory, through funding by the EPSRC (DARTS reference 98E14).

References

1. T. W. Dekleva and D. Forster, *Adv. Catal.*, 1986, **34**, 81; D. Forster, *Adv. Organomet. Chem*, 1979, **17**, 255.
2. M. J. Howard, M. D. Jones, M. S. Roberts and S. A. Taylor, *Catal. Today*, 1993, **18**, 325.
3. P. M. Maitlis, A. Haynes, G. J. Sunley and M. J. Howard, *J. Chem. Soc., Dalton Trans.*, 1996, 2187.
4. G. J. Sunley and D. J. Watson, *Catal. Today*, 2000, **58**, 293.
5. M. S. Jarrell and B. C. Gates, *J. Catal.*, 1975, **40**, 255.
6. A. Haynes, B. E. Mann, G. E. Morris and P. M. Maitlis, *J. Am. Chem. Soc.*, 1993, **115**, 4093.
7. R. S. Drago, E. D. Nyberg, A. El A'mma and A. Zombeck, *Inorg. Chem.*, 1981, **3**, 641; R. S. Drago and A. El A'mma, (University of Illinois), U.S. 4328125, 1982.
8. N. De Blasio, M. R. Wright, T. E, C. Mazzocchia and D. J. Cole-Hamilton, *J. Organomet. Chem.*, 1998, **551**, 229; D. Z. Jiang, X. B. Li and E. L. Wang, *Macromol. Symp.*, 1996, **105**, 161.
9. C. R. Marston and G. L. Goe, (Reilly Tar), E.P. 0 277 824, 1988; Y. Shiroto, K. Hamato, S. Asaoka and T. Maejima, (Chiyoda), E.P. 0 567 331, 1993; D. J. Watson, B. L. Williams and R. J. Watt, (BP Chemicals), E.P. 0 612 712, 1994; M. O. Scates, R. J. Warner and G. P. Torrence, (Hoechst Celanese), U.S. 5466874, 1995; T. Minami, K. Shimokawa, K. Hamato, Y. Shiroto and N. Yoneda, (Chiyoda), U.S. 5364963, 1994.
10. N. Yoneda, T. Minami, J. Weiszmann and B. Spehlmann, *Stud. Surf. Sci. Catal.*, 1999, **12**, 93.
11. P. H. Findlay, S. M. Leinonen, M. G. J. T. Morrison, E. E. A. Shepherd and D. C. Sherrington, *J. Mater. Chem.*, 2000, **10**, 2031.
12. N. A. Cruise and J. Evans, *J. Chem. Soc., Dalton Trans.*, 1995, 3089.
13. G. W. Adamson, J. J. Daly and D. Forster, *J. Organomet. Chem.*, 1974, **71**, C17.
14. A. Adams, N. A. Bailey, B. E. Mann, C. P. Manuel, C. M. Spencer and A. G. Kent, *J. Chem. Soc., Dalton Trans.*, 1988, 489.
15. D. C. Sherrington, in *Polymer-supported Reactions in Organic Synthesis*, ed. P. Hodge and D. C. Sherrington, Wiley, Chichester, U.K., 1980, ch. 1, p. 59-74.
16. A. Fulford, C. E. Hickey and P. M. Maitlis, *J. Organomet. Chem.*, 1990, **398**, 311.
17. T. R. Griffin, D. B. Cook, A. Haynes, J. M. Pearson, D. Monti and G. E. Morris, *J. Am. Chem. Soc.*, 1996, **118**, 3029.
18. P. R. Ellis, J. M. Pearson, A. Haynes, H. Adams, N. A. Bailey and P. M. Maitlis, *Organometallics*, 1994, **13**, 3215.

AN ORIGINAL BEHAVIOUR OF COPPER(II)-EXCHANGED Y FAUJASITE IN THE RUFF OXIDATIVE DEGRADATION OF CALCIUM GLUCONATE

Gwénaëlle Hourdin, Alain Germain, Claude Moreau and François Fajula

Laboratoire de Matériaux Catalytiques et Catalyse en Chimie Organique
UMR-CNRS 5618, Ecole Nationale Supérieure de Chimie de Montpellier
8, Rue de l'Ecole Normale, 34296 Montpellier Cedex 5, France

1 INTRODUCTION

Starting from a salt of an aldonic acid, the Ruff oxidative degradation reaction[1] leads to an aldose with loss of one carbon atom. Known since 1898, the original process used aqueous hydrogen peroxide as oxidant, in the presence of catalytic amounts of ferric salts.

GLUCONATE   **ARABINOSE**

This reaction is particularly appropriate to produce pentose from a readily accessible and cheap material such as glucose, and has received some improvements in connection with the preparation of D-arabinose from D-gluconate[2-6]. Copper(II) has been proved a better catalyst than iron(III)[5], but the catalyst is always a soluble salt of a transition metal. However, in spite of the well-known complexant character of the D-gluconate anion for multivalent cations, it was an interesting challenge to investigate this reaction in the presence of heterogeneous catalysts. Indeed, solid catalysts possess several advantages compared to their homogeneous counterparts, such as ease of handling, ease of recovery and recycling, and amenability to continuous processing, all things that provide technical, economical and environmental benefits.

The aim of this paper is to relate our attempts to perform heterogeneous catalysis of Ruff degradation of calcium D-gluconate to D-arabinose, using copper(II)-exchanged Y zeolite. Leaching of the transition metal is studied with the greatest care. From this study, a peculiar behaviour of the reactive system will be discovered.

2 EXPERIMENTAL

2.1 Reactants

Hydrogen peroxide (30 wt% aqueous solution) was purchased from Prolabo. Calcium gluconate was a gift from Roquette Frères SA. Peroxide tests were obtained from Merck. Copper(II) chloride and copper(II) sulphate were purchased from Aldrich.

2.2 Catalysts

Cu(70)/FAU(2.4) stands for a faujasite with Si/Al ratio equal to 2.4 (Y zeolite) and 70 % exchanged by copper(II).

Na/FAU (1 g), obtained from Süd Chemie (CBV 100), was stirred at room temperature for 24 h in a 1 M copper(II) chloride solution. The solid was filtered, washed repeatedly with water until obtaining a neutral filtrate, dried overnight at 353 K and calcined under flowing dry air (250 mL.min^{-1}) at 723 K for 6 hours.

Preservation of the zeolite structure was verified by X-ray powder diffraction (XRD) patterns recorded on a CGR Theta 60 instrument using Cu Kα_1 filtered radiation. The chemical composition of solids was determined at the Service Central d'Analyse CNRS (Solaize, France). Copper in the zeolite was characterised by DR-UV-visible spectroscopy using a Perkin-Elmer Lambda 14 apparatus, equipped with a reflectance sphere, and by temperature programmed reduction (TPR), using a Micromeritics Autochem 2910, equipped with a katharometer (3% H$_2$/Ar gas mixture at 30 mL.min^{-1} and 10 K.min^{-1}).

2.3 Catalytic tests

The reactions were carried out in an open 50 mL glass reactor, thermostated at 293 K and equipped with a magnetic stirrer. A mixture of calcium gluconate (2.30 g, 10.2 mmol of carboxylate) and catalyst (180 µmol of copper) was prepared in 20 mL of water. The pH of the solution was adjusted to 6.5 with a sodium hydroxide solution (0.5 M). 2.6 mL (25 mmol) of 30% aqueous hydrogen peroxide (Prolabo) was gradually added with a peristaltic pump over a period of one hour. During the oxidation, the pH was kept constant by adding sodium hydroxide solution, using an automatic burette (Metrohm 718 Stat-Titrino). After complete consumption of hydrogen peroxide, as determined by peroxide test, the stirring was stopped and the mixture was filtered and analysed. After appropriate derivatization of the major reaction products: aldoses (arabinose, erythrose, glyceraldehyde) and aldonic acids (gluconic, arabinonic, erythronic et glyceric acids), the composition of the final solution was determined by GC equipped with a capillary DB-1 column (J & W Scientific). Carbon dioxide (or its hydrogenocarbonate form), as well as formic acid, resulting from the deep oxidation reaction, were not analysed. For the blank experiment, using Na-FAU, the reaction mixture was acidified to pH 2.5 before filtration.

Copper in solution was determined by plasma atomic absorption spectroscopy (CIRAD Montpellier).

3 RESULTS AND DISCUSSION

Copper(II)-exchanged faujasite was tested in the Ruff degradation of calcium gluconate and compared, under the same reaction conditions, with homogeneous catalysis by copper(II) sulphate. The amount of copper was the same in both cases and the reactions were stopped when hydrogen peroxide was totally consumed. D-arabinose and D-erythrose were the main products, glyceraldehyde was formed in small amount ($\leq 2\%$ yield). Blank experiment was carried out with the starting Na-faujasite, for the same reaction time as the one required for the oxidation in presence of copper zeolite.

Table 1 *Yield of arabinose and erythrose in the oxidative degradation of calcium gluconate by hydrogen peroxide catalysed by Cu(II)-exchanged faujasite. Comparison with homogeneous catalysis,*

Catalyst	Conversion (%)	Arabinose (%)	Erythrose (%)	Reaction time (h)	Leaching* (%)
CuSO$_4$	100	63	8	1.5	100
Na/FAU(2.4)	0	0	0	3.5	0
Cu(70)/FAU(2.4)	100	62	9	3.5	0.23
Cu(70)/FAU(2.4)**	100	64	9	3.5	2
Cu(70)/FAU(2.4)**	100	63	9	4.5	0.15

Gluconate: 10.2 mmol; H$_2$O$_2$: 25.4 mmol; Cu(II): 180 μmol; H$_2$O: 20 mL; 293 K; pH = 6.5
* *Relative amount of copper dissolved in the filtrate at the end of the reaction*
** *Successive reuses of the solid recovered at the end of the previous experiment*

According to the results given Table 1, the Cu(II) exchanged faujasite Cu(70)/FAU(2.4) behaved as a good heterogeneous catalysis for the Ruff reaction: the arabinose yield was high, equivalent to the yield obtained in homogeneous catalysis. Only the rate of the reaction was lower. Moreover, the amount of dissolved copper at the end of the reaction was very low. The solid recovered at the end of the reaction was reused twice without loss of efficiency. Such results were in agreement with a real heterogeneous catalysis, but some precautions must be taken. Effectively, it was recently shown[7-10] that leaching of very small amounts of transition metal (Co, Cr, V) could be responsible of the catalytic oxidative activity of some "redox molecular sieves", because of the existence, in these cases, of a maximum of activity at low concentration of metal[7]. In order to clarify this point, the homogeneous catalysis by cupric salts was studied at low concentration. The reaction was stopped after 3.5 h which corresponds to the necessary time for the complete consumption of hydrogen peroxide when Cu(70)/FAU(2.4) was used. The obtained arabinose yield is shown in Figure 1 as a function of the amount of cupric sulphate used. The catalytic activity varied monotonically with the amount of copper up to an amount of ca 75 μmol. Using the same amounts of copper than the ones dissolved at the end of the reactions with copper faujasite (0.3 to 3.6 μmol), the yield did not exceed 3%. So, such an amount of dissolved copper could not be responsible for the catalytic activity of Cu(70)/FAU(2.4). On the other hand, the highest amount of copper shown figure 1 corresponded to that generally used in homogeneous catalysis. It can be observed that the arabinose yield had reached a maximum around 70%, which was the result of the formation of secondary products (erythrose, glyceraldehyde and the indeterminate formic acid).

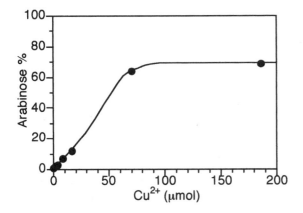

Figure 1

Arabinose yield as a function of Cu(II) in solution.

* Same conditions as in Table 1, except the reaction time set to 3.5 h

Though such a study could constitute an argument in favour of the heterogeneous nature the catalysis, it cannot be considered as an absolute proof: because the metal dissolved during the reaction could precipitate when the reaction was complete. According to Sheldon[11] "Rigorous proof of heterogeneity can be obtained by filtering the catalysts at the reaction temperature before completion of the reaction and testing the filtrate for activity". Unfortunately, in the present case, such a method was not possible because calcium gluconate is not totally soluble. In order to circumvent this difficulty, we have studied the variation of the amount of dissolved copper as a function of the reaction time. For that purpose, the reaction was stopped after various time periods and copper in solution was dosed, each time, after filtration.

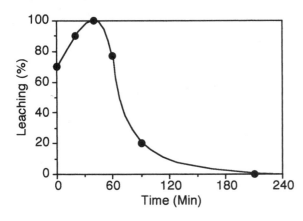

Figure 2

Leaching of copper from Cu(70)/FAU(2.4) as a function of the reaction time

* Same conditions as in Table 1

The results reported in Figure 2 show unambiguously that copper leaching occurred from the very initial mixing of zeolite and gluconate, before starting the addition of hydrogen peroxide. Copper was completely dissolved during the reaction, but was not recovered in the solution at the end of the reaction. So, the reaction behaved apparently as being heterogeneously catalysed, but, in fact, it was homogeneously catalysed.

In order to determine the cause of copper leaching, various components of the reaction (reagents and products) were tested in the conditions of the reaction (concentration, temperature and time). The results are given in Table 2.

Table 2 *Effect of various components of the reaction on the leaching of Cu*

Cu(70)/FAU(2.4): 175 mg (Cu: 180 µmol); H_2O: 20 mL; pH = 6.5; 293 K; 3.5 h

* *H_2O_2: 25.4 mmol*
** *Calcium gluconate: 10.2 mmol*
*** *Arabinose: 10.2 mmol*

Component	Cu leaching (%)
None	0.1
H_2O_2*	0.1
Calcium gluconate**	100
Arabinose***	0.1

First of all, the acidity could be a favourable factor for the leaching of copper from the solid by exchange with protons, but the pH of the Ruff reaction was strictly controlled at 6.5 and leaching was negligible at this pH value. Hydrogen peroxide, which was already claimed for promoting leaching in reactions using chromium molecular sieves[8], was not responsible for the leaching of copper in the present case. It was the same with arabinose, but, on the contrary, calcium gluconate caused the complete leaching of copper from the zeolite. This fact was assigned to the high complexing power of gluconate anion towards multivalent transition metal cations[12,13]. Thus, copper was dissolved and was able to react in solution.

Temperature programmed reduction measurements demonstrated that the nature of the copper recovered in the solid at the end of the reaction was different from that in the exchanged zeolite. Indeed, copper in exchanged faujasite exhibited two reduction steps[14] : the low temperature peak was attributed to the reduction of Cu^{2+} into Cu^+ and the high temperature peak was attributed to the reduction of Cu^+ into Cu^0. In opposite, copper in the final solid exhibited only one reduction peak at an intermediate temperature. Moreover, X-ray powder diffraction revealed the pattern of calcium carbonate, beside that of the faujasite, but this technique was unable to detect any species containing copper. Taking into account that the amount of copper in the recovered solid was smaller than 2 weight %, the absence of detection was attributed to the low sensitivity of the method. However, copper was observable by UV spectroscopy, and, by comparison with the possible species ($CuSO_4$, $CuCO_3$, Cu gluconate, CuO, $Cu(OH)_2$), it was concluded that the final solid (λ_{max} = 728 nm) contained copper hydroxide (λ_{max} = 733 nm).

When sodium gluconate was used in place of calcium gluconate, copper was always remaining in solution at the end of the reaction. Such a difference of behaviour let to think that the precipitation of calcium carbonate led the precipitation of copper hydroxide when calcium gluconate was used .

4 CONCLUSIONS

Considering a classical criterion like catalyst recycling, copper(II)-exchanged Y faujasite appeared as a good heterogeneous catalyst for the Ruff degradation of calcium D-gluconate into D-arabinose. However, this was an illusion. Following the evolution of the concentration of copper in solution during the reaction, it was possible to show that the gluconate leached the metal from the zeolite at the beginning of the reaction and that the dissolved copper disappeared from the solution at the end of the reaction. So, homogeneous copper was responsible for the reaction. But, when the aldonic acid was completely consumed, copper precipitated as an hydroxide, certainly helped by the surface of the zeolite and the concomitant crystallisation of calcium carbonate. As with a true heterogeneous catalysis, recycling was possible on one hand, but, on the other hand, such reaction could not permit a continuous process.

It was previously stressed that apparent heterogeneous catalysis could be the result of the leaching of minute amount metal [7-10]. By this work, we have demonstrated the existence of a new kind of "pseudo heterogeneous" catalysis: a reaction in which the effective catalyst is dissolved to be active, but precipitates when the conversion is complete. These results constitute a new demonstration that recycling is not an appropriate proof of the heterogeneity of catalysis in liquid phase oxidation.

G. H. gratefully acknowledges Roquette Frères for a scholarship.

References
1.   O. Ruff, *Ber.*, 1898, **31**, 1573.
2.   R.C. Hockett and C.S. Hudson, *J. Am. Chem. Soc.*, 1934, **56**, 1632.
3.   H.G. Fletcher, H.W. Diehl and C.S. Hudson, *J. Am. Chem. Soc.*, 1950, **72**, 4546.
4.   R.G.P. Walon, US Patent 3 755 294 (1973)
5.   V. Bilik, CZ Patent 6749-83 (1986)
6.   M. Rosenberg, J. Svitel, E. Sturdik, J. Kocan, P. Magdolen and J. Kubala, CZ Patent 4458-90 (1990)
7.   I. Belkhir, A. Germain, F. Fajula and E. Fache, *J. Chem. Soc., Faraday Trans.*, 1998, **94**, 1761.
8.   H.E.B. Lempers and R.A. Sheldon, *J. Catal.*, 1998, **175**, 62.
9.   M.J. Haanepen, A.M. Elemans-Mehring and J.H.C. van Hooff, *Appl. Catal. A*, 1997, **152**, 203.
10. E.V. Spinacé, U. Schuchardt and D. Cardoso, *Appl. Catal. A*, 1999, **185**, L193.
11. R.A. Sheldon, M. Wallau, I.W.C.E. Arends and U. Schuchardt, *Acc. Chem. Res.*, 1998, **31**, 485.
12. D.T. Sawyer, *Chem. Rev.*, 1964, **64**, 633.
13. Y.E. Alekseev, A.D. Garnovskii and Y.A. Zhdanov, *Uspekhi Khim.*, 1998, **67**, 723.
14. S. Kieger, G. Delahay and B. Coq, *Appl. Catal. B.*, 2000, **25**, 1.

POLYMER-BOUND ORGANOMETALLIC COMPLEXEX AS CATALYSTS FOR USE IN ORGANIC SYNTHESIS

Nicholas E. Leadbeater

Department of Chemistry
King's College London
Strand
London WC2R 2LS
United Kingdom

1 INTRODUCTION

Transition metal complexes have, for many years, found applications in synthetic organic chemistry both as reagents and catalysts. Although there are many advantages in using metal complexes in synthesis, the problems of metal extraction and product purification make them less than ideal for use in applications such as synthesis of fine chemicals where contamination of the product with heavy metals is highly undesirable. There is now increasing interest in the development of polymer-bound metal catalysts and reagents for organic synthesis that maintain high activity and selectivity.[1] Advantages of attaching a catalyst to a polymer support include ease of separation from the product mixture at the end of a reaction and the fact that attaching a metal complex to a polymer can reduce the toxicity and air sensitivity of the species considerably. In addition, as the catalyst is easily removed from the reaction mixture, there is the possibility that it can be re-used in subsequent reactions.

In the first part of this article, focusing attention on polymer-supported cobalt phosphine complex **1** and arene ruthenium complex **2**, we review contributions from our laboratory that show how organometallics can be efficiently attached to derivitised polystyrene and we outline their synthetic versatility.[2,3] Following this, we discuss the preparation of a supported ruthenium complex, **3**, and its use in oxidation and transfer hydrogenation catalysis.

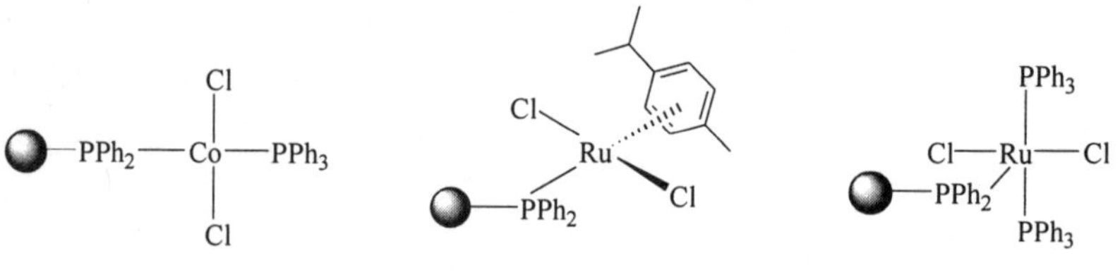

2   PREPARATION OF POLYMER-SUPPORTED COMPLEXES **1** AND **2**

The polymer support chosen for immobilisation of the both complexes was commercially available 'polymer-supported triphenylphosphine' (polystyrene crosslinked with 2% divinylbenzene, 3 mmol P per g of resin). The immobilised cobalt complex **1** was prepared by exchange of a phosphine ligand on the homogeneous analogue $CoCl_2(PPh_3)_2$ (**4**) for a phosphine moiety on the support (Figure 1).[2] This is achieved by agitating a solution of **4** with the functionalised resin overnight. Filtration, washing and drying of the polymer gives **1** which, like **4**, is bright blue in colour. The catalyst loading was optimised at 0.9 mmol Co per gram of resin.

The immobilised ruthenium complex, **2**, was prepared by thermolysis of the dimeric ruthenium complex $[Ru(p\text{-cymene})Cl_2]_2$ with the functionalised resin (Figure 1).[3] The choice of solvent in which the reaction is performed proves to be very important. Highest yields of the homogeneous phosphine complex $Ru(p\text{-cymene})Cl_2(PPh_3)$ (**5**) are obtained using toluene as the solvent however the swelling properties of the resin in the solvent are very low. On the other hand, dichloromethane is a good solvent for swelling the resin but is not suitable for forming the desired phosphine ruthenium complex. Other solvents such as thf and dioxane led to significant by-product formation and so we found that the reaction is best performed in a mix of dichloromethane and toluene (1:2). Filtration, washing and drying of the polymer gave a deep red powder which was characterised as **2** with a loading of approximately 1.1 mmol Ru per gram of resin.

Both the polymer bound complexes **1** and **2** are stable in air, no decomposition being noted over the period of three months at room temperature. This compares favourably to the homogeneous analogues, especially **5**, where decomposition occurs if the materials are not stored under an inert atmosphere and at low temperature.

Figure 1   *The preparation of polymer-supported complexes 1 and 2*

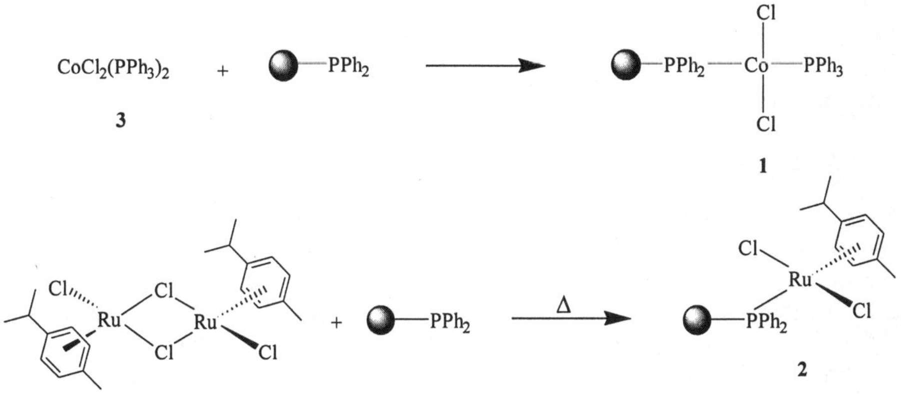

3   USE OF SUPPORTED COMPLEXES **1** AND **2** IN CATALYSIS

3.1 Use of 1 as a catalyst for the oxidation of alcohols

Using a catalytic amount of **1**, we have shown that the efficient oxidation of primary and secondary benzylic alcohols using *t*-butyl hydroperoxide as oxidant can be effected. Our results show that the attachment of the metal complex to the polymeric support has little effect on the yields of reaction compared to the homogeneous analogue.[4] Using both **1** and **4** the oxidation is selective for benzylic alcohols, simple aliphatic alcohols being unaffected. One interesting difference between the homogeneous and supported cobalt catalysts is the amount of acid formed in the case of primary alcohol oxidation. Using **2**, equal yields of benzaldehyde and benzoic acid are reported on oxidation of benzyl alcohol after 1.5 h.[4] Using the supported catalyst, formation of the acid is greatly reduced although not fully inhibited (86% aldehyde and 9% acid after 1.5 h). Pure aldehyde can be obtained if the reaction is performed for a shorter time but this is at the expense of yield, 80% of aldehyde being formed after 1 h, the remaining 20% being unreacted alcohol.

3.2 Use of 2 as a catalyst for enol formate synthesis

Arene ruthenium complexes are used frequently in metal-mediated organic synthesis for a wide range of reactions.[5] For the purposes of our studies we have focused attention mainly on enol formate synthesis as a representative reaction for comparing the activity of **2** with its non-supported analogue **5**. As with the supported cobalt complex, we find that attachment of **5** to a polymer support has little effect in its catalytic activity with a range of enol formates being prepared in high yield.

3.3 Leaching and recyclability

Together with high activity, two of the key factors for consideration when determining whether a polymer-supported metal catalyst is viable are whether it can recycled and whether the metal leaches off the support during the reaction. Our results have illustrated that both **1** and **2** can be recycled a number of times without loss of catalytic activity. In order to assess the leaching of the metal complex from the support, the crude reaction mixture in each case was analysed by ^1H, ^{13}C and ^{31}P NMR spectroscopy. There were no peaks characteristic for the presence of **4** or **5** observed, this suggesting that there was no observable catalyst leaching to this level of detection. In the case of **2** we recorded the UV-VIS spectrum of the product mixture and this showed no absorptions due to Co(II) complexes, this being taken as another indication that there was no significant leaching of the catalyst from the polymer support.

4 PREPARATION AND USE OF A RESIN-BOUND RUTHENIUM OXIDATION AND TRANSFER HYDROGENATION CATALYST

The ruthenium complex $RuCl_2(PPh_3)_3$ (**6**) can be easily immobilised in an analogous manner to that of **2** by agitating a dichloromethane solution of **6** with resin-bound triphenylphosphine overnight using a mechanical shaker (Figure 2). Filtration, washing and drying of the polymer gave a black powder which was characterised as resin-bound $RuCl_2(PPh_3)_3$ (**3**) by elemental analysis and by comparison of spectroscopic data with that of **6** which is also black in colour. The polymer bound complex formed is stable in air, but is better kept under an atmosphere of nitrogen for prolonged storage.

Like its homogeneous analogue, **3** shows high activity in both the oxidation of unsaturated hydrocarbons and the transfer hydrogenation of ketones, representative examples being shown in Table 1. For comparative purposes, reported yields for the analogous reactions using **6** are also shown.

Table 1 The use of **3** in transfer hydrogenation[§] and oxidation catalysis[¶]

Substrate	Product	Yield using 3 (%)	Yield using 6 (%)
(cyclohexanone)	(cyclohexanol)	85	89 [6]
(cyclopentanone)	(cyclopentanol)	64	60 [6]
(cyclohexane)	(cyclohexanone)	45	47 [6]
(ethylbenzene)	(acetophenone)	87	91 [7]
(diphenylmethane)	(benzophenone)	70	75 [7]

[§] *Typical procedure for catalytic transfer hydrogenation of ketones.* To solid **3** (100 mg) was added degassed propan-2-ol (10 ml). The mixture heated to reflux then the ketone (10 mmol) was added. The resulting mixture was stirred for 15 min and then a solution of NaOH in propan-2-ol (10 mg, in 2 ml) was added dropwise. After 1 h at reflux the mixture was cooled, the supported ruthenium complex filtered off and the product analysed.

[¶] *Typical procedure for the oxidation of hydrocarbons.* To a stirred mixture of hydrocarbon (2 mmol) and **3** (100 mg) in 1,2-dichloroethane/ethyl acetate (7:1) was added a 30% solution of peracetic acid in ethyl acetate (6 mmol in 4 ml) dropwise at reflux over the period of 2 h. After 2 h at reflux the mixture was cooled, the supported ruthenium complex filtered off and the product analysed.

Our results show that, as before, the attachment of the metal complex to the polymeric support has little effect on the yields of reaction compared to the homogeneous analogue, any small decrease in yield being more than compensated by the ease of removal of the catalyst from the product mixture. To show that **3** can be recycled a number of times, the oxidation of 1-phenylethanol to benzaldehyde was repeated five times using the same batch of supported catalyst. As seen in Table 2, the yields remain around 85% clearly illustrating the re-usability of the catalyst.

Table 2   The reuse of **3** in the oxidation of 1-phenylethanol

Experiment	Yield (%)
1	87
2	84
3	85
4	84
5	83

A further objective of our studies was to assess the leaching of the catalyst from the support and also to determine whether the catalysis was due to **3** or to a homogeneous ruthenium complex that comes off the support during the reaction and then returns to the support at the end. To test for leaching, we focused on the oxidation of 1-phenylethanol to benzaldehyde. We filtered off the resin after 30 minutes of reaction time and allowed the filtrate to react further. The catalyst filtration was performed at the reaction temperature (80°C) in order to avoid possible re-coordination or precipitation of soluble ruthenium upon cooling. We found that, after this hot filtration, the reaction continued thereby indicating that at least some of the catalyst comes off the resin during the course of the reaction. In contrast, when the mixture was allowed to cool before filtration, little further reaction was observed. To continue our studies we performed the reaction with the loaded resin **3** encapsulated in a MicroKan (**A**).[*] We placed some resin-bound triphenylphosphine in a separate MicroKan (**B**) and added this to the reaction vessel also. We then performed the reaction and, at the end, collected the two MicroKans and analysed the resin in each for ruthenium to see whether any metal complexes had transferred from **A** to **B** during the course of the reaction. We found this was the case thereby confirming our initial postulation that a proportion of the catalyst comes off the resin during the course of the reaction. Of interest was that there was only a trace of ruthenium in the product mixture. This suggests that, although the catalyst comes off the support at elevated temperatures during the reaction, when cooled, the catalyst returns to the support. Work is ongoing to determine whether the ruthenium complex at the end of the reaction is the same as that at the beginning.

---

[*] A MicroKan is very much like a tea bag in that the resin sample is placed inside the Kan trough a cap and the and the cap is sealed holding the sample within the can. When placed in the reaction flask, solvent and reactants can penetrate the Kan but the resin cannot escape.

5 CONCLUSIONS

In conclusion, we have shown that attachment of transition metal complexes to polymer supported triphenylphosphine leads to air stable, versatile immobilised catalysts that are as active as their homogeneous analogues and have the advantage that they can be re-used numerous times. Work is currently underway to exploit the activity of other polymer-supported organometallic complexes in metal-mediated organic synthesis.

Acknowledgements

The author is grateful to K.A. Scott and L.J. Scott for their dedicated experimental work, to the Royal Society for a University Research Fellowship and to AstraZeneca, Pfizer and Novartis for financial assistance.

References

1. For an introduction to the area see: Pomogailo, A.D. *Catalysis by Polymer-Immobilized metal Complexes*, 1998, Gordon and Breach, Amsterdam.
2. N.E. Leadbeater, K.A. Scott and L.J. Scott, *J. Org. Chem.*, 2000, **65**, 4770.
3. N.E. Leadbeater, K.A. Scott and L.J. Scott, *J. Org. Chem.*, 2000, **65**, 3231.
4. S. Iyer and J.P. Varghese, *Synth. Comm.,* 1995, **25**, 2261.
5. (a) Y. Jiang, Q. Jiang and X.J. Zhang, *J. Am. Chem. Soc.,* 1998, **120**, 3817; (b) D.L. Davies, J. Fawcett, S.A. Garratt and D.A. Russell, *J. Chem. Soc., Chem. Commun.,* 1997, 1351; (c) F. Simal, A. Demonceau, and A.F. Noels, *Tetrahedron Lett.*, 1998, **39**, 3493.
6. R.L. Chowdhury and J.-E. Bäckvall, *J. Chem. Soc., Chem. Commun.*, 1991, 1063.
7. S.-I. Murahashi, Y. Oda, N.Komiya and T. Naota, *Tetrahedron Lett.*, 1994, **35**, 7953.

DEHYDROISOMERISATION OF n-BUTANE INTO ISOBUTENE OVER Ga-CONTAINING ZEOLITE CATALYSTS

D. B. Lukyanov and T. Vazhnova

Department of Chemical Engineering
University of Bath
Bath BA2 7AY, UK

1 INTRODUCTION

Direct dehydroisomerisation (DHI) of n-butane into isobutene over bifunctional zeolite-based catalysts represents a potential new route for the generation of isobutene utilising cheap n-butane feedstock. Isobutene is used worldwide for production of methyl tert-butyl ether (MTBE) and polyisobutylene. It is currently obtained via extraction from refinery/cracker C_4 streams or via conversions of isobutane (in one step) or n-butane (in two steps).[1,2] Isobutene can also be produced via the isomerisation of n-butenes,[3] although there is no evidence that this is practised commercially.[2,3]

Two possible process options have been considered for direct conversion of n-butane into isobutene. These were based either (a) on a two-bed reactor with a dehydrogenation catalyst in combination with an isomerisation catalyst or (b) on a single-bed reactor with a bifunctional catalyst that combined both the dehydrogenation and isomerisation functions. These two options have been briefly reviewed in a recent paper by Pirngruber et al.,[4] who performed investigation of n-butane DHI into isobutene over a number of Pt-containing zeolite catalysts.[4-6] The results obtained by this research group have demonstrated that medium-pore zeolites modified by Pt can be considered as promising bifunctional catalysts for n-butane dehydroisomerisation.

The main objective of the present work was to investigate the possibilities of direct (and selective) n-butane dehydroisomerisation into isobutene over Ga-containing zeolites. Another objective was to evaluate the role played by Ga and acid sites in this reaction. For this work such medium pore zeolites, as ferrierite (FER) and theta-1, were chosen because of their superior performance in n-butene isomerisation reaction.[3,7] The modifying metal, Ga, was chosen due to the known high dehydrogenation activity of Ga-ZSM-5 catalysts in propane and n-butane conversions.[8-10] However, Ga-ZSM-5 catalysts were not used in this study because of their high aromatisation activity,[8,9] which would not allow to stop the reaction at the stage of formation and isomerisation of butenes.

2 EXPERIMENTAL

Two medium-pore zeolites, namely, ferrierite (Si/Al = 6.3) and theta-1 (Si/Al = 30) were used in this work. They were used as catalysts in the H-form or were modified by gallium before catalytic experiments. In the latter case, Ga was introduced into the zeolites by an incipient wetness impregnation method, using aqueous solutions of $Ga(NO_3)_3$. In this work, catalysts with a Ga content of 2.2 wt.% were investigated.

Kinetic studies of dehydroisomerisation of n-butane were performed at 530°C. The reaction was carried out at atmospheric pressure in a continuous flow microreactor with 100% n-butane as feed. Reaction products were analysed by on-line GC equipped with two detectors: TCD (analysis of H_2) and FID (analysis of hydrocarbons). Activities and selectivities of the fresh (not deactivated) catalysts were determined after 7 mins on stream. The activities of the catalysts in the different initial reaction steps (n-butane dehydrogenation and cracking steps) were estimated by extrapolation of the rate data to zero conversions following the procedure described previously.[11] Different levels of conversions were obtained by performing experiments at different space velocities. Prior to the catalytic experiments, the catalyst samples were activated under N_2 flow at 530°C for 4 h.

FTIR spectra of the self-supported zeolite discs were collected at a resolution of 2 cm^{-1} using a Nicolet Magna 550 FTIR spectrometer. Pyridine was used as a probe molecule to characterise Broensted and Lewis acid sites. Prior to all FTIR experiments the samples were activated under vacuum at 400°C for 16 h. The detailed experimental procedure is described elsewhere.[12] The IR spectra of the reduced samples were collected at ambient temperature after reduction of the catalysts by hydrogen at 500°C (P_{H2} = 50 torr, duration of treatment = 1 h).

3 RESULTS AND DISCUSSION

3.1. FTIR Examination of the Parent Zeolites (H-Form)

The H-forms of the ferrierite and theta-1 zeolites were examined by FTIR spectroscopy, and the IR spectra (OH region) showed the presence of two bands at ~3745 and ~3600 cm^{-1} corresponding to terminal silanol groups ($\equiv$SiOH) and bridging hydroxyls ($\equiv$Al(OH)Si$\equiv$), respectively. Practically no hydroxyls associated with the extralattice Al species (band at ~3680 cm^{-1}) were observed in these spectra. This result reflected the retention of Al atoms in the zeolite framework positions after calcination and ion exchange of the as-synthesised zeolites with an aqueous solution of NH_4NO_3. Pyridine adsorption revealed that the ratio between the intensities of the bands at 1550 and 1450 cm^{-1} was around 10 (or higher), indicating that the amount of the Lewis acid sites was much less than the amount of the Broensted acid sites present in the zeolites.

3.2. Conversion of n-Butane over Parent and Ga-Containing Zeolites

Investigation of n-butane conversion over H-forms of the ferrierite and theta-1 zeolites demonstrated that the isobutene selectivities were similar (and low) for these catalysts. The maximum selectivities (7-8 %) were obtained at low n-butane conversions (5-10 %) and decreased with increasing conversion of n-butane due to olefin interconversion and aromatisation reactions. Isobutene was in equilibrium with the other butene isomers due to the high isomerisation activity of the parent zeolites. The maximum selectivity to butenes, which was observed at low conversions, was around 20 %. This value reflects a moderate contribution of the dehydrogenation steps in n-butane transformation over H-forms of the ferrierite and theta-1 zeolites and indicates an important role of the n-butane protolytic cracking steps over these two catalysts.

Figures 1 and 2 compare the initial product distributions as functions of n-butane conversion over parent and Ga-containing zeolites. Data presented clearly demonstrate that insertion of Ga in the zeolites has resulted in redistribution of the initial products in favour of the products of the n-butane dehydrogenation steps (hydrogen and butenes). Consequently, a decrease in the concentrations of the products of the n-butane protolytic cracking steps (methane and ethane) is observed. Changes in the product distributions are essential for both zeolites (see Figures 1 and 2), although for the ferrierite catalysts these changes are smaller than in the case of the theta-1 catalysts. Consequently, the highest increase in the isobutene selectivity was observed for the Ga-containing theta-1 catalyst. This is reflected in Table 1, which shows that the highest selectivity to isobutene, obtained in this work over Ga-theta-1 catalyst, was around 27 %, and the corresponding selectivity to butenes was around 70 %. These selectivities were observed

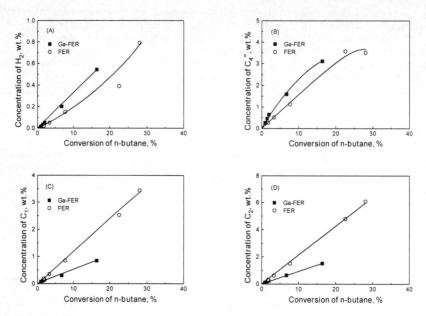

Figure 1   *Concentrations of the initial reaction products as functions of n-butane conversion over H-FER and Ga-FER catalysts.*

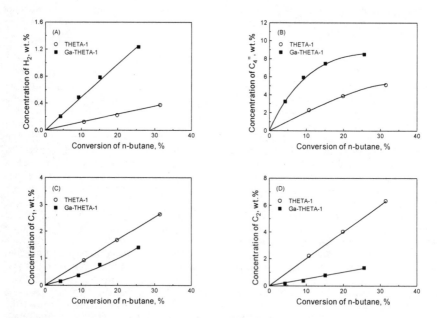

Figure 2   *Concentrations of the initial reaction products as functions of n-butane conversion over H-theta-1 and Ga-theta-1 catalysts.*

Table 1   *Maximum Selectivities to Butenes and Isobutene Observed During n-Butane Conversion over Ga-Theta-1 and Ga-FER Catalysts*

Catalyst	$C_4^=$ selectivity, wt.%	$i\text{-}C_4^=$ selectivity, wt.%
Ga-Theta-1	69.7	27.2
Ga-FER	30.5	11.9

at low n-butane conversions (5-10 %) and decreased with increasing conversion of n-butane due to olefin aromatisation. It should be noted that the latter reaction was enhanced considerably by the presence of gallium species in the catalysts.

Thus, based on the data presented above, we can conclude that further improvement of the Ga-containing zeolite catalysts for direct DHI of n-butane into isobutene requires supression of the catalyst aromatisation activity. Further reduction of the cracking activity of the catalysts is also necessary.

3.3 On the Role of Ga and Acid Sites in n-Butane Conversion over Ga-Theta-1 Catalysts

In order to get better understanding of the role of gallium and acid sites in n-butane transformation over Ga-containing catalysts, we have considered the rate data obtained over H- and Ga-theta-1 catalysts. These catalysts were chosen, since they produced much better results when compared with the ferrierite-based catalysts. Consequently, the activities of the theta-1 catalysts in the initial n-butane dehydrogenation and cracking steps were determined. This was done by the extrapolation of the rate data on formation of the primary reaction products (hydrogen, methane and ethane) to zero n-butane conversions, as shown in Figure 3.

The obtained estimates for the catalyst activities in the different initial reaction steps of n-butane conversion are collected in Table 2. Data presented indicate that insertion of gallium into H-theta-1 zeolite results in enhancement of the dehydrogenation activity and reduction of the cracking activity of the catalyst. Based on the literature data,[8-10,13-15] it can be concluded that the observed enhancement of the dehydrogenation activity is due to gallium, which is reduced to Ga^{1+} species under reaction conditions.[14,15] The observed decrease in the cracking activity (see Table 2) can be explained (i) by the decrease in the number of the acid sites (bridging OH groups) or/and (ii) by the decrease in the acid strength of these sites by gallium species present in the Ga-theta-1 catalysts. Obviously, in both cases, one should expect the same decrease in the rates of formation of methane and ethane (assuming that both reactions proceed on the same active sites). However, Table 2 shows that the decrease in the methane formation rate (~3 times) is significantly smaller than the decrease in the rate of ethane formation (~6 times). Such a difference can be understood if ethane formation proceeds over acid sites only, while the formation of methane involves both the acid and Ga active sites. The latter assumption is directly supported by the results of kinetic studies of propane transformation over Ga-ZSM-5 catalysts,[10] which have demonstrated that Ga species are active in formation of methane from propane. Hence, based on the assumption that ethane is produced on the acid sites only, we arrive at conclusion that the acidity of the Ga-theta-1 zeolite should be considerably lower than the acidity of the parent H-theta-1 zeolite.

In order to check this conclusion, we collected the IR spectra of the H-theta-1 and Ga-theta-1 zeolites after their calcination in vacuum. These spectra are shown in Figure 4 (spectra a and b), and demonstrate that there is practically no difference in the number of the acid sites in these two catalysts. However, this result does not prove that the number of the acid sites in these catalysts should be similar under reaction conditions. Indeed,

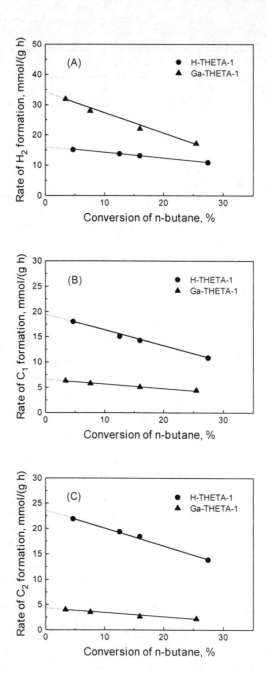

Figure 3   *Rates of formation of the initial reaction products as functions of n-butane conversion over H-theta-1 and Ga-theta-1 catalysts.*

Table 2   *Rates (mmol/(g h)) of Formation of the Initial Products of n-Butane Conversion over H-Theta-1 and Ga-Theta-1 Catalysts*

Catalyst	R_{H2}	R_{C1}	R_{C2}
H-Theta-1	16.1	19.5	23.7
Ga-Theta-1	34.1	6.7	4.1

according to the literature,[14,15] Ga species are reduced under reaction conditions, and this process should result in the disappearance of the acidic OH groups, as strongly suggested by the experimental data published by two research groups.[15,16] Consequently, we decided to obtain the IR spectrum of the reduced Ga-theta-1 catalyst. The reduction was carried out in the IR-cell in hydrogen at 500°C, and the IR spectrum of the reduced Ga-theta-1 zeolite is shown in Figure 4 (spectrum c). Quantitative data on the intensity of the 3603 cm^{-1} band (corresponding to the acidic OH groups) in the reduced catalyst vs parent catalyst are included in Table 3. Comparison of these data with the data on C_2 formation rate over parent and Ga-theta-1 catalysts (see Table 3) leads to the conclusion that the observed decrease in the catalyst cracking activity is due to the decrease in the number of the acid sites in the working Ga-theta-1 catalysts. Moreover, data in Table 3 suggest that measuring of the C_2 formation rate during n-butane conversion over Ga-containing zeolite catalysts could provide a quantitative estimate for the acidity of these catalysts under reaction conditions. More information concerning this suggestion could be found elsewhere.[17]

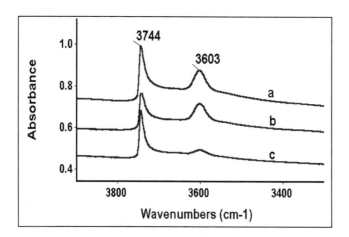

Figure 4 *IR spectra (OH region) of the parent H-theta-1 catalyst (a) and Ga-theta-1 catalysts after calcination (b) and reduction by hydrogen (c).*

Table 3   *Reduction of Cracking Activity of the Ga-Theta-1 Catalyst vs Decrease in Intensity of the 3603 cm^{-1} Band in the Same Catalyst Reduced in Hydrogen*

Catalyst	H-Theta-1	Ga-Theta-1
C_2 formation rate, rel.un.	1	0.17
Band intensity, rel.un.	1	0.19

4 CONCLUSIONS

Conversion of n-butane into isobutene over theta-1 and ferrierite zeolites was studied in a continuous flow microreactor at 530°C and 100% n-butane as a feed. The zeolites were used as catalysts in the H- and Ga-forms. Insertion of Ga into the zeolites resulted in improved isobutene selectivities due (i) to an increase in the dehydrogenation activities and (ii) to a decrease in the cracking activities of the catalysts. The highest selectivities to isobutene (~27%) and butenes (~70%) were obtained with the Ga-theta-1 catalyst at n-butane conversions around 10%. These selectivities decreased with increasing conversion due to olefin aromatisation, which was enhanced considerably by the Ga species present in the catalysts.

The rates of formation of the primary reaction products (C_1, C_2 and H_2) over H- and Ga-theta-1 catalysts were determined. These data were analysed together with the results of the FTIR studies of as-prepared and reduced (in H_2) Ga-theta-1 catalysts. Based on this analysis, the role of Ga and acid sites in the initial steps of n-butane transformation over Ga-theta-1 catalysts was clarified, and it was shown that the number of the acid sites in these catalysts was reduced significantly under reaction conditions.

Summarising the results of this study, it can be concluded that further development of the highly selective Ga-containing zeolite catalysts for direct dehydroisomerisation of n-butane into isobutene requires suppression of the catalyst aromatisation activity. Further reduction of the cracking activity is also necessary.

5 ACKNOWLEDGEMENTS

A part of the experimental work described in this paper was done in the UMIST Centre for Microporous Materials. Discussions with John Dwyer and other members of this Centre are gratefully acknowledged. DBL thanks EPSRC for the award of an Advanced Research Fellowship.

6 REFERENCES

1. M. R. Sad, C. A. Querini, R. A. Comelli, N. S. Figoli, J. M. Parera, *Appl. Catal. A: General,* 1996, **146**, 131.
2. M. L. Morgan, *Chemistry and Industry,* 1998 (Feb 2) 90.
3. H. H. Mooiweer, K. P. de Jong, B. Kraushaar-Czarnetzki, W. H. J. Stork, B. C. H. Krutzen, *Proc. of the 10th Intern. Zeolite Conference (Eds. J.Weitkamp et al.), Stud. Surf. Sci. Catal.,* 1994, **84**, 2327.
4. G. D. Pirngruber, K. Seshan, J. A. Lercher, *J. Catal.,* 1999, **186**, 188.

5. G. D. Pirngruber, O. P. E. Zinck-Stagno, K. Seshan, J. A. Lercher, *J. Catal.*, 2000, **190**, 374.
6. G. D. Pirngruber, K. Seshan, J. A. Lercher, *J. Catal.*, 2000, **190**, 396.
7. M. W. Simon, S. L. Suib, C. L. O'Young, *J. Catal.*, 1994, **147**, 484.
8. M. Guisnet, N. S. Gnep, F. Alario, *Appl. Catal.*, 1992, **89**, 1.
9. Y. Ono, *Catal. Rev.-Sci. Eng.*, 1992, **34**, 179.
10. D. B. Lukyanov, N. S. Gnep, M. R. Guisnet, *Ind. Eng. Chem. Res.*, 1995, **34**, 516.
11. D. B. Lukyanov, V. L. Zholobenko, J. Dwyer, S. A. I. Barri, W. J. Smith, *J. Phys. Chem. B*, 1999, **103**, 197.
12. V. L. Zholobenko, M. A. Makarova, J. Dwyer, *J. Phys. Chem.*, 1993, **97**, 5962.
13. P. Meriaudeau, C. Naccache, *J. Mol. Catal.*, 1990, **59**, L31.
14. G. L. Price, V. Kanazirev, *J. Catal.*, 1990, **126**, 267.
15. G. D. Meitzner, E. Iglesia, J. E. Baumgartner, E. S. Huang, *J. Catal.*, 1993, **140**, 209.
16. El-M. El-Malki, R. A. van Santen, W. M. H. Sachtler, *J. Phys. Chem. B*, 1999, **103**, 4611.
17. D. B. Lukyanov, T. Vazhnova, *J. Phys. Chem. B*, submitted.

GUANIDINE CATALYSTS SUPPORTED ON SILICA AND MICELLE TEMPLATED SILICAS. NEW BASIC CATALYSTS FOR ORGANIC CHEMISTRY

Duncan J Macquarrie*[a] James E G Mdoe[a] Daniel Brunel[b] Gilbert Renard[b] and Alexandre Blanc[b]

(a) Centre for Clean Technology, Department of Chemistry, University of York, Heslington, YORK, YO10 5DD, England
(b) Laboratoire des Matériaux Catalytiques et Catalyse en Chimie Organique, ENSCM, UMR-5618-CNRS, 8, rue de l'Ecole Normale, 34296 Montpellier, cédéx 5 France

The development of efficient solid base catalysts for the synthesis of fine chemicals is an important target, with many processes being catalysed by homogeneous amine bases, or by inorganic species such as hydroxide, which are generally neutralised, with the formation of considerable quantities of salt waste. Recently, work has been done on a range of basic catalysts, including hydrotalcites,[1] modified natural phosphates,[2,3] and amine-containing silica based materials, with both amorphous silica [4-6] and the newer Micelle Templated Silicas[7-9] being used as catalyst supports[10-14]. These materials show promise as novel basic catalysts for a range of reactions.

The attachment of guanidines to the surfaces of inorganic supports is an area of great interest. Guanidines are significantly stronger bases than simple amines, and have a basicity similar to that of simple metal hydroxides, one of the most commonly used inorganic bases. Thus, heterogeneous bases derived from guanidines will be of great interest as hydroxide replacements.

Two approaches have been employed to prepare supported guanidines – a ring-opening of the glycidyl-functionalised silica by 1,5,9-triazabicyclodecane (TBD)[15], and the nucleophilic displacement of a chloro substituent again by TBD [16]

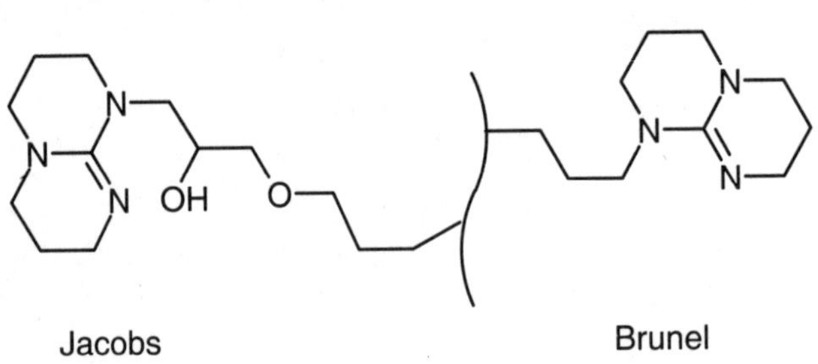

Jacobs   Brunel

Scheme 1 *Existing TBD-derived supported catalysts.*

The catalyst derived from chlorine displacement was effective in the base catalysed ring-opening of epoxides to form monoglycerides.[16] The other TBD-derived catalyst (from epoxide ring-opening) was also found to be active in the Knoevenagel reaction, the Michael addition and the base catalysed epoxidation of alkenones.[15] Results in the epoxidation reaction were of particular interest, as this leads to highly labile and valuable intermediates in the synthesis of eg. the manumycin series of antibiotics.[17] However, conversions were moderate, and selectivity towards hydrogen peroxide was relatively low (ca. 20%). Nevertheless, this indicates promise for such catalysts in an important reaction type.

We now present our results on the preparation and use of supported 1,1,3,3,- tetramethylguanidine (TMG) as a novel base catalyst.

Preparation of catalysts.

Initial results using the epoxide route proved the ring-opening to be difficult, reaction being only partially achieved after long periods at elevated temperatures. This was probably due to steric difficulties caused by the methyl groups on N-1 and N-3 which are very close to the imine N. Better results were obtained by ring-opening of the epoxide silane before grafting onto the support surface. However, it was found that the optimum route using TMG was chlorine displacement from chloropropyl groups, either before attachment to the silica, or after incorporation of the chloropropyl unit into the material. Three routes were used. The first involved attachment of the chloropropyl unit to the surface of a silica material by either (a) grafting of chloropropylsilane onto an amorphous silica (Merck Kieselgel 60) or a MTS silica or (b) templated sol-gel synthesis of a chloropropyl-MTS. Subsequent to this step, the surface was silylated to minimise the number of silanols, and to provide a hydrophobic environment. Finally, the chlorine atom was displaced by TMG in the presence of a stronger base to remove HCl (Scheme 2).

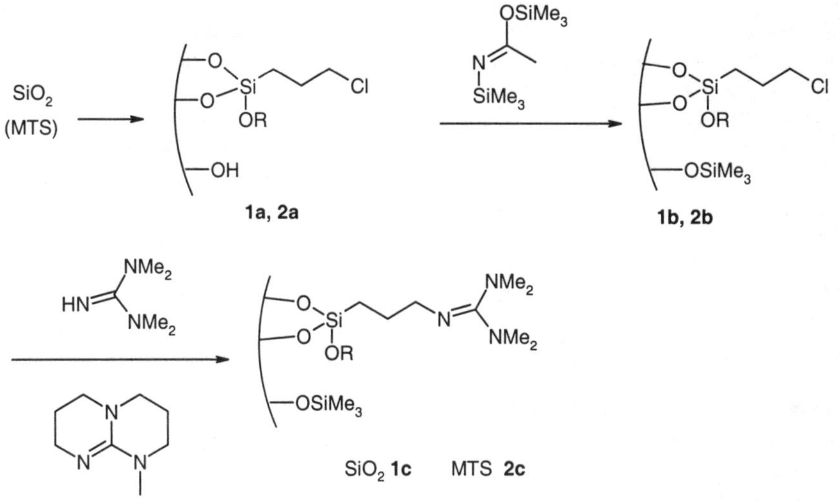

Scheme 2 *Preparation of catalysts by nucleophilic displacement of chloropropyl silicas.*

The second route involved the preparation of the TMG-silane[18] followed by attachment to the surface of silica or MTS. These catalysts thus have exposed surface hydroxyls and are significantly more hydrophilic than those prepared by the first route (Scheme 3).

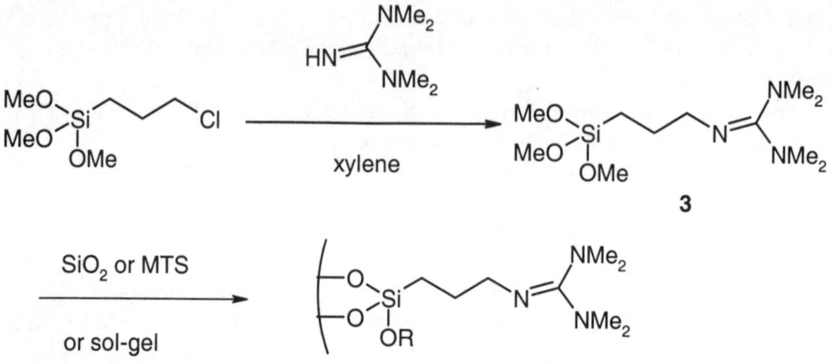

Scheme 3 *Preparation of supported guanidine materials using preformed guanidine silane*

A third route was also investigated. This involved the attempted templated sol-gel synthesis of the guanidine silane with TEOS. This route failed to give structured catalysts, although an amorphous material with reasonable surface area and activity was produced. It is likely that the guanidine becomes protonated under the synthesis conditions, and this is thought to disrupt the self-assembly process, leading to amorphous materials.[19] However, a templated synthesis of a 1.1mmol g^{-1} chloropropyl material was successfully achieved [20] and this material was successfully transformed into a supported guanidine **6** using the methodology outlined in Scheme 2.

Analysis of the catalysts was carried out by elemental analysis, thermal analysis, nitrogen porosimetry, and in selected cases, ^{13}C CPMAS spectroscopy and DRIFT infra-red spectroscopy. The spectroscopic characterisation indicated that the guanidine unit had indeed been built up on the surface, and that the grafting of the preformed material had been successfully accomplished. Within the limitations of the technique, nitrogen porosimetry gave the expected results, with the very high surface area of the parent MTS supports being reduced upon grafting and subsequent functionalisation. Similarly, the tight pore size distribution of the materials is broadened by the (partial) lining of the pore walls (and by the changes in the wall-nitrogen interaction this will cause). These changes are consistent with the grafting and functionalisation reactions taking place in the pores of the material. Similar changes were noted for the amorphous silica materials (Table 1)

Catalytic activity

The catalysts were evaluated in two reactions – the base catalysed epoxidation of electron deficient alkenes),[15] and in the Linstead variation of the Knoevenagel condensation to give 3-nonenoic acid. This reaction utilises malonic acid, and leads to an unusual dehydration, giving the β,γ-unsaturated acid, rather than the more typical α,β-enoic acid.[21-24] The product can be used as a precursor to the lactone, which is a flavour component of coconut oil.

Table 1 *Physical characteristics of the catalysts.*

Material	SSA(m^2g^{-1})	Pore diameter (nm)	loading (mmol g^{-1})
Kieselgel 100	424	6.0 (broad)	-
4	244	6.0 (broad)	1.1
MTS	972	3.0	-
5	221	1.9	1.7
2a	912	2.8	
2b	741	2.4	
2c	807	2.5	0.4
6	1281	2.4	1.4

The epoxidation of cyclohexenone was chosen as a test reaction to evaluate the catalysts. Initial experiments[25] have shown that simple base catalysts such as aminopropyl-silica and variations (e.g. in situ sol-gel versions, and aminoethyl-aminopropyl systems [14]) are active in this reaction, but are very rapidly deactivated, presumably by oxidation. Typical conditions involve the use of methanol as solvent and 20°C as reaction temperature. Hydrogen peroxide (30% v/v) was used throughout, and was added slowly over the period of the reaction. Three parameters were used to evaluate the performance of the catalysts: Conversion of reactant enone was measured; Organic Selectivity being the ratio of epoxide formed : enone used; Inorganic Selectivity is the ratio of epoxide formed : hydrogen peroxide added. This latter parameter is important in that the volumes of oxidant can be high if low selectivity is obtained. This leads to poor vessel occupancy, and to substantial amounts of water being present in the reaction system, leading to separation difficulties for a water-sensitive product.

Table 2 *Results of the epoxidation of cyclohexenone*

Catalyst	time (h)	conversion	Selectivity Organic	Inorganic
4	4	55%	42%	4%
6	12	85%	65%	7%
5	18	82%	85%	10%
2c	2	40%	89%	46%
6	2	65%	89%	56%

What can be seen from Table 2 is that the catalysts with unpassivated silica surfaces (i.e. hydrophilic, silanol-rich materials) work well in terms of conversion and Organic Selectivity, but their Inorganic Selectivity is rather poor, with the vast majority of peroxide being decomposed before reaction was possible. The materials whose surfaces were passivated by trimethylsilylation are generally much better, giving high conversions, and excellent Organic and Inorganic Selectivities. The selectivites obtained are higher than seen with earlier supported guanidines,[15] especially when the good degree of conversion is also taken into account. These catalysts thus represent a significant advance towards genuinely useful epoxidation catalysts for enones.

The second reaction type investigated is the Linstead-Knoevenagel condensation of malonic acid with heptanal (Scheme 4). The product from this is a precursor to the lactone, a component of coconut oil.

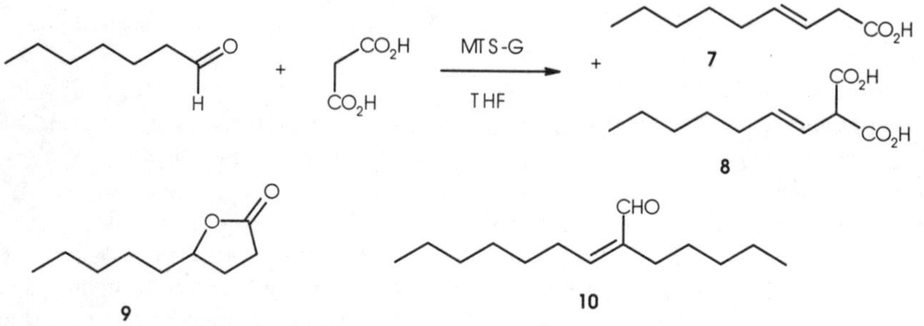

Scheme 4 *The preparation of non-3-enoic acid*

Previous work on this reaction has included the use of triethanolamine as catalyst, as well as triethylamine as catalyst and solvent.[21-24] The use of elevated temperatures (>75°C) can lead to uncontrolled decarboxylation of malonic acid before condensation, giving acetic acid, which is then too weak a carbon acid to condense. This difficulty means that often up to 3 equivalents of the malonic acid need to be used to achieve good conversion. Our aim in this work was therefore to find a catalyst which would cause the condensation to occur efficiently, but at low enough temperatures to avoid decomposition of the malonic acid. Using THF as solvent and a 1:1 ratio of malonic acid to aldehyde, with 15g of catalyst per mole of reagent, we obtained high levels of conversion of aldehyde in a reasonable time (Table 3).

Table 3 *Conversion of heptanal to non-3-enoic acid using supported guanidine catalysts.*

Catalyst	Time (h)	Conversion of heptanal (%)	Yield of 7 (%)	Yield of 10 (%)
2c	48	95	58	14
5	48	78	40	5

As for the epoxidation reaction, the passivated catalyst **2c** outperformed the non-passivated analogue **5**, giving higher conversion and yield of product. In both cases, a significant quantity of the diacid was also formed, indicating that the condensation did indeed occur very readily at relatively low temperatures (refluxing THF, 65°C), but the subsequent decarboxylation was slow. In both cases, significant quantities of diacid could be detected. Work is in progress to attempt to solve this problem, but the choice of solvents is limited by the poor solubility of malonic acid. Initial attempts to directly convert the crude mixture from the condensation step into the lactone under acid catalysis have met with some success, but considerably more will need to be done to make this a genuinely viable option. The major by-product of the process derives from the condensation of the aldehyde to give a dimer, **10**. This occurs to a greater extent (14%) with **2c** than with **4** (5%). Thus, initial work has shown that the guanidine catalysts are active enough to cause excellent conversions of malonic acid, without significant losses due to decarboxylation. Fine tuning of the reaction system is now required to ensure efficient decarboxylation of the product diacid.

CONCLUSIONS

Novel supported guanidine catalysts, based on 1,1,3,3-tetramethylguanidine have been prepared using a variety of synthetic routes. The catalysts have been shown to be efficient catalysts for the base-catalysed epoxidation of electron-deficient alkenes, and show promise in the Linstead variation of the Knoevenagel condensation.

ACKNOWLEDGEMENTS

DJM thanks the Royal Society for a Fellowship, and for travel funds for a Research Visit to Montpellier, JEGM thanks the NORAD scheme for a studentship. We also thank Professor François Fajula for hospitality and support, Dr Annie Finiels for NMR analysis and Dr Patrick Graffin for GC-MS analysis.

References

1. D. Tichit and F. Fajula, Stud. Surf. Sci., *Catal.*, 1999, **125,** 329.
2. S. Sebti, R. Nazih and R. Tahir, App. Cat. A, *Gen.*, 2000, **197,** L187.
3. S. Sebti, H. Boukhal, N. Hanafi and S. Boulaajaj, *Tet. Lett.*, 1999, **40,** 6207.
4. E. Angeletti, , C. Canepa, G. Martinetti and P. Venturello, *Tet. Lett.*, 1988, **29,** 2261.
5. E. Angeletti, , C. Canepa, G. Martinetti and P. Venturello, *J. Chem. Soc.*, Perkin 1, 1989, 105.
6. D. J. Macquarrie, J. H. Clark, A. Lambert, J. E. G. Mdoe and A. Priest, *React. Funct. Polym.,* 1997, **35,** 153.
7. J. S. Beck, J. C. Vartuli, W. J. Roth, M. E. Leonowicz, C. T. Kresge, K. D. Schmitt, C. T. W Chu, D. W. Olson, E. W. Sheppard, S. B. McCullen, J. B. Higgins, J. L. Schlenker and J. Amer. *Chem. Soc.*, 1992, **114,** 10834.
8. P. T. Tanev and T. J. Pinnavaia, *Science*, 1995, **267,** 865.
9. J. Y. Ying, C. P. Mehnert and M. S. Wong, Angew. Chem., *Int Ed. Engl.*, 1999, **38,** 56.
10. D. Brunel, Microporous, *Mesoporous Mater.*, 1999, **27,** 329.
11. M. Laspéras, T. Lloret, L. Chaves, I. Rodriguez and D. Brunel, Stud, Surf. Sci. *Catal.*, 1997, **108,** 75.
12. D. J. Macquarrie, D. B. Jackson, *Chem. Commun.*, 1997, 1781.
13. D. J. Macquarrie, *Green Chem.*, 1995, **1,** 195.
14. B. M. Choudary, M. L. Kantam, P. Sreekanth, T. Bandopadhyay, F. Figueras and A. Tuel, *J. Mol. Cat. A*, 1999, **142,** 361.

15. Y. V. Subba Rao, D. E. de Vos and P. A. Jacobs, Angew. Chem., Int. Ed., *Engl.*, 1997, **36,** 2661.
16. A. Derrien, G. Renard and D. Brunel, Stud. Surf. Sci., *Catal.*, 1998, **117,** 445.
17. C. L. Dwyer, C. D. Gill, O. Ichihara and R. J. K. Taylor, *Synlett*, 2000, 704.
18. T. Takago (to Shin-Etsu Chemical Industry Co. Ltd.), Ger. Offen. 2 827 293 (1979) [US Patent, 4 248 992 (1981)]
19. R. J. P. Corriu, A. Mehdi and C. Reyé, C. R. Acad. Sci, Paris, *t.2, Série IIc* 35, 1999.
20. D. J. Macquarrie, D. B. Jackson, J. E. G. Mdoe and J. H. Clark, *New J. Chem.*, 1999, **23,** 539.
21. S. E. Boxer and R. P. Linstead, *J. Chem. Soc.*, 1931, 740.
22. R. P. Linstead and E. G. Noble, *J. Chem. Soc.*, 1933, 557.
23. N. Ragoussis, *Tetrahedron Lett.*, 1987, **28,** 93.
24. H. M. S. Kumar, B. V. S. Reddy, E. J. Reddy, J. S. Yadav, *Tet. Lett.*, 1999, **40,** 2401.
25. J. E. G. Mdoe, D Phil Thesis, University of York, 1999.

ORGANIC MODIFICATION OF HEXAGONAL MESOPOROUS SILICA

Dominic B. Jackson, Duncan J. Macquarrie*, James H. Clark
Department of Chemistry
University of York
YORK YO10 5DD, UK

1 INTRODUCTION

Tightening environmental legislation and rising costs of waste disposal require the fine chemicals industry to consider cleaner production methods as the 21st century dawns.[1] Of particular interest is the heterogeneous catalysis of liquid-phase organic reactions which affords the immediate advantage of easy recoverability and recyclability of the catalyst, often with the further advantage of increased selectivity to the desired product over traditional homogeneous systems. Much work has been performed on supported acids to replace systems such as H_2SO_4 and $AlCl_3$ in industrial processes but less work is available in the field of supported bases. However there is an equally pressing need to supersede materials such as RNH_2 or metal hydroxides/alkoxides in the synthesis of fine organic chemicals.

Base-catalysed reactions such as the Knoevenagel, Michael and aldol reactions continue to be of importance in industrial routes to synthetic chemicals and are often inherently clean, with water (or nothing) as the by-product. Traditional homogeneous methods of catalysis often require upwards of 40 mol% catalyst (such as piperidine) with the attendant difficulties in recovery and reuse of the catalyst. They often offer extremely poor selectivity to the desired products, either due to competing processes (side reactions) or further reaction of the first-formed product.

Bases we have investigated include phenolates[2] and guanidines supported on either amorphous K-100 silica (a chromatographic silica, average pore size 10 nm, surface are c. 250 m^2g^{-1}, available from Merck) or structured hexagonal mesoporous silica (HMS). We have also investigated simple amines and methyl/dimethyl amines supported on K-100 silica.[3,4] We now wish to discuss analogues of the amine-silica materials supported on HMS, prepared either via an in-situ sol-gel method[5] or grafting onto a pre-prepared HMS support.[6]

The amorphous silica-supported amine systems show promising selectivity and recyclability for the heterogeneous catalysis of the Knoevenagel reaction (scheme 1). However they also demonstrate distinct limitations on the choice of solvent for the reaction and moderate turnover numbers.[3] Materials prepared via grafting of HMS or in-situ preparation of organo-functionalised HMS will hopefully overcome these limitations.

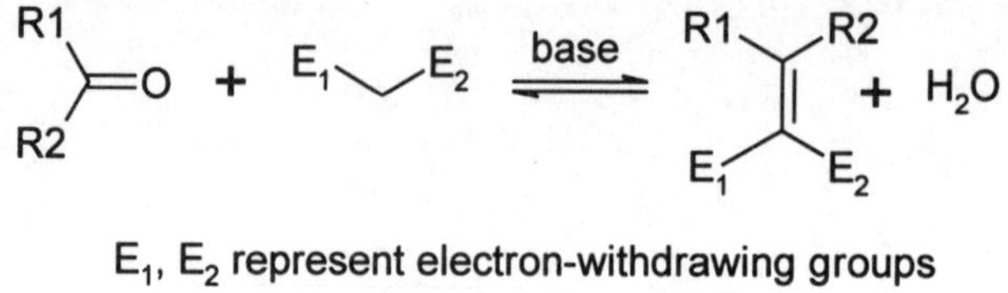

E_1, E_2 represent electron-withdrawing groups

Scheme 1: Outline of the Knoevenagel reaction

2   ROUTES TO ORGANIC FUNCTIONALISATION OF HMS

Scheme 2 depicts the two routes to introduction of organic functionality under discussion. The template in each case is a long-chain amine such as *n*-dodecylamine (we have also used *n*-decylamine and *n*-octylamine and other workers have used block copolymer materials[7] as templates) in a solvent system of c. 50:50 ethanol: water by volume. Addition of silanes and ageing for c. 18 h at room temperature affords the crude product from which the template can be removed by refluxing in a suitable solvent.

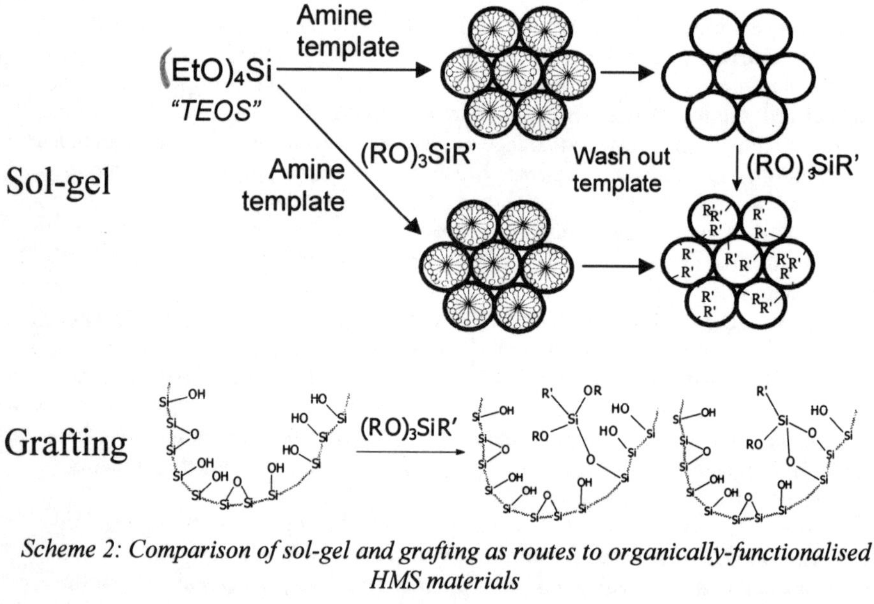

Scheme 2: Comparison of sol-gel and grafting as routes to organically-functionalised HMS materials

Both methods are easy to perform in experimental terms and offer a route to the incorporation of useful groups such as γ-aminopropyl and 2-cyanoethyl functionalities. In the case of the sol-gel method high loadings can be achieved whilst still maintaining quantitative incorporation of organic groups, although this is at the expense of the ordered nature of the product material. Furthermore in the case of high-loading γ-aminopropyl materials produced by this method, less solid than expected is recovered at the end of the sol-gel process – suggesting formation of water-soluble polymers. The sol-gel method also has the obvious disadvantage that organic groups susceptible to hydrolysis cannot be incorporated, and it is also found that charged groups (or groups liable to become charged during the sol-gel process, such as guanidines) cannot be introduced via this method without the templating mechanism breaking down.[8]

Grafting is traditionally performed by refluxing the silica and the group to be attached in toluene for 18-24 h – we have found no advantage in removing the alcohol by-product during the progress of the reaction. Loadings tend to be limited (a practical maximum of c. 1 mmolg[-1] is often found when grafting K-100 silica with γ-aminopropyltrimethoxy-silane for example) and the stability of the organic groups is found to be inferior to that of groups introduced by the in-situ sol-gel method, as discussed in 2.2.

2.1 Choosing a support: HMS versus amorphous silica

Until the discovery of the MCM class of materials in 1992[9] and the development by Pinnavaia of the HMS class of neutral-templated analogues in 1995 the only highly structured porous support available was the zeolite family. The pore sizes of these materials are, at <1 nm, too small to accommodate large organic molecules and hence their use as supports is limited. Amorphous silicas with pore sizes in the mesoporous range (1 – 10 nm) continue to be of interest as supports due to their open structure allowing easy diffusion of substrates to and from the active catalytic sites. Typical properties of a γ-aminopropyl-grafted silica such as K-100 and the sol-gel aminopropyl-HMS analogue are compared in table 1.

2.2 Stability of attached organic groups

We have performed ^{29}Si NMR studies on both materials functionalised with γ-aminopropyl groups by grafting and sol-gel methods in order to quantify the number of bonds between the silicon atom directly attached to the organic group and the rest of the silica in the materials in question.[10] The results are shown in table 2 and are consistent with the results from thermal analysis (figure 1) in that they indicate a more robustly-bound silane is obtained via the sol-gel process. Figure 1 shows that the weight loss due to the organic groups is clearly defined over the range 400-620 °C. On continuing the analysis up to 1000 °C continued weight loss is minimal and may be due to loss of residual organic groups and condensation of surface hydroxy groups to give siloxanes.

The corresponding weight loss from the grafted material is much less well-defined – the weight loss beginning at c. 300 °C and continuing up to 625 °C represents the loss of the organic groups. However the poorer definition indicates a much wider spread of organic site binding strengths.

3 CATALYSIS BY SUPPORTED BASES

The model reaction of interest is the Knoevenagel reaction between cyclohexanone and ethyl cyanoacetate (ECA). Table 3 shows the activity of various aminopropyl-functionalised materials in this reaction, under the optimum conditions for each catalyst.[3] In the case of the grafted silica materials the loading can be taken as c. 1mmolg[-1]. In the case of the HMS-based catalysts the ratios in the first column indicate the molar ratio of TEOS to organosilane used during their preparation: 9:1 indicates 10% organic organosilane corresponding to a loading of 1.1 mmolg[-1]; 4:1 corresponds with a loading of 2.5 mmolg[-1].

	Sol-gel AMP-HMS	Grafted K-100 AMP-silica
Surface area/m^{2}g^{-1} (typical, from BET model)	750	350
Pore size/nm (BET model)	3.5 (i)	Broad distribution centred on 10nm
N loading/mmolg^{-1}	≤ c. 2.5 normal, >4 (not templated) (ii)	Typical maximum 1.1 through grafting (iii)
E_N^T (Surface polarity) [H$_2$O=1, Me$_4$Si=0][11]	0.90 for typical 1.1mmolg^{-1} loaded material	0.56 for typical 1mmolg^{-1} grafted (iv)

Table 1: Comparison of properties of HMS and amorphous silica

NOTES:

(i)   A geometrical model is used to calculate pore sizes (*d*) of *d*=4*V*/*S* where *V* is the specific pore volume and *S* the specific surface area of a sample. The pore size of a typical *n*-dodecylamine-templated HMS reduces to c. 2.5 nm if nitrogen adsorption at higher partial pressures (P/P$_O$ > 0.5) is discounted – this is due predominantly to textural porosity.

(ii) Incorporation of organic groups is still quantitative at these high loadings and the product behaves very like an amorphous silica, albeit composed of >40% organic character.

(iii) We have produced higher loading materials via grafting onto HMS: see later.

(iv) The difference in surface polarity is at least partially due to the larger surface area of HMS leaving more exposed surface silanol groups, for which $E_N^T \approx 1.00$.

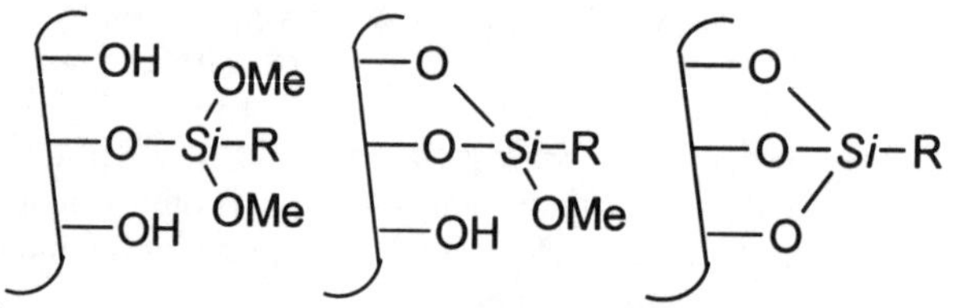

^{29}Si (ppm)			
AMP-HMS	N/A	-63.1	-68.3
AMP-silica	-57	-66.2	N/A

% present			
AMP-HMS	0	24	76
AMP-silica	49	51	0

Table 2: ^{29}Si NMR data for sol-gel γ-aminopropyl HMS and grafted γ-aminopropyl K-100 silica

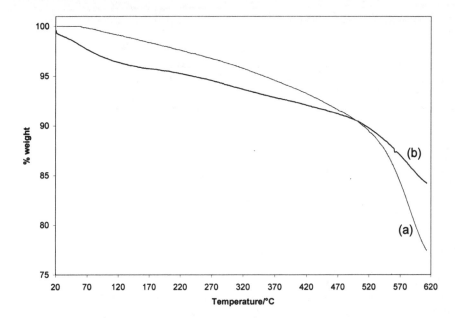

Figure 1: Thermal analysis of (a) 1.1 mmolg^{-1} sol-gel and (b) 1.0 mmolg^{-1} grafted aminopropyl-K100 materials over the range 20-620 °C.

Each catalyst is evaluated catalytically under its optimum conditions, these conditions differ for the HMS-based catalysts and the amorphous silica-based systems. The temperature column in table 3 indicates the reflux temperature of the solvent in use: for AMPS this is cyclohexane whereas for the HMS-based catalysts it is toluene. For AMPS it is found that the higher boiling point of toluene confers no advantage upon the rate of reaction – indeed the reaction proceeds more slowly in toluene when catalysed by AMPS.

This implies latent differences in surface chemistry between the two types of catalyst. Also of note is that, whilst the rates of reaction are very similar for the two types of catalyst operating under their optimum conditions, the turnover numbers for the HMS-based catalysts are much higher. This suggests that a different poisoning mechanism may be in operation and again points to differences in surface chemistry. Also notable is the efficiency of these environmentally-benign catalysts in the reaction of ketones which is traditionally achieved by systems such as TiCl$_4$/pyridine. Homogeneous amine-based systems give very poor selectivity even at up to 40 mol% catalyst.

Catalyst	Description	R1	R2	Time/h	Temp/°C	Yield/%	Turnover
AMPS	Grafted K-100	c-C$_5$H$_{10}$		1	82	98	650
9:1 AMP	Sol-gel HMS	c-C$_5$H$_{10}$		2	110	92	2450
4:1 AMP	Sol-gel HMS	c-C$_5$H$_{10}$		0.5	110	95	2650
AMPS	Grafted K-100	Et	Et	2	82	97	1127
9:1 AMP	Sol-gel HMS	Et	Et	18	110	95	1250
4:1 AMP	Sol-gel HMS	Et	Et	4	110	97	265

Table 3: Catalytic activities of grafted amorphous silica and sol-gel HMS

3.1 Deactivation mechanisms

Scheme 3 shows the accepted poisoning mechanism for grafted silica systems in our model reaction. This is characterised by amide and nitrile bands in the infra-red spectrum of used catalysts which cannot be displaced by washing. We have also produced a similar spectrum by refluxing fresh catalyst with ethyl cyanoacetate and, as expected, the treated material is not active if used in a model reaction.

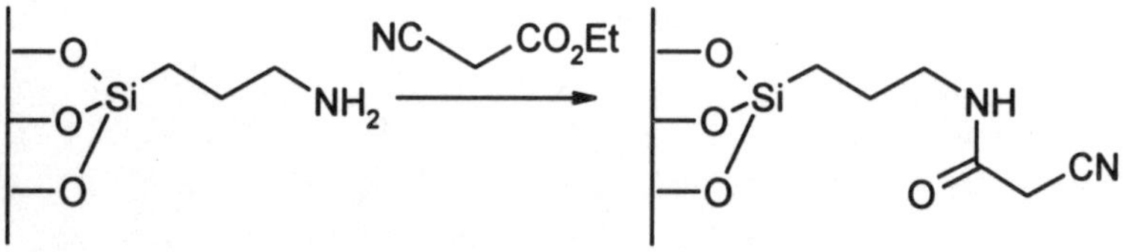

Scheme 3: Deactivation mechanism of aminopropyl-grafted silicas in the Knoevenagel reaction

Less is known about the sol-gel HMS materials. Unlike grafted silica-based materials, which react rapidly with benzaldehyde to form a surface-bound imine, the fresh sol-gel HMS materials do not react (or react very slowly) under similar conditions. The infra-red spectrum of these catalysts after use shows only adsorbed organics. We have performed some regeneration experiments by refluxing deactivated catalysts in polar solvents such as ethanol and after stripping off the solvent a small quantity of liquid is recovered. Upon GC analysis this appears to consist of the components present in the original reaction mixture in various proportions, plus other unidentified species. In addition, the surface area of a typical catalyst drops from 687 m^2g^{-1} (fresh catalyst) to 564 m^2g^{-1} and 480 m^2g^{-1} after complete deactivation (by refluxing with 15 ml each ECA and cyclohexanone) and further re-use respectively. We suggest therefore that deactivation of these HMS-based materials may occur via pore blockage with substrates/products or their oligomers during the reaction.

Figure 2 shows the results of an experiment undertaken to determine the pKa of the surface. The inflexion obtained is typical of a silanol indicating that the surface may actually be composed of an SiO^-/NH_3^+ ion pair with the active basic site being the silanol. This would fit with the evidence from the reactivities of the fresh catalysts towards benzaldehyde (see above).

4   SURFACE MODIFICATION

HMS materials made via the sol-gel system without the presence of an additional organosilane (i.e. TEOS only) still exhibit a significant weight loss at high temperatures (>500 °C) in thermal analysis. This is attributed to the large number of surface-bound ethoxy groups derived from unhydrolysed OEt ligands on the TEOS monomer. Treatment of the as-synthesised material with dilute acid or base completes the hydrolysis of these groups to surface-bound hydroxyl groups – acid treatment is preferable as the silica structure is very vulnerable to attack by even dilute base and partial collapse can result. Figures 3 and 4 show the thermal analysis and diffuse reflectance infra-red (recorded at 120 °C) respectively of a typical material before and after acid treatment – of note is the sharp weight loss from c. 520 °C upwards in the original material which is much reduced after acid treatment and the strong C-H bands in the IR of the original between c. 2750 and 3000 cm^{-1} which again disappear after acid treatment, to be replaced by a broad, less-structured OH band.

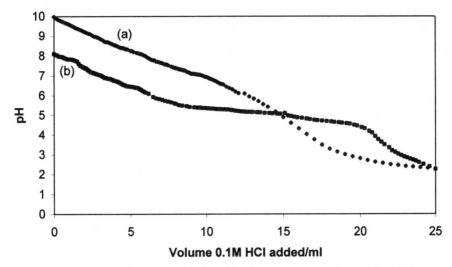

Figure 2: Variation of pH with added acid for (a) γ-aminopropyl-grafted K-100 silica and (b) sol-gel aminopropyl-HMS

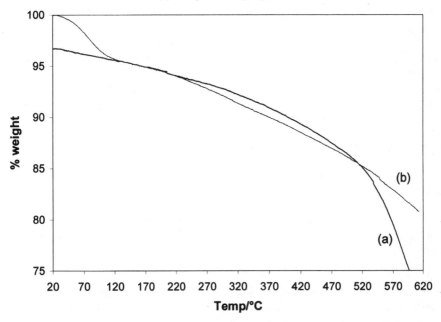

Figure 3: Thermal analysis of (a) as-synthesised HMS material and (b) the same material after acid treatment. The sharp weight loss at high temperatures of (a) is much reduced in (b).

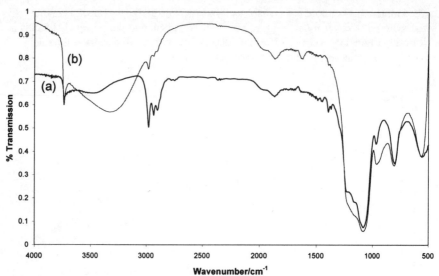

Figure 4: *Diffuse reflectance infra-red spectra recorded at 120 °C of (a) as-synthesised HMS material and (b) the same material after acid-treatment. Note the intense C-H stretching bands in (a) have largely disappeared in (b).*

This "tuneability" of the number of ethoxy groups on the surface of the material is potentially of great value in preparing supports with specific hydrophilicity/ hydrophobicity properties to suit specific applications. It is also of interest to us from the point of view of optimising conditions for grafting the surface with organic groups, increasing the loading of such groups and ultimately making more efficient supported base catalysts. In theory such treatment can alter the organophilicity of the surface and influence adsorption/desorption from the surface under catalytic conditions (possibly shedding light on the reaction kinetics along the way).

Preliminary results indicate that ethanol is the superior solvent for grafting in terms of obtaining high surface area, well-structured products. Toluene grafting can produce higher loadings (indeed, we have produced a material with nitrogen loading >5 mmolg^{-1} by grafting an acid-treated material in toluene, albeit with a surface area of 25 m^{2}g^{-1} compared to 670 m^{2}g^{-1} for the parent material) often at the expense of HMS structured character. We have also established that ethanol reacts with the silica surface during template extraction (and possibly during grafting with a basic silane), while toluene does not. We believe this arises from amine-catalysed nucleophilic attack of ethanol on silicon centres with effective substitution of ethoxy for hydroxyl on the silicon centre in question.

We have also performed basic catalyst order studies upon aminopropyl-grafted HMS materials produced via various methods of surface modification in our chosen model reaction. Results of one such study are shown in figure 5 – the important result is that the order of all the catalysts studied is approximately 1, indicating that adsorption/ desorption is not a rate-limiting step in our model systems.

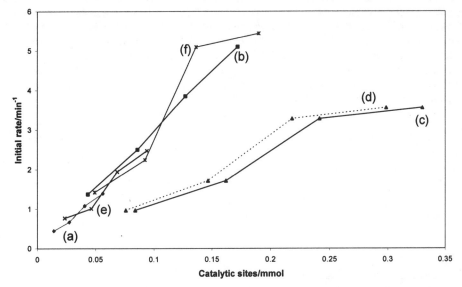

Figure 5: Catalyst order plot for five aminopropyl-functionalised HMS materials produced via surface modification: (a) as-synthesised HMS grafted in ethanol (loading 0.45 mmolg⁻¹ nitrogen), (b) as-synthesised HMS grafted in toluene (1.4 mmolg⁻¹), (c) acid-treated HMS grafted in ethanol (2.6 mmolg⁻¹), (d) correction for residual Cl content of (c) (- 0.25 mmolg⁻¹), (e) base-treated HMS grafted in ethanol (0.75 mmolg⁻¹), (f) base-treated HMS grafted in toluene (1.5 mmolg⁻¹)

As can be seen from figure 4 the catalysts are mostly well-behaved and have activities correlating well with their respective loadings, with the exception of (c). This material is of a significantly higher loading than the other materials but is less reactive than would be expected for this loading.[12] No explanation is readily forthcoming for this behaviour: correction for residual chlorine content (assuming this derives from residual HCl and that some amine sites are consequently protonated) yields line (d) which does not relate much better with the data for the other four catalysts. Doping a more typical catalyst such as (b) with an equivalent molar quantity of HCl results in an enhanced catalytic activity – this is explainable as certain steps of the mechanism to the Knoevenagel reaction are acid-catalysed (which also accounts for the usefulness of silica-supported bases as catalysts for the reaction).

5   LARGER ORGANIC FUNCTIONALITY

We have reported previously that addition of a second organosilane in the sol-gel system results in quantitative incorporation of both organic functionalities into the isolated product, and that the catalytic activities of, for example, HMS materials functionalised with both phenyl and aminopropyl groups is superior to that of analogous materials containing solely aminopropyl groups.[13] We now wish to report the results we have obtained in the preparation and characterisation of analogous materials prepared via grafting methodology.

As noted above, ethanol will react with the silica surface via a substitution mechanism during the extraction of the amine template at elevated temperatures. Our investigations into this phenomenon have revealed that heating a silica with an alcohol (to between 80 and 100 °C for c. 6 hours) and a liquid amine will add alkoxy groups from the alcohol to the silica surface. We have utilised this methodology to introduce groups such as *n*-hexoxy and phenethoxy onto the surface to act as spectator groups during the catalytic process. The materials produced by this alkoxylation method are then grafted with aminopropyl groups – data from the characterisation of the materials including loading of spectator groups is shown in table 4.

Material	Preparation method	Specific surface area/m^2g^{-1}	Loading N/mmolg^{-1}	Loading C/mmolg^{-1}	Loading OC$_6$H$_{13}$/ mmolg^{-1}	Loading OCH$_2$CH$_2$Ph/ mmolg^{-1}	Ratio OR:N
1	Heating HCl-treated HMS w. PhCH$_2$CH$_2$OH/Et$_3$N	575	0.00	17.55	N/A	2.19	N/A
2	Heating HCl-treated HMS w. C$_6$H$_{13}$OH/Et$_3$N	591	0.00	15.18	2.53	N/A	N/A
3	Grafting 1 in Na-dry toluene	456	1.21	10.70	N/A	0.88	0.73
4	Grafting 2 in Na-dry toluene	545	0.86	10.25	1.28	N/A	1.48

Table 4: Characterisation of bifunctional grafted HMS materials

As can be seen from table 4 the loadings of spectator groups achieved in the materials are respectable. Dry toluene is used for the grafting reaction as it is important to exclude water: our expectation is that the C–O–Si linkage in these materials would not be very robust. Also worthy of note are both the respectable nitrogen loadings achieved in the final products and the retention of some of the spectator groups after the grafting process is complete. Catalytic order studies have been performed on these materials and the results fit well with the data shown in figure 4. This further implies that in our model systems the accessibility of the catalyst surface is not an important rate-determining factor, even in the presence of such large spectator groups.

Conclusions

The one-pot sol-gel route to organically-functionalised HMS materials is of great use for synthesising HMS-supported amine materials. The resulting materials are catalytically active in the Knoevenagel reaction with higher turnover numbers than comparable aminopropyl-grafted amorphous silica-based materials. Both classes of catalyst afford excellent selectivity to the desired single condensation product. They exhibit different solvent dependencies and a different deactivation mechanism to the amorphous silica-based catalysts due to the higher surface area of the HMS support having a larger number of exposed surface silanols and hence a higher surface polarity. The active basic site is probably not a true amine but an SiO$^-$/NH$_3^+$ ion pair. The poisoning mechanism is not yet clearly defined but appears to be due to slow pore blockage by the reaction substrates or products and their oligomers. The loading of organic groups attainable is higher than that obtainable by grafting amorphous silica, albeit at the expense of structured character in the product material.

Grafting a pre-formed HMS support is also a useful method for introducing organic functionality however NMR and thermal analysis studies reveal that the organic group is not as firmly bound to the surface in this case compared to the sol-gel method. Various methods are available to modify the surface before grafting, the principal one being acid treatment to convert residual ethoxy groups into hydroxy groups. It is also possible to add larger organic functionality to the surface by heating with a suitable alcohol and a liquid base catalyst. Toluene appears to be a superior solvent to ethanol for carrying out the grafting reaction in terms of achieving higher loadings of organic groups however it can lead to structural damage. Using ethanol consistently affords high surface area, well-structured products. In our chosen model reaction all the grafted catalyst materials evaluated appear to have a catalyst order of approximately 1, indicating that adsorption/desorption from the catalyst surface is not a rate-determining step in our systems even with large organic spectator groups present.

Acknowledgements

JHC thanks the Royal Academy of Engineering/EPSRC for a fellowship. DJM thanks the Royal Society for a University Research Fellowship and DBJ is grateful to EPSRC for a studentship.

References

[1] J. H. Clark and D. J. Macquarrie, *Chem. Soc. Rev.*, 1996, 303
[2] D. J. Macquarrie, *Chem. Commun.*, 1997, 601; D. J. Macquarrie, *Tet. Lett.*, 1998, 4125
[3] D. J. Macquarrie, J. H. Clark, A. Lambert, A. Priest and J. E. G. Mdoe, *React. Func. Polym.*, 1997, **35**, 153
[4] J. E. G. Mdoe, D. J. Macquarrie and J. H. Clark, *Syn. Lett.*, 1998, 625
[5] D. J. Macquarrie and D. B. Jackson, *Chem. Commun.*, 1997, 1781
[6] P. T. Tanev and T. J. Pinnavaia, *Science*, 1995, **267**, 865
[7] R. Richer and L. Mercier; *Chem. Commun.*, 1998, 1775
[8] R. J. P. Corriu, A. Mehdi, C. Reyé, *C. R. Acad. Sci.*, 1992, Paris, t.2, Série IIc 35
[9] C. T. Kresge, M. E. Leonowicz, W. K. J. Roth, J. C. Vartuli and J. S. Beck, *Nature*, 1992, **359**, 710
[10] D. J. Macquarrie, D. B. Jackson, J. E. G. Mdoe and J. H. Clark, *New J. Chem.*, 1999, **23**, 539
[11] D. J. Macquarrie, S. J. Tavener, G. W. Gray, P. A. Heath, J. S Rafelt, S. I. Saulzet, Jeffrey J. E. Hardy, J. H. Clark, P. Sutra, D. Brunel, F. di Renzo and F. Fajula, *New J. Chem.*, 1999, **23**, 725
[12] M. Laspéras, T. Llorett, L. Chaves, I. Rodriguez, A. Cauvel and D. Brunel, *Stud. Surf. Sci. Catal.*, 1997, **108**, 75
[13] D. J. Macquarrie, *Green Chemistry*, 1999, **1**, 195

TOWARDS PHTHALOCYANINE NETWORK POLYMERS FOR HETEROGENEOUS CATALYSIS

Neil B. McKeown, Hong Li and Saad Makhseed

Department of Chemistry
University of Manchester
Manchester M13 9PL, UK

1 INTRODUCTION

Certain transition metal derivatives of phthalocyanine (Pc, Figure 1) are well established catalysts which mediate a wide range of chemical reactions.[1] In particular, the ability of Pcs to activate oxygen and enable it to become a useful chemical reagent is especially significant in the context of the toxic and environmentally harmful nature of common transition metal-based oxidising agents. For example, iron, cobalt and manganese phthalocyanine (FePc, CoPc and MnPc) are useful catalysts for the industrially important 'sweetening' of crude petrochemicals by the aerobic oxidation of mercaptan impurities (the *Merox* process).[2] These compounds are also valuable catalysts for the laboratory scale oxidation of phenols to quinones,[3] alkenes to epoxides or carbonyls (e.g. the Wacker reaction),[4] and alkanes to alcohols or ketones.[5] More recently there has been a growing interest in the use of Pcs for the treatment of industrial effluent waste, especially the oxidative degradation of chlorinated waste[6] and polycyclic aromatics hydrocarbons (PAHs).[7] A related area of interest is the use of Pcs as photosensitisers for the production of singlet oxygen which is a versatile oxidising and cytotoxic reagent.[8]

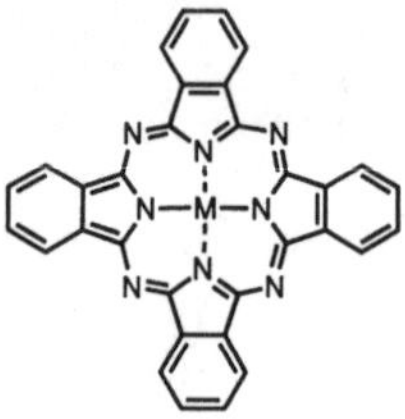

M = Fe, Co, or Mn for oxidation reactions

M = H_2 or Zn for singlet oxygen production

Figure 1 *The structure of phthalocyanine (Pc)*

1.1 The Problems Associated with Pc Catalysts.

In order for the Pc catalyst to function efficiently it is necessary for both oxygen (or an oxygen donor) and the reactant to have access to the active metal centres. Unsubstituted Pcs are very insoluble and, therefore, have to be used as heterogeneous catalysts with activity restricted to particulate surfaces. However, it is apparent that Pcs are unreliable heterogeneous catalysts with activity varying widely from batch to batch even when obtained from the same supplier.[9] Sulphonation or carboxylation of Pcs produces water-soluble derivatives with consistent homogeneous catalytic activity, however, cofacial aggregation of the hydrophobic Pc unit limits their activity and they are incompatible with non-aqueous soluble reagents.[10] Placing suitable substituents (e.g. alkyl chains) on the Pc ring provide good solubility in organic solvents and these derivatives provide reliable homogeneous catalysis.[11] In addition, such compounds are excellent photosensitisers for singlet oxygen formation.[12] Unfortunately, homogeneous Pc catalysts suffers from self-oxidation[13] and problems associated with retrieval of the catalyst from the product - this is a particular problem as even a very small amount of residual Pc catalyst produces a highly coloured product.

1.2 Supported Pc Catalysts

With the aim of providing heterogeneous Pc catalytic systems with reliable activity, Pcs have been immobilised onto polymer substrates and encapsulated within zeolites. In particular, zeolite encapsulation (Figure 2), achieved by the *in-situ* cyclotetramerisation of phthalonitrile to form the Pc within the zeolite voids, demonstrates beautifully the benefit of enforcing isolation of the Pc molecules.[14] For example, each encapsulated FePc molecule can mediate the oxidation of 30 000 molecules of ethylbenzene at room temperature.[15] In addition, the enforced isolation of zeolite encapsulation avoids the problem of self-oxidation encountered with homogeneous Pc catalysis. Unfortunately, the loading of Pc within zeolites is poor with, at best, only 1 in 6 pores containing a Pc molecule. Other problems associated with zeolite encapsulated Pcs include hindered access of large reactant molecules (e.g. PAHs) to the Pcs and a poor compatibility of the host with organic reactants although this can be improved by embedding the zeolite in a polymer membrane. Such organic/inorganic hybrid materials are reported to mimic the catalytic activity of cytochrome-P450.[16] In comparison with zeolite encapsulated Pcs polymer supported Pcs show poor activity. Partially, this failure is a result of using poorly defined systems prepared using crude synthetic methodology.[17] In addition, it appears a common feature of polymer supported Pcs that high surface loading results in poor efficiency due to the formation of aggregates.[18]

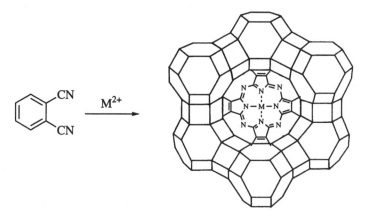

Figure 2 *Zeolite encapsulated phthalocyanine*

1.3 Pc Network Polymers

Pcs are best prepared *via* the cyclotetramerisation of a phthalonitrile. It has been demonstrated that subjection of a compound containing two phthalonitrile units to conditions which are favourable to Pc formation (e.g. high temperature with the addition of a base catalyst or metal-ion template) provides an insoluble network polymer which contains a high concentration of Pc functionality (Figure 3).[19] However, due to the free rotation about the linking groups (L) between the Pc residues during network formation, a random structural arrangement is obtained in which the natural tendency of the Pc to aggregate is encouraged. This produces densely-packed materials with no advantage for catalysis due to the restriction of access to the active metal-ion sites.

The aim of an on-going research programme at Manchester is the synthesis of rigid Pc networks in which spirocyclic linking groups prohibit the rotation of the Pc rings during network formation. Ultimately, it is hoped that such materials will provide nanoporous materials suitable for heterogeneous catalysis.

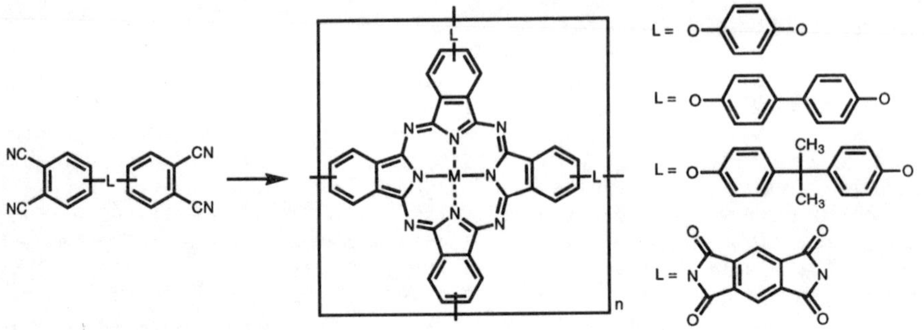

Figure 3 *Pc network formation from bis(phthalonitrile)s*

2 RESULTS AND DISCUSSION

It is anticipated that Pc formation derived from a bis(phthalonitrile) with a linking group containing two spiro-centres will result in a network in which each Pc unit is co-planar with its neighbours to provide a 2-D sheet polymer. In contrast, a network polymer derived from a bis(phthalonitrile) with a linking group containing a single spiro-centre will possess a structure in which each Pc is orthogonal to its neighbours.

2.1 Preparation of a 2-D Pc Network

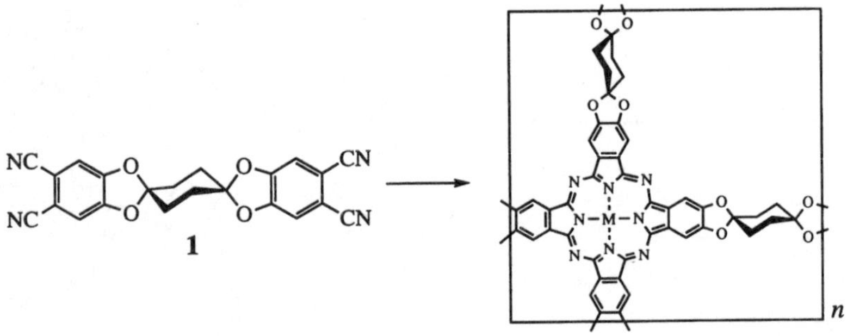

Figure 4 *Preparation of a rigid 2-D Pc network*

Bis(phthalonitrile) 1 with a linking group containing two spiro-centres is readily prepared from the bis(catechol) ketal of 1,4-cyclohexanedione *via* bromination followed by the Rosamond von Braun reaction with copper (I) cyanide (Figure 4).[20] The reaction of 1 with lithium pentyloxide in refluxing pentanol produces a highly insoluble, darkly coloured material. Reaction of 1 with a suitable mono(phthalonitrile), such as the 4,5-di(1,4,7,10-tetraoxaundecyl)phthalonitrile,[20] gives soluble hyperbranched polymers of a wide mass distribution but containing a large proportion of high mass polymer, as measured by GPC. This demonstrates that the Pc formation from 1 is of sufficient efficiency to produce the required 2-D network.

2.1 Preparation of a 3-D Pc Network

Using similar synthetic methods the bis(phthalonitrile) **2** is prepared from 2,2'-spiro-bis(indane) (Figure 5). The action of lithium pentoxide on **2** produces a highly insoluble material. UV/visible absorption spectroscopy of the material as a dispersion in chloronapthalene shows that the Pc units are not cofacially aggregated - indeed the Q-band absorption, centred at 680 nm, resembles that of an isolated Pc molecule in dilute solution. This is a significant result as it implies the possibility of unrestricted access of reagents to a metal-ion held in the central cavity of the Pc ring.

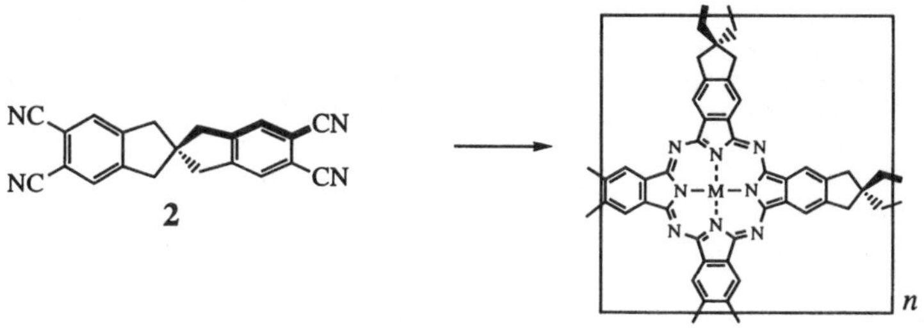

Figure 5 *Preparation of a rigid 3-D Pc network*

3 CONCLUSIONS

This encouraging preliminary synthetic work needs to be followed by a detailed structural analysis of the Pc network materials. In particular, the potential nanoporosity of the 3-D system is to be explored using nitrogen adsorption measurements. Should a high surface area be obtained, these materials would offer an interesting alternative to zeolite encapsulated Pcs for efficient heterogeneous catalysis.

ACKNOWLEDGEMENTS

We are grateful to the EPSRC (HL) and Kuwait University (SM) for funding this work and the Leverhulme Trust for providing a Research Fellowship to NBM.

References

1. N. B. McKeown, *Phthalocyanine Materials: Synthesis, Structure and Function*, CUP, Cambridge, 1998, Chapter 7, p. 140.
2. (a) S. A. Borisenkova, *Petroleum Chem.*, 1991, **21**, 379; (b) J. J. Alcaraz, B. J. Arena, R. D. Gillespie and J. S. Holmgren, *Catalysis Today*, 1998, **48**, 89.
3. (a) A. Zsigmond, F. Notheisz and J. E. Backvall, *Catalysis Let.*, 2000, **65**, 135; (b) T. Schmidt, W. Hartung and F. Wasgestian, *Inorgan. Chim. Acta*, 1998, **274**, 126.
4. (a) K. Kasuga, K. Tsuboi, M. Handa, T. Sugimori and K. Sogabe, *Inorg. Chem. Commun.*, 1999, **2**, 507; (b) E. M. Gaigneaux, R. Maggi, P. Ruiz and B. Delmon, *J. Mol. Catalysis.*, 1996, **109**, 67; (c) M. Kimura, T. Nishigaki, T. Koyama, K. Hanabusa and H. Shirai, *Reactive Funct. Polym.*, 1996, **29**, 85.
5. (a) R. F. Parton, P. E. Neys, P. A. Jacobs, R. C. Sosa and P. G. Rouxhet, *J. Catalysis*, 1996, **164**, 341; (b) G. Langhendries, G. V. Baron, P. E. Neys and P. A. Jacobs, *Chem. Eng. Sci*, 1999, **54**, 3563; (c) F. Thibaultstarzyk, R. F. Parton and P. A. Jacobs, *Stud. Surf. Sci. Catal.*, 1994, **84**, 1419.
6. (a) M. Sanchez, A. Hadasch, A. Rabion and B. Meunier, *Comptes Rendue Acad. Sci., C*, 1999, **2**, 1387; (b) L. Ukrainczyk, M. Chibwe, T. J. Pinnavaia and S. A. Boyd, *Environ. Sci. Tech.*, 1995, **29**, 439; (c) M. Bressan, N. Celli, N. D' Alessandro, L. Liberatore, A. Morvillo and L. Tonucci, *J. Organomet. Chem.*, 2000, **594**, 416.

7.   A. Sorokin and B. Meunier, *Eur. J. Inorg. Chem.*, 1998, **1**, 1269.
8.   A. A. Gorman and M. A. J. Rodgers, *Chem. Soc. Rev.*, 1981, **10**, 205.
9.   H. Grennberg and J.E. Backvall, *Acta Chem. Scand.*, 1993, **47**, 506.
10. H. Shirai, H. Tsuiki, E. Masuda, T. Koyama, K. Hanabusa and N. Kobayashi, *J. Phys. Chem.*, 1991, **95**, 417.
11. K. Kasuga, K, Tsuboi, M. Handa, T. Sugimori and K. Sogabe, *Inorg. Chem. Commun.*, 1999, **2** 507; (b) T. Schmidt, W. Hartung and F. Wasgestian, *Inorg. Chim. Acta*, 1998, **274**, 126.
12. P. M. Baigel, A. A. Gorman, I. Hamblett and T. J. Hill, *J. Photochem. Photobiol.*, 1998, **43**, 229.
13. K. W. Hampton and W. T. Ford, *J Mol. Catalysis*, 1996, **113**, 167.
14. (a) R. Raja, C. R. Jacob and P. Ratnasamy, *Catalysis Today*, 1999, **49**, 171; (b) N. S. Zefirov and A. N. Zakharov, *Can. J. Chem.*, 1998, **76**, 955; (c) K. J. Balkus, A. K. Khanmamedova, K. M. Dixon and F. Bedioui, *Appl. Catalysis*, 1996, **143**, 159; (d) K. J. Balkus, A. G. Gabrielov, S. L. Bell, F. Bedioui and J. Devynck, *Inorg. Chem.*, 1994, **33**, 67.
15. S. Ernst, Y. Traa and U. Deeg, *Stud. Surf. Sci. Catal.*, 1994, **84**, 925.
16. (a) R. F. Parton, I. F. J. Vankelecom, M. J. A. Casselman, C. P. Bezoukhanova and P. A. Jacobs, *Nature*, 1994, **370**, 541; (b) G. Langhendries, G. V. Baron, I. F. J. Vankelecom, R. F. Parton and PA. Jacobs, *Catalysis Let.*, 2000, **65**, 131; (c) G. Langhendries, G. V. Baron, P. E. Neys and P. A. Jacobs, *Chem. Eng. Sci*, 1999, **54**, 3563; (d) R. F. Parton, P. E. Neys, P. A. Jacobs, R. C. Sosa and P. G. Rouxhet, *J. Catalysis*, 1996, **164**, 341; (e) I. F. J. Vankelecom, R. F. Parton, M. J. A. Casselman, J. B. Uytterhoeven and P. A. Jacobs, *J. Catalysis*, 1996, **163**, 457.
17. e.g. M. Gebler, *J. Inorg. Nucl. Chem.*, 1981, **43**, 2759.
18. S. A. Borisenkova, E. G. Girenko, B. G. Gherassimov, L. M. Mazyarkina, V. P. Erofeeva, V. M. Derkacheva, O. L. Kaliya and E. A. Lukyanets, *J. Porph. Phthalo.*, 1999, **3**, 355.
19. (a) C.S. Marvel and J. H. Rassweiler, *J. Am. Chem. Soc.*, 1958, **80**, 1197; (b) A. W. Snow, J. R. Griffith and N. P. Marullo, *Macromolecule*, 1981, **17**, 1614; (c) D. Wöhrle and B. Schulte, *Makromol. Chem.*, 1988, **189**, 1167; (d) D. Wöhrle and B. Schulte, *Makromol. Chem.*, 1988, **189**, 1229.
20. G.J. Clarkson, N.B. McKeown and K.E. Treacher, *J. Chem. Soc., Perkin 1*, 1995, 1817.

SUZUKI COUPLING USING Pd(0) AND KF/Al$_2$O$_3$

G. W. Kabalka, R. M. Pagni, C. M. Hair, L. Wang, and V. Namboodiri

Department of Chemistry
University of Tennessee
Knoxville, TN 37996-1600

1 INTRODUCTION

Among the most important reactions in organic synthesis are those that form carbon-carbon bonds. Classically this was accomplished using base-catalyzed reactions such as the aldol, Claisen, and Knoevenagel condensation reactions. Modern versions of these reactoins often display remarkable stereoselectivity.[1] Unfortunately, many types of carbon atoms cannot be joined together by these reactions, for example, an aryl carbon to another aryl carbon. It took the development of transition metal-catalyzed reactions before new types of carbon-carbon and carbon-heteroatom bonds could be created. Of particular note in this regard are the reactions catalyzed by palladium, usually in its 0 or +2 oxidation state.[2,3,4,5]

Of the close to 100 types of palladium-catalyzed carbon-carbon bond forming reactions known, the Heck reaction of aryl halides with alkenes, the Suzuki reaction of boronic acids with aryl halides, and the Stille reaction of vinyl triflates with vinylstannanes are the best known and most widely used. The catalysts are usually of two types: Pd(0) bonded to a phosphine ligand [e.g. Pd(PPh$_3$)$_4$] or a palladium salt such as PdCl$_2$ or Pd(OAc)$_2$ in the presence of a phosphine. Buchwald and coworkers, for example, recently reported a very active catalyst for the Suzuki reaction consisting of Pd(OAc)$_2$ and (o-biphenyl)P(cyclohexyl)$_2$.[6] It is generally believed that when Pd(+2) salts are used, the palladium is reduced to Pd(0) before the coupling commences. Reetz and Westermann in fact have recently shown that PdCl$_2$(PhCN$_2$)$_2$ yields colloidal Pd(0) during the Heck coupling of styrene with PhBr in N-methylpyrrolidone at 130°C.[7]

Many of the coupling reactions require a base such as OAc⁻ or NEt$_3$ in addition to the palladium. In the Heck reaction, for example, the base is used to effect elimination of hydrogen halide from an intermediate palladium complex, thus regenerating the palladium for use in further catalytic cycles. In the Suzuki reaction, on the other hand, the base binds to the boron atom of the boronic acid which activates the carbon-carbon bond for further reaction.

Because of our extensive experience in carrying out organic reactions on alumina[8] (a potential base), we were intrigued with the possibility of carrying out the coupling reactions under heterogeneous conditions, with all the potential benefits that entails. Although heterogeneous examples on resins,[9] zeolites,[10] silica,[11] glass beads[12] and clay[13] are known, we have found a superior method for carrying out the Suzuki reaction using two solids, ligandless palladium black [Pd(0)] and KF/Al$_2$O$_3$, a very strong base.[14] We have also successfully carried out the Sonogashira reaction, i.e., the coupling of an aryl halide with a terminal alkyne, using Pd black and KF/Al$_2$O$_3$.

1.1 Results and Discussion

Based on our experience in using Al$_2$O$_3$ as base in the iodination[15] and bromination[16] of vinylboronic acids, we hoped that this solid would likewise function as base in the Suzuki reaction. To simplify the reaction, we hoped to use ligandless palladium black [Pd(0)] as catalyst, thus alleviating the need for expensive or custom-made phosphine ligands.

To test these ideas the reaction of *p*-tolylboronic acid and iodobenzene (eq. 1) was studied.[17-19] As seen in Table 1, the reaction does not occur with Pd(0) and basic alumina, but does to varying degrees using the transition metal and bases adsorbed on alumina. Because KF/Al$_2$O$_3$ was the most effective base in inducing the Suzuki reaction, it was used in all subsequent reactions. It is worth noting that the Pd(0) - KF/Al$_2$O$_3$ system could be reused for further Suzuki reactions by addition of more KF to the alumina.

Table 1 *Coupling of p-Tolylboronic Acid with Iodobenzene using Pd(0) and Solid Bases*

Base[a]	Reaction Time[b]	Yield (%)
Al$_2$O$_3$ (basic)	6 hr	0
NaOH	3 hr	23
K$_2$CO$_3$	3 hr	27
K$_3$PO$_4$	3 hr	71
NaF	4 hr	5
KF	3 hr	86

[a]40% by weight. [b]Run with 5% Pd(0) at 80°.

The amount of Pd(0) used in the Suzuki coupling of *p*-tolylboronic acid with iodobenzene is critical in determining the optimal reaction time. In otherwise identical reactions where the amount Pd(0) was varied from 1 wt% to 5 wt% on 40% KF/Al$_2$O$_3$, 4 wt% Pd(0) gave the highest yield of 4-methylbiphenyl (Table 2); higher concentrations of the metal did not improve yields. Lower concentrations of Pd(0) could be used in the reaction but longer reaction times were required to obtain comparable yields. As noted earlier, the expensive Pd(0) could be used in subsequent Suzuki couplings.

Table 2. *The Effect of Palladium Concentration on the Yield of 4-Methylbiphenyl*[a]

Palladium (%)	Average Yield (%)
1	70.3
2	76.1
3	83.3
4	95.6
5	96.0

[a]Reaction run on 40% KF/Al$_2$O$_3$ at 80° for 4 hr.

To see how general the solventless Suzuki reaction is, the coupling of *p*-tolylboronic acid with several aryl, vinyl, and alkyl halides was attempted (eq. 2) (Table 3). In the reactions of the halobenzenes, the order of reactivity was observed to be: PhI>PhBr>PhCl, with PhF being unreactive; this order of reactivity is identical to what is seen in solution. Although the coupling worked well with iodo- and bromobenzene, the results with chlorobenzene were disappointing, especially in light of the recent successes of coupling of chloroarenes in solution by Buckwald[6] and Fu.[20] Interestingly, the vinyl bromide, (*E*)-1-bromopropene, is unreactive in the Suzuki coupling on Pd-KF/Al$_2$O$_3$. All of the alkyl halides that we investigated, 1-iodobutane, 1-iodohexane, 2-iodopropane, iodocyclohexane, and 1-bromobutane, did not couple as well. The allylic halides, 3-iodopropene and (*E*)-1-bromo-2-butene, on the other hand, did successfully couple. Surprisingly, even in those cases where no Suzuki coupling was observed, no starting halide was recovered in most instances on reaction workup.

$$RX \quad + \quad \underset{CH_3}{\underset{|}{\bigcirc}}\text{—}B(OH)_2 \quad \xrightarrow[\text{40\% KF/Al}_2\text{O}_3]{\text{5\% Pd(0) on}} \quad \underset{CH_3}{\underset{|}{\bigcirc}}\text{—}R \qquad (2)$$

Table 3. *Coupling of Organic Halides with p-Tolylboronic Acid on Pd-KF/Al$_2$O$_3$*

R-X[a]	Yield (%)	Recovered Halide (%)
Fluorobenzene	0	0
Chlorobenzene	5	0
Bromobenzene	60	0
Iodobenzene	99	<1
3-Iodopropene	39	0
(E) - 1-Bromo-2-butene	66	0
(E) - 1-Bromopropene	0	0
Alkyl Halide	0	0

[a]Run at 100° for 4 hr.

A less extensive study of the reaction of boronic acids with different structural characteristics - aryl, vinyl, allyl, and alkyl- with iodobenzene on Pd-KF/Al$_2$O$_3$ was also undertaken (eq. 3) (Table 4), Only the allylic boronic acid, 3-propenylboronic acid, did not undergo the Suzuki coupling although it was largely consumed.

$$RB(OH)_2 \quad + \quad \underset{}{\text{PhI}} \quad \xrightarrow[\text{40\% KF/Al}_2\text{O}_3]{\text{5\% Pd(0) on}} \quad \underset{}{\text{Ph-R}} \qquad (3)$$

Table 4. *Suzuki Coupling of Boronic Acids with Iodobenzene[a]*

Boronic Acid[a]	Product Yield %	Recovered PhI (%)
p-Tolylboronic Acid	98	0
(E) - 1-Pentenylboronic Acid	79	0
3-Propenylboronic Acid	0	29
1-Butylboronic Acid	26	trace

[a]Reaction run at 100° for 4 hr.

The majority of Suzuki couplings on Pd-KF/Al$_2$O$_3$ took approximately 4-5 hours at 80°-100° to complete. A considerable savings of time could be obtained by running the reactions under microwave irradiation.[21] In the coupling of *p*-tolylboronic acid with iodobenzene (eq. 1), for example, 4-methylbiphenyl was obtained in 80% yield after two minutes of irradiation. Although the reaction times were considerably reduced under microwave irradiation, the reaction characteristics were identical to those observed to the thermal reactions. For instance, in the coupling of *p*-tolylboronic acid with halobenzenes the order of reactivity was found to be: iodobenzene>bromobenzene>>chlorobenzene, with fluorobenzene being unreactive. Other representative examples of Suzuki coupling using Pd-KF/Al$_2$O$_3$ under microwave irradiation (2 minutes of full power) are shown in equation 4.

$$\text{Pd-KF/Al}_2\text{O}_3 \quad \text{microwaves (2 min)} \tag{4}$$

X = F	86%
X = Cl	87%
X = Br	80%
X = OCH$_3$	84%

As noted earlier, the Sonogashira coupling of aryl (and vinyl) iodides and terminal alkynes was successfully carried out using Pd(0) and KF/Al$_2$O$_3$, also under microwave irradiation.[22] In addition to Pd(0) and KF/Al$_2$O$_3$, this coupling reaction required CuI, presumably to activated the terminal alkyne for reaction. Interestingly no reaction occurred here in the absence of a phosphine ligand (PPh$_3$). Thus the coupling of iodobenzene worked best when Pd(0), KF/Al$_2$O$_3$, CuI and PPh$_3$ were all present (eq. 5). As shown in equation 6, the heterogeneous Sonogashira reaction was tolerant of several functional groups on the iodobenzene.

$$\text{CH}_3(\text{CH}_2)_6\text{C}\equiv\text{CH} \quad \frac{\text{Pd/CuI/PPh}_3}{\text{KF/Al}_2\text{O}_3} \quad \text{microwaves (150 sec)}$$

All reagents	94%
No Pd	0%
NoCuI	0%
No PPh$_3$	0%
No KF/Al$_2$O$_3$	0%

$$\text{—(CH}_2)_6\text{CH}_3 \tag{5}$$

$$X = p\text{-Acetyl} \quad 82\%$$
$$X = p\text{-NO}_2 \quad 67\%$$
$$X = m\text{-F} \quad 96\%$$
$$X = o\text{-NMe}_2 \quad 82\%$$

1.2   Conclusions

We have discovered a heterogeneous, solventless version of the Suzuki coupling reaction using Pd(0) and KF/Al_2O_3 which is quite competitive with the solution version of the reaction and does not require the use of expensive or custom-made phosphine ligands. The reaction is even more attractive when carried out under microwave activation because the reactions occur in two or three minutes. We have also shown that the Sonogashira coupling of aryl iodides with terminal alkynes occurs on Pd-KF/Al_2O_3 (+ CuI) under microwave activation. In this instance the inexpensive triphenylphosphine is needed for the reaction to occur.

ACKNOWLEDGEMENT

This research was supported by the U. S. Department of Energy and the Robert H. Cole Foundation.

References

1. M. B. Smith, *Organic Synthesis*, McGraw-Hill: New York, 1999.
2. J.-L. Malleron, J.-C. Fraud, and J.-Y. Legros, *Handbook of Palladium-Catalyzed Organic Reactions:* Academic Press: San Diego, 1997.
3. P. Theil, *Angew. Chem. Int. Ed.*, 1999, **38**, 2345.
4. R. Stürmer, *Angew. Chem. Int. Ed.*, 1999, **38**, 3307.
5. J. Tsuji, *Palladium Reagents and Catalysts: Innovations in Organic Synthesis*, Wiley: Chichester, 1995.
6. J. P. Wolfe, R. A. Singer, B. H. Yang, and S. L. Buchwald, *J. Am. Chem. Soc.*, 1999, **121**, 9550.
7. M. T. Reetz and E. Westermann, *Angew. Chem. Int. Ed.*, 2000, **39**, 165.
8. G. W. Kabalka and R. M. Pagni, *Tetrahedron*, 1997, **53**, 7999 and references therein.
9. T. Y. Zhang and M. J. Allen, *Tetrahedron Lett.*, 1999, **40**, 5813.
10. S. E. Sen, S. M. Smith, and K. A. Sullivan, *Tetrahedron*, 1999, **55**, 12657.
11. M. S. Anson, A. M. Mirza, L. Tonks, and J. M. J. Williams, *Tetrahedron Lett.*, 1999, **40**, 7147.

12. M. P. Leese and J. M. J. Williams, *Synlett*, 1999, 1645.

13. C. Waterlot, D. Couturier and B. Rigo, *Tetrahedron Lett.*, 2000, **41**, 317.

14. (a) J. Yamawaki and T. Ando, *Chem. Lett.*, 1979, 755. (b) L. M. Weinstock, J. M. Stevenson, S. A. Tomellini, S.-A. Pan, and T. Uthe, *Tetrahedron Lett.*, 1986, **27**, 8345. (c) T. Ando, S. J. Brown, J. H. Clark, D. G. Cork, T. Hanafusa, J. Ichihara, J. M. Miller and M. S. Robertson, *J. Chem. Soc. Perkin Trans. II*, 1986, 1133.

15. W. R. Sponholtz III, R. M. Pagni, G. W. Kabalka, J. F. Green, and L. C. Tan, *J. Org. Chem.*, 1991, **56**, 5700.

16. D. A. Willis, M. B. McGinnis, G. W. Kabalka, and R. M. Pagni, *J. Organometal. Chem.*, 1995, **487**, 35.

17. G. W. Kabalka, R. M. Pagni, and C. M. Hair, *Organic Lett.*, 1999, **1**, 1423.

18. M. Hair, R. M. Pagni, and G. W. Kabalka, in *Contemporary Boron Chemistry*, in press.

19. G. W. Kabalka, R. M. Pagni, C. M. Hair, J. L. Norris, L. Wang, and V. Namboodiri, in *ACS Symposium Series*, in press.

20. A. F. Littke, C. Dai, and G. C. Fu, *J. Am. Chem. Soc.*, 2000, **122,**, 4028.

21. G. W. Kabalka, R. M. Pagni, L. Wang, V. Namboodiri, and C. M. Hair, *Green Chem.*, in press.

22. G. W. Kabalka, L. Wang, V. Namboodiri, and R. M. Pagni, *Tetrahedron Lett.*, in press.

UNUSUAL REGIOSELECTIVITIES OBSERVED IN THE OLIGOMERIZATION OF PROPENE ON NICKEL(II) ION-EXCHANGED SILICA-ALUMINA CATALYSTS

Christakis P. Nicolaides# and Michael S. Scurrell*

#Chemiçal Process Engineering Research Institute/National Centre for Research and Technology-Hellas, P O Box 361, 570 01Thermi-Thessaloniki, Greece

*Applied Chemistry and Chemical Technology Centre, Department of Chemistry, University of the Witwatersrand, P O Wits, Johannesburg, 2050, South Africa

1 INTRODUCTION

Catalysts based on nickel(II) ion exchanged silica-alumina have recently been shown [1,2] to exhibit very high activities and stabilities in the oligomerization of ethene under both fixed-bed and slurry operating conditions. The essential characteristics of the system are summarized in the following manner: the conversion of ethylene results in oligomers containing an even number of carbon atoms per molecule, with a high degree of linearity in the products up to the hexenes. A very high activity in excess of 11 mol gcat^{-1} h^{-1} at 120°C (equivalent to an estimated TOF (turnover frequency) of 0.42 s^{-1}) is found in slurry-phase conditions.[2] The high activity is associated with the use of reaction temperatures in the range 100 –120°C and lower activities result from attempts to use higher temperatures. Most of our work has been carried out using an ethene pressure of 3.5 MPa.

The behaviour of these catalysts in the oligomerization of higher alkenes, propene and butene has not previously been reported. This paper addresses this aspect, and in addition focuses on the nature of the structure of the oligomers produced by propene dimerization. In their pioneering work, Hogan et al [3] indicated some formation of linear hexenes from propene dimerization, but their catalyst probably did not contain nickel in an exclusively ion-exchanged form. Various studies [4-7] conducted more recently on solid catalysts containing nickel give little or no detailed information on the nature of the C$_6$ dimer produced and again, many of the catalysts appear to contain nickel in a form other than comparatively isolated cations. Parallel studies of homogeneous catalysts are numerous. Much of the literature has been reviewed [8,9] and suggests that apart from a few specific systems, regioselectivities to linear hexenes rarely exceed about 20-23 mass%[9] to about 42 mass%[9]. Keim et al[10] have reported 75% linearity for very specifically designed complexes containing cyclooctenyl and and oxygen containing bidentate ligands incorporating CF$_3$ groups. Studies of Ni-phosphine complexes suggest that regioselectivity is controlled by steric factors[8] with a 22% selectivity being seen with the Ni(PPh$_3$)$_3$ complex itself.

It is noted that there have been a few, largely unsuccessful attempts to produce solid nickel catalysts for the dimerization of butene to linear octenes.[11,12]

In summary the detailed nature of the hexene isomers produced on various solid nickel catalysts by the dimerization of propene is still largely unknown. This paper attempts to provide new knowledge in this area.

2 EXPERIMENTAL

The catalysts studied were prepared and purified by methods previously described[1]. Briefly, the silica-alumina support was prepared by a coprecipitation method on to the support and possessed a Si/Al ratio of 25. Ni^{2+} ions were introduced from an aqueous solution of nickel chloride by ion exchange at reflux. The final catalysts contained about 1.5 mass% Ni, with some residual sodium and chlorine, corresponding to a Cl:Ni mol ratio of <0.1, and a Ni:Al mol ratio close to 1.0 (calculated with the assumption that all residual sodium is associated with some of the Al sites in the material (Na/Al mol ratio typically ca. 0.5)).

The experimental arrangement for conducting the oligomerization reaction has been described elsewhere[1]. Propene was fed at a pressure of 3.5MPa and the MHSV was 2. At temperatures of ca 110°C conversion was close to 90%. In order to gain detailed information on the nature of the dimer products obtained, the on-line gas chromatograph arrangement was supplemented with an off-line analysis system used for the analysis of the condensed liquid-phase products. Full separation of all C_6 alkene isomers was made possible via the use of two capillary columns designated OV-101 and BP1. A simpler optional analysis of the degree of linearity of the hexenes was sometimes carried out by first subjecting the alkene oligomers to a catalytic hydrogenation treatment prior to analysis in order to eliminate the positional isomers and simplify the resulting chromatogram. Such an approach becomes a necessity rather than an option when attempting to analyze C_8 products because of the very much larger number of alkene isomers that exist.

An indication of the state of the Ni present in the prepared catalysts was obtained by studies of the temperature programmed reduction of the solids. TPR spectra were recorded using catalyst samples that had been pre-treated at 150°C for 2h in a flow of nitrogen (UHP grade), before being reduced in hydrogen (UHP grade).

3 RESULTS

3.1 Conversion and Selectivity as a Function of Temperature.

The temperature dependence of the percentage conversion of reactant is depicted in Figure 1 which shows data for both propene and butene. At low temperatures butene is significantly less reactive than propene, but at higher temperatures (essentially above ca 210°C) these differences in reactivities are much less pronounced and both alkenes react at similar rates. With propene as feed, the major oligomers are C_6, C_9 with a small fraction of C_{12}.

3.2 Nature of the Oligomers Produced.

The gas chromatographic data (Figure 2) reveals clearly that the conversion process is one of true oligomerization. This was as found with ethene[1]. There is a great contrast here with acid-catalyzed conversion since in the latter case oligomerization is usually accompanied by cracking and re-oligomerization of cracked products leading to a product that contains all possible numbers of carbon atoms per molecule within the applicable product range[13]. The detailed nature of the dimers formed from propene is shown in Table 1. 4-Methylpentenes and 2-methylpentenes together constitute about 47 mass% of the products, with a minor amount of 2,3-dimethylbutenes (ca 5 mass%). Most significantly, the major products are the n-hexenes at ca 45 mass%. The distribution of isomers within the 4-methypentenes and

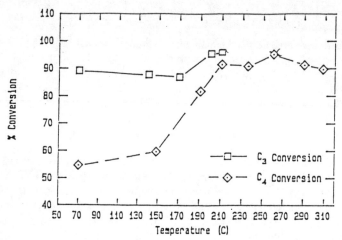

Figure 1   *Conversion of propene and butene as a function of temperature*

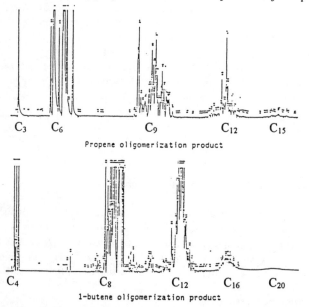

Figure 2   *Gas-chromatographic analysis of the oligomerization products from propene and butene*

2-methylpentenes are significantly different from values calculated for equilibrium conditions (see Table 1). Similar differences are also found within the n-hexene isomers produced. It is noted, however, that very little of the n-hexene product is 1-hexene, indicating that double-bond-shift is a relatively fast process with the reaction conditions and catalysts used (see below).

Table 1 *Isomer distributions obtained for the dimer fraction produced in the oligomerization of propene*
Reaction temperature = 110 °C

Isomer	mass% in $C_6^=$	Relative mass% within each geometric isomer group	
		Found	Equilibrium*
hex-1-ene	1.5	3.3	1.0
cis-hex-3-ene	1.2	2.7	4.0
trans-hex-3-ene	7.8	17.4	20.2
cis-hex-2-ene	9.4	20.9	41.6
trans-hex-2-ene	25.2	55.7	33.2
	45.1	100.0	100.0
4-Me-pent-1-ene	3.1	8.8	4.2
4-Me-cis-pent-2-ene	6.9	19.6	31.5
4-Me-trans-pent-2-ene	25.2	71.6	64.3
	35.2	100.0	100.0
2-Me-pent-1-ene	2.1	16.9	25.2
2-Me-pent-2-ene	10.3	83.1	74.8
	12.4	100.0	100.0
2,3-diMe-but-1-ene	0.7	15.2	45.9
2,3-diMe-but-2-ene	3.9	84.8	54.1
	4.6	100.0	100.0
other minor products	2.7	n.a.	n.a.

*Equilibrium distribution calculated from thermodynamic data

4 DISCUSSION

For both ethene and propene the overall nature of the products formed can be rationalized in terms of the widely accepted sequence of steps depicted in Figure 3. The relative amounts of isomeric hexenes produced from propene indicate that specific features of the catalyst probably determine the regioselectivity. The selectivity to n-hexenes is considerably higher than that found by Hogan et al with their catalyst[3] and may well reflect differences in composition and preparation method, and perhaps detailed nature of the support. The observation that the isomer distribution differs from that expected for the equilibrium situation indicates that kinetic factors exert a strong influence. It may well be that the steric factors held by some to operate in homogeneous systems are also of significance in the present work. However, more work is required to pursue this aspect. Distributions closer to the equilibrium values are found for the acid-catalyzed case[13] because these catalysts allow sequential cracking and oligomerization to occur at very high rates relative to those of the primary oligomerization step.

It is also clear that double-bond-shift is relatively facile on the nickel-silica-alumina catalysts. Double-bond shift may occur on the nickel cation centers or on the silica-alumina support or on both[14]. The hexene products formed from ethene are as expected for a reaction sequence involving (1) dimerization of ethene to but-1-ene etc, (2) double-bond-shift of but-1-ene to a but-2-ene mixture, and (3) reaction of but-2-ene with a further ethene

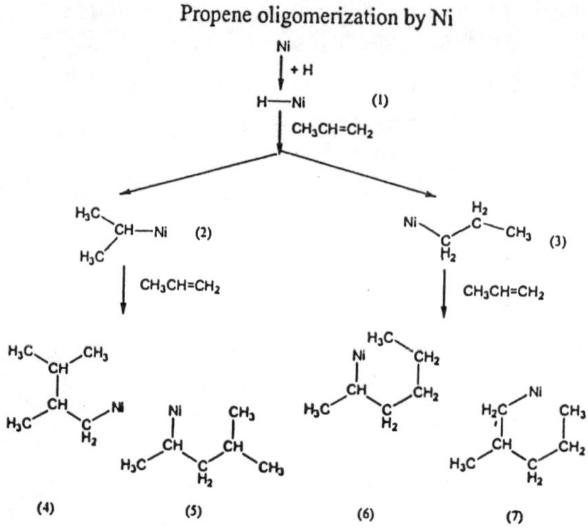

Propene oligomerization by Ni

Figure 3   *Outline reaction scheme for propene oligomerization by Ni-based catalysts. For clarity only alkyl product structures are depicted.*

molecule. This sequence of steps is also consistent with models developed for these catalysts operating in a slurry-phase reactor system[2]. The important conclusion to be drawn from the ease with which double-bond shift occurs on such catalysts is that attempts to produce linear octenes from butene dimerization will be difficult, though apparently not completely unsuccessful[11]. The initial double-bond shift reaction of butene-1 guarantees that the first alkene insertion step results almost exclusively in the formation of a branched alkyl unit.

The TPR data depicted in Figure 4 showing a single large peak at ca 710°C contrasts markedly with TPR data obtained by others for Ni-Al$_2$O$_3$ systems prepared by impregnation or coprecipitation methods.[15] The much lower temperatures at which peak maxima were encountered in those studies in the range typically 200-450°C suggests that the Ni cation centres in our catalyst are much more resistant to reduction, probably because they are present in an ion-exchanged form and are probably essentially isolated from each other due to the relatively high Si/Al ratio of the support. Similar high reduction temperatures (705°C) were observed by Sachtler and co-workers in their Co-ZSM-5 systems prepared by wet ion exchange[16].

It might be felt that any regioselective effects experienced with the present catalyst could perhaps be enhanced by choosing supports having a greater potential for imposing structure related effects such as steric constraints (assuming that steric effects are largely responsible for the observed regioselectivities observed in the present work). An obvious thought arising here involves the use of zeolites or molecular sieves. At present there appears to be little or no suitable data obtained with oligomerization catalysts based on nickel cations present in a zeolite environment. What little evidence there is[17] suggests that such systems tend to be much less stable than those based on the use of amorphous silica-alumina as the exchange host and catalyst deactivation is relatively rapid. Detailed product distributions were not reported and therefore no conclusions about the regioselectivities observed can be drawn.

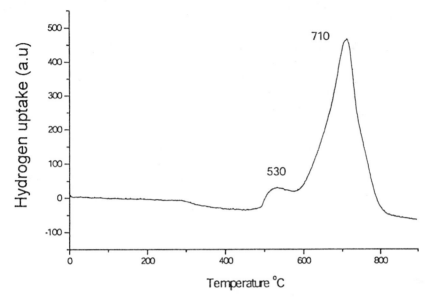

Figure 4 *Temperature programmed reduction of Ni(II) ion exchanged silica-alumina oligomerization catalyst. Catalyst was pretreated in nitrogen at 150 °C for 2h.*

It is however clear that the use of carefully ion-exchanged amorphous supports can lead to catalysts that have superior regioselectivities in propene dimerization compared with many homogeneous systems. The approach may well be more generally applicable to, for example, other areas of transition metal mediated catalysis and this aspect is now attracting our attention.

3 ACKNOWLEGEMENTS

We acknowledge fruitful discussions with J. Heveling and R.D. Forrester. Thanks are given to S. Kauchali, University of the Witwatersrand, for assistance with the calculations of the equilibrium distributions within the $C_6^=$ isomers using ASPEN procedures.

References

1. J. Heveling, C.P. Nicolaides and M.S. Scurrell, *Appl. Catal. A*, 1998, **173**, 1.
2. M.D. Heydenrych, C.P. Nicolaides and M.S. Scurrell, *J. Catal.*, in press.
3. J.P. Hogan, R.L. Banks, W.C. Lanning and A. Clark, *Ind. Eng. Chem.*, 1955, **47**, 752.
4. T. Cai, *Catal. Today*, 1999, **51**, 153.
5. W. Skupinski and S. Malinowski, *J. Mol. Catal.*, 1978, **4**, 95.
6. D. Kiessling, G. Wendt, M. Jusek and R. Schoellner, *React. Kinet. Catal. Lett.*, 1991, **43**, 255.
7. W. Skupinski and S. Malinowski, *J. Organometallic Chem.*, 1976, **117**, 183.
8. B. Bogdnovic, *Adv. Organometallic Chem.*, 1979, **17**, 105.
9. S.M. Pillai, M. Ravindranathan and S. Sivaram, *Chem. Rev.*, 1986, **86**, 353.

10. W. Keim, B. Hoffmann, R. Lodewick, M. Peuckert and G. Schmitt, *J. Mol. Catal.*, 1979, **6**, 39.
11. P. Beltrame, L. Forni, A. Talamini and G. Zuretti, *Appl. Catal. A*, 1994, **110**, 39.
12. B. Nkosi, F.T.T. Ng and G.L. Rempel, *Appl. Catal. A.*, 1997, **161**, 153.
13. R.J. Quanu, L.A. Green, S.A. Tabak and F.J. Krambeck, *Ind. Eng. Chem. Res.*, 1988, **27**, 565.
14. C.P. Nicolaides and M.S. Scurrell, unpublished observations.
15. J. Zielinski, *Appl. Catal. A.*, 1993, **94**, 107.
16. X. Wang, H-Y Chen and W.M.H. Sachtler, *Appl. Catal. B*, 2000, **26**, L227.
17. J. Heveling, A. van der Beek and M. de Pender, *Appl. Catal.*, 1988, **42**, 325.

SELECTIVITY THROUGH THE USE OF HETEROGENEOUS CATALYSTS

Keith Smith

Centre for Clean Chemistry
Department of Chemistry
University of Wales Swansea
Swansea SA2 8PP
UK

1 INTRODUCTION

Increasingly stringent environmental regulations and political pressure demand that many traditional chemical processes be conducted in a cleaner manner.[1] Aromatic substitution reactions are of considerable importance for production of fine chemicals, but in this field the traditional processes are notorious, suffering a number of disadvantages, such as low selectivity towards the desired product and the requirement for large quantities of mineral or Lewis acid activators. In turn these acids cause plant corrosion problems and generate large volumes of waste. Major efforts are therefore needed to reduce these problems.

The ability to control the selectivity of the substitution process is fundamental to achieving success in the area, but often there are few features to give assistance. Therefore, it is necessary to gain control through the use of external agents. Heterogeneous catalysts such as zeolites have in a number of cases proved to be central to this strategy. For example, in previous studies, we have shown that zeolites can be used successfully for cleaner organic synthesis in reactions such as bromination,[2] chlorination,[3] acylation[4] and methanesulfonylation[5] of aromatic compounds.

In these liquid phase reactions there are two principal roles for the zeolites. (1) They provide sites capable of enhancing the rate of the reaction in either a stoichiometric or catalytic manner. (2) By having these sites located within the rigid pores of the inorganic matrix they also impose additional constraints on the reacting partners, favouring production of one isomer (usually the most linear one) over other possible ones. Often, the active sites are acidic, created by the presence of aluminium atoms located in a silica framework. Typically, the topographical constraints are exerted most acutely on the transition state of the reactions concerned, which means that substrate, reagent and solvent may all be important in determining the level of selectivity achieved.

Two areas of current interest where we have been investigating the use of zeolites to gain much needed selectivity enhancements are the nitration of aromatic substrates and the alkylation of naphthalene. Some of our progress is reported in this paper.

2 NITRATION OF AROMATIC COMPOUNDS

Aromatic nitro compounds represent versatile chemical feedstocks for a wide range of industrial products, such as pharmaceuticals, agrochemicals, dyestuffs and explosives.

Traditionally, nitration has been performed with a mixture of nitric and sulfuric acids (mixed acid method). However, the method is highly unselective for nitration of substituted aromatic compounds and disposal of the spent acid reagents presents a serious environmental issue. In order to address these problems several alternative methods for aromatic nitration have been developed recently. For example, lanthanide triflates catalyse nitration with nitric acid, which avoids the use of large volumes of sulfuric acid but provides no enhancement of selectivity.[6] Selectivity of nitrations with alkyl nitrates,[7] acyl nitrates,[8] or even nitric acid itself[9,10] can, however, be enhanced by zeolites.

We have reported the use of zeolite Hβ in conjunction with a mixture of acetic anhydride and nitric acid, which currently offers the best combination of yield and *para*-selectivity for nitration of simple aromatic compounds.[11] For example, this system allows quantitative mononitration of toluene with selectivity for the *para* isomer of around 80 %. For commercial production of 4-nitrotoluene this would require less than half the toluene or nitric acid needed for the conventional mixed acid method and would generate less than one third of the waste. In general, the system offers excellent possibilities for nitration of substrates of moderate activity. However, it is not very successful with deactivated substrates. One aspect of the work reported here is therefore the modification of this approach so that it accommodates deactivated substrates.

2.1 Selective Nitration of Deactivated Substrates

Normally, nitration of deactivated compounds (and therefore polynitration of toluene) is carried out using aggressive nitric acid – oleum mixtures. Dinitration of toluene with mixed acids produces a 4:1 ratio of 2,4- and 2,6-dinitrotoluenes, from which the former is isolated for manufacture of toluenediisocyanate (TDI) and toluenediamine, both of which are used in the manufacture of polyurethanes. Zirconium and hafnium derivatives catalyse nitration of *o*-nitrotoluene, but ratios of 2,4-:2,6-dinitrotoluene are modest (66:34).[12] Dinitration of toluene using Claycop (copper nitrate on K10 clay), acetic anhydride and nitric acid in the presence of carbon tetrachloride produced dinitrotoluenes in a yield of 85% with a ratio of 2,4-:2,6-dinitrotoluene of 9:1.[13] This method, however, requires a large excess of nitric acid, the use of an unacceptable solvent and long reaction times. The direct nitration of toluene to 2,4-dinitrotoluene using nitric acid over a zeolite β catalyst, with azeotropic removal of water, is reported to give a 2,4:2,6 ratio of 14, but full results are yet to be published.[14]

After our success in nitrating moderately active monosubstituted benzenes with acetic anhydride and nitric acid over zeolite β,[11] we decided to try the use of trifluoroacetic anhydride and nitric acid over zeolite β for nitration of deactivated substrates. Although trifluoroacetyl nitrate is known to be more active than acetyl nitrate, it has not been widely used in nitration reactions.[15] Nitrobenzene has been successfully nitrated using fuming nitric acid and trifluoroacetic anhydride in equimolar proportions at 45-55 °C.[16] However, no dinitration of toluene was reported.

Our initial studies involved nitration of *p*-nitrotoluene, which nitrates exclusively to give 2,4-dinitrotoluene. We found that the trifluoroacetic anhydride - nitric acid mixture was active enough to nitrate this substrate in high yield at room temperature without the need for a zeolite. For nitration of *o*-nitrotoluene (**1**), there were two main products, 2,4-dinitrotoluene (**2**) and 2,6-dinitrotoluene (**3**, Scheme 1). Again, we found that the trifluoroacetic anhydride - nitric acid mixture alone was active enough to effect nitration. However, the presence of zeolite Hβ improved regioselectivity for the 2,4-isomer slightly (2,4:2,6 ratio 3:1 compared to 2:1 without the zeolite), although it did not affect the overall reaction rate or product yield much. It appeared that the free solution reaction was so fast that most of the reactants did not have time to diffuse into the pore system and so reaction occurred primarily in free solution or at the external surface of the zeolite.

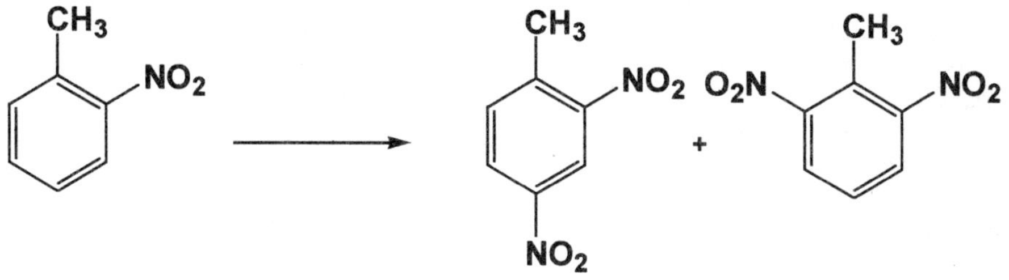

Scheme 1

In order to slow down the overall reaction we added acetic anhydride as a diluent, thereby giving the zeolite a better chance to exert an influence over the reaction. Indeed, this led to a slower reaction and zeolite Hβ then exerted a greater influence over both the rate and selectivity. The yield after 2 h at –10 °C in the absence of the zeolite was only 16 % and the **2:3** ratio was the usual 2:1, but with the zeolite present the yield increased to 99 % and the ratio to 17:1. *p*-Nitrotoluene could also be nitrated with the optimised system used for *o*-nitrotoluene, but the reaction was much slower. Therefore, for direct dinitration of toluene it would be necessary to minimise the amount of the diluent used.

2.1.1 Dinitration of Toluene. We attempted to carry out a single step nitration of toluene (Scheme 2) under conditions similar to those developed for selective nitration of *o*-nitrotoluene, but using two equivalents of nitric acid and a smaller quantity of acetic anhydride. The reaction produced dinitrotoluenes in almost quantitative yield. For a 17.5 mmol reaction the selectivity was exceptional (**2:3** = 25:1) using 1 g zeolite, and as high as has been reported heretofore (**2:3** = 14:1) even when only 0.5 g of zeolite was used.

Excellent though these results were, the ratio of **2:3** was less than expected if the first step were to give *o*- and *p*-nitrotoluenes in the published ratio for nitric acid / acetic anhydride / Hβ (*ca.* 4:1), and *p*-nitrotoluene gave only **2** while *o*-nitrotoluene gave a 17:1 ratio of **2:3**. It seemed likely that the problem was a less selective first step using trifluoroacetic anhydride. Therefore, we investigated a two step sequence in which trifluoroacetic anhydride was introduced only during the second step. When we also added extra zeolite for the second step (Scheme 3), the regioselectivity improved to a **2:3** ratio of 70:1, with quantitative yields, after 2 hours at around -10 °C. Pure 2,4-dinitrotoluene could be isolated in 90 % yield from this system simply by filtering the zeolite, concentrating the mother liquor, and recrystallisation from acetone.

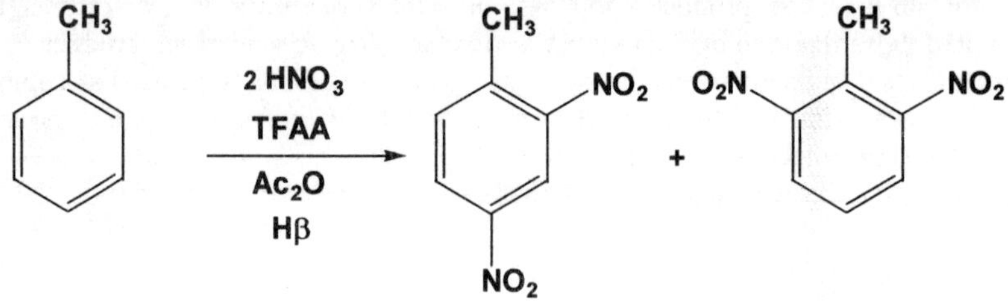

Scheme 2

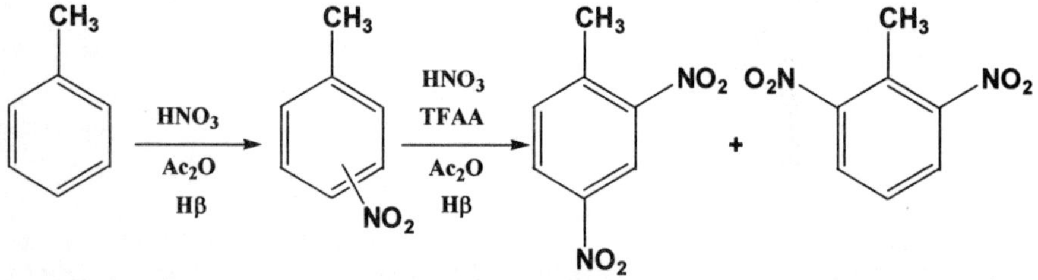

Scheme 3

This new approach to the nitration of deactivated aromatics has been successfully applied to a range of substrates and details have recently been published.[17] As indicated above, the approach was modelled on the successful nitration of moderately activated substrates using nitric acid, acetic anhydride and zeolite β that we had developed earlier.[11] However, commercial organisations appear reluctant to adopt that new technology despite its greatly superior selectivity and potential environmental impact. Therefore, we continue to look for alternative methods of achieving selective nitrations.

2.2   Nitration with Dinitrogen Tetroxide

Another approach towards clean nitration involves use of dinitrogen tetroxide with oxygen or ozone as an oxidant.[18] The ozone method most likely involves dinitrogen pentoxide, known to be a highly active nitrating agent. The method utilising oxygen is less clear cut and requires use of tris(pentane-2,4-dionato)iron(III) (Fe(acac)$_3$) as catalyst in an organic solvent.[19] In principle, this could lead to a highly atom-efficient process, but it is not regioselective. Therefore, we decided to study dinitrogen tetroxide in the presence of zeolites to see if they could catalyse the process and impart *para*-selectivity.

Zeolites have been used before in the vapour phase nitration of aromatic compounds using nitrogen dioxide.[20] However, the conditions were harsh and there was no regioselectivity. Initially, therefore, we attempted to reproduce the mild conditions of Suzuki for nitration of chlorobenzene. Liquid N$_2$O$_4$ (approx. 10 ml) was condensed into a trap at -78 °C and was then warmed to 0 °C. Fe(acac)$_3$ (0.355 g) and chlorobenzene

(10 mmol) were added, the system was flushed with oxygen and the mixture was stirred at 0 °C for 48 h. The product contained nitrochlorobenzenes in proportions that approximated those reported by Suzuki, but also an acetylacetone nitration product.

We then carried out similar reactions in which various zeolites were used as catalysts instead of Fe(acac)$_3$ in an attempt to determine which zeolite, if any, would be most applicable to *para*-selective aromatic nitration. Zeolites Hß and Naß produced the greatest selectivity for *para*-chloronitrobenzene (85 %) and the highest yields (90 and 96 %). Therefore, zeolite Hß was tested with a range of other substrates (Scheme 4). The results are shown in Table 1.

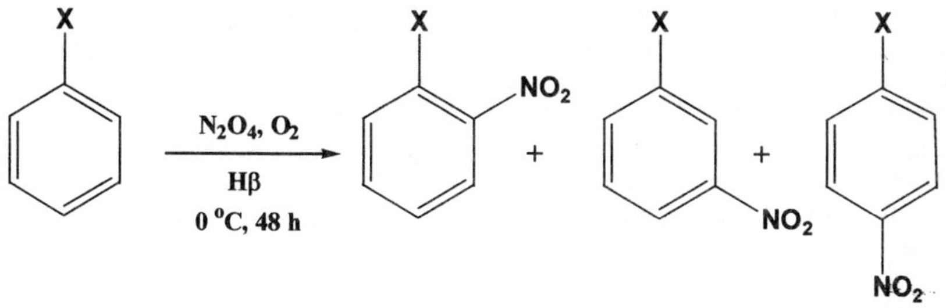

Scheme 4

The results demonstrate that zeolite ß is an efficient catalyst for nitration of halogenobenzenes with dinitrogen tetroxide and oxygen and gives *para*-selectivities that are high compared to classical nitration methods. The method represents a potentially clean synthesis of halogenonitrobenzenes using an easily recycled catalyst system.[21] However, at this preliminary point the reactions are not as selective as we would like, involve a large excess of dinitrogen tetroxide and long reaction times, and use an undesirable chlorinated solvent. Our future efforts in this area will be concentrated on minimising these factors.

Table 1 *Nitration reactions according to Scheme 4[a]*

Substrate (5)	X	Yield (%)	Proportions (%) ortho (6)	meta (7)	para (8)
toluene	CH$_3$	85	53	2	45
benzene	H	50	-	-	-
fluorobenzene	F	95	7	0	93
chlorobenzene	Cl	95	14	<1	85
bromobenzene	Br	94	22	<1	77
iodobenzene	I	95	37	1	62

[a] To a mixture of 1,2-dichloroethane (30 ml), the substrate (10 mmol) and zeolite Hß (1 g) at 0 °C was added liquid N$_2$O$_4$ (approx. 10 ml), the system was flushed with oxygen and the mixture was stirred at 0 °C for 48 h.

3 DIALKYLATION OF NAPHTHALENE

Regioselective dialkylation of naphthalene is another reaction of considerable interest as 2,6-dialkylnaphthalenes can be oxidised to naphthalene-2,6-dicarboxylic acid, which is used in the synthesis of the commercially valuable polymer, poly(ethylene naphthalenedicarboxylate) (PEN).[22] PEN has properties that are generally superior to those of poly(ethylene terephthalate) (PET) and has become the polymer of choice for a variety of applications such as in films, industrial fibres, packaging, liquid crystalline polymers, coatings, inks and adhesives.   However, the high cost of naphthalenedicarboxylic acid has been a major hindrance to widespread application.

Amoco chemicals have commercialised a procedure that uses *o*-xylene as starting material. However, this synthesis of naphthalene-2,6-dicarboxylic acid involves five steps, including isomerisation of 1,5-dimethylnaphthalene to 2,6-dimethylnaphthalene. The direct regioselective alkylation of naphthalene (**9**, Scheme 5) is therefore more attractive, and has been investigated by, amongst others, NKK, Chiyoda and Catalytica.[22] Synthesis of **10** by this approach is complicated, however, by the possible formation of nine other dialkyl isomers, which can be difficult to separate from the 2,6-isomer.

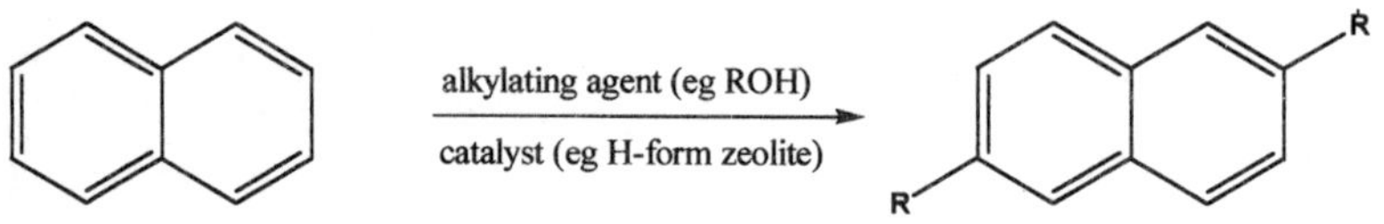

	alkylating agent (eg ROH)	
	──────────────────────►	
	catalyst (eg H-form zeolite)	

9   **10**

Scheme 5

Initial studies of methylation gave mainly ß-dialkylation over HZSM-5, but with 2,6/2,7 ratios of approximately 1. Subsequently, *iso*-propylation of naphthalene over HM provided higher 2,6/2,7 ratios,[23,24] while *tert*-butylation over HY gave even higher ratios and the added advantage of easy purification of 2,6-dialkylnaphthalene by crystallisation.[25] Examples of the best 2,6-dialkylnaphthalene yields and 2,6/2,7 ratios for *iso*-propylation[24] and *tert*-butylation[25] are given in common form in Table 2. The most promising approach seemed to be *tert*-butylation, as the highest 2,6/2,7 ratio was obtained as well as an easy separation of the desired product. Therefore, we undertook a detailed study of the *tert*-butylation reaction to see if further improvements could be made in the yield and selectivity for the 2,6-isomer.

Table 2   *Literature results for iso-propylation[24] and tert-butylation[25] of naphthalene*

Alkylation type	*iso-Propylation*	*tert-Butylation*
Catalyst	HM	HY
Naphthalene converted (%)	96.6	52.4
2,6-Dialkylnaphthalene (%)	54.2	23.3
Ratio of 2,6- to 2,7-isomer	4.0	5.9

3.1 Preliminary screening of zeolites

Initially, a range of different solids was screened (Table 3) for efficacy in the *tert*-butylation reaction, under conditions similar to those used by Moreau.[25] Where direct comparison was possible, the trends in the results were consistent with the findings of Moreau. Minor variations could be due to use of different zeolite samples and a different autoclave with different dimensions and different temperature control. HY was the most active solid while Hβ was essentially non-selective for the 2,6- over the 2,7-isomer. HZSM-5 gave no reaction. With mesoporous HMS the naphthalene conversion was similar to that with Hβ, but with a higher yield of **10** and a surprisingly higher 2,6/2,7 ratio. However, our attention was attracted most by the result with HM. Although quite inactive, it was potentially the most promising for selectivity, as only the 2-alkyl and 2,6-dialkyl products were detected under these conditions. Therefore, we undertook a more detailed study of the reaction with HM as catalyst in the hope of enhancing the conversion into dialkylnaphthalene while retaining high selectivity.

Table 3   *tert-Butylation of naphthalene over solid catalysts according to Scheme 5[a]*

Catalyst	HZSM-5	HM	Hβ	HY	HMS[b]
Naphthalene converted (%)	0	22	49	89	43
2,6-Dialkylnaphthalene (%)	0	2	2	33	6
Ratio of 2,6- to 2,7-isomers	-	-	1.1	2.7	1.9

[a] Catalyst (0.5 g), naphthalene (10 mmol), *tert*-BuOH (20 mmol), cyclohexane (100 ml), 2 h autoclave reactions at 160 °C. [b] A synthetic mesoporous material.

3.2 Optimisation of the conditions using HM as catalyst

We attempted to increase the conversion into desirable product by using more forcing conditions (increasing the time and temperature) and by adding more catalyst and/or alkylating agent. Doubling the amount of catalyst to 1.0 g under the original conditions increased the conversion by 13 % and the dialkylnaphthalene yield by 5 %. Also, it was then possible to identify the 2,7-isomer in the mixture and a 2,6/2,7 ratio of 17.3 could be seen, already a significant improvement over all previously reported selectivities. Further studies were conducted using different times; temperatures; pressures; *tert*-butylating species; solvents; stoichiometry; Si/Al ratio of the catalyst; and modes of addition.

Following this study, we found that the optimum conditions for maximising the yield while retaining selectivity involved two successive 1 h autoclave reactions at 180 °C, using HM (4.0 g, Si/Al ratio 17.5), *tert*-butanol (80 mmol) and cyclohexane (10 ml) for an initial 10 mmol of naphthalene. Under these conditions, a 2,6-di-*tert*-butylnaphthalene yield of 50 % and a 2,6/2,7 ratio of 58.5 were achieved after the first reaction, and a 60 % yield with a ratio of 50.6 after the second stage. The optimised result is compared with the most selective literature result in Table 4.

It can be seen that we have successfully improved the yield of 2,6-di-*tert*-butylnaphthalene (**10**, R = *tert*-butyl) to over 60 % and simultaneously increased the 2,6/2,7 ratio to over 50. The product is easily isolated in over 50 % yield by direct crystallisation from ethanol, with the remainder obtainable by Kugelrohr distillation.[26]

Table 4   *Comparison of the best literature results for tert-butylation of naphthalene[25] with the optimum new results[26]*

	Literature results	The new work
Catalyst	HY	HM
Naphthalene converted (%)	52.4	96
2,6-Dialkylnaphthalene (%)	23.3	60
Ratio of 2,6- to 2,7-isomer	5.9	50.6

4 CONCLUSIONS

Zeolites offer possibilities for both catalysis and selectivity enhancement in aromatic substitution reactions, where such effects are much needed to overcome the many problems caused by the traditional methods.   Two reaction types where superior methodologies are particularly needed, namely selective nitration of simple benzenoids and 2,6-dialkylation of naphthalene, have been addressed in the present study.

Our earlier nitration system involving nitric acid and acetic anhydride in conjunction with zeolite Hβ has been adapted to include trifluoroacetic anhydride, which enhances the reactivity considerably and allows the nitration of deactivated aromatics. Optimisation of the process has enabled the most selective double nitration of toluene yet attained, giving 2,4-dinitrotoluene in high yield.

In an alternative approach to mononitration of simple aromatics dinitrogen tetroxide has been used in conjunction with zeolite Hβ.   This provides selectivity for *para*-substitution that is higher than in the traditional approach, though there is still need for further improvement of the system.

Finally, highly regioselective dialkylation of naphthalene has been achieved over zeolite HM using *tert*-butanol as alkylating agent.   By optimisation of the reaction parameters, 2,6-di-*tert*-butylnaphthalene has been obtained in a yield of 60 % with a 2,6/2,7 ratio of over 50.   This is the highest yield of a 2,6-dialkylnaphthalene and easily the highest 2,6/2,7 ratio yet reported from a direct dialkylation of naphthalene.

We continue to explore the possibilities for exploitation of zeolites in selective aromatic substitution reactions.

ACKNOWLEDGEMENTS

Gratitude is due to the students who conducted the work reviewed herein.   Tracy Gibbins developed nitration of deactivated aromatics, Saeed Almeer is developing nitration using dinitrogen tetroxide, and Simon (now Dr.) Roberts developed naphthalene dialkylation. My colleague Dr. Steve Black and all members of my research group are also thanked for creating the general research environment that enabled the work to be accomplished.

The financial sponsors are warmly thanked.   DERA at Fort Halstead funded the work on nitration of deactivated aromatics, and particular thanks are due to our industrial colleagues, Dr. Ross Millar and Dr. Rob Claridge, who made substantial contributions to the development of the ideas.   The Government of Qatar is funding the studentship of Saeed Almeer and the EPSRC funded the studentship of Simon Roberts.

Thanks are also due to the EPSRC Mass Spectrometry Centre for mass spectra and the EPSRC and the University of Wales for grants enabling purchase of NMR equipment.

References

1. J.H. Clark, Ed., *Chemistry of Waste Minimisation*, Chapman and Hall, London, 1995.
2. K. Smith and D. Bahzad, *J. Chem. Soc., Chem. Commun.*, 1996, 467.
3. K. Smith, M. Butters and B. Nay, Synthesis, 1985, 1157; K. Smith, M. Butters, W. E. Paget, D. Goubet, E. Fromentin and B. Nay, *Green Chemistry*, 1999, **2**, 83.
4. K. Smith, Z. Zhenhua and P. K. G. Hodgson, *J. Mol. Catal. A: Chemical*, 1998, **134** 121.
5. K. Smith, G. M. Ewart and K. R. Randles, *J. Chem. Soc., Perkin Trans. 1*, 1997, 1085.
6. F. J. Waller, A. G. M. Barrett, D. C. Braddock and D. Ramprasad, *Chem. Commun.*, 1997, 613.
7. T. J. Kwok and K. Jayasuriya, *J. Org. Chem.*, 1994, **59**, 4939.
8. A. Cornelis, A. Gerstmans and P. Laszlo, *Chem. Letters*, 1988, 1839.
9. K. Jayasuriya and R. Damavarapu, *US pat. Appl. No 5946638*, 1999.
10. B. M. Choudary, M. Sateesh, M. L. Kantam, K. K. Rao, K. V. R. Prasad, K. V. Raghavan and J. A. R. P. Sarma, *Chem. Commun.*, 2000, 25.
11. K. Smith, A. Musson and G. A. DeBoos, *J. Org. Chem.*, 1998, **63**, 8448.
12. F. J. Waller, A. G. M. Barrett, D. C. Braddock and D. Ramprasad, *Tetrahedron Lett.*, 1998, **39**, 1641.
13. B. Gigante, A. Prazeres, M. Marcelo-Curto, A. Cornelis and P. Laszlo, *J. Org. Chem.*, 1995, **60**, 3445.
14. D. Vassena, A. Kogelbauer, R. Prins and J. Armor, *Abstracts of EUROPACAT 4*, Rimini, September 5-10, 1999, p 222.
15. (a) G. A. Olah, *Nitration: Methods and Mechanisms*, VCH, New York, 1989; (b) K. Schofield, *Aromatic Nitration*, Cambridge University Press, Cambridge, 1980; (c) R. Taylor, *Electrophilic Aromatic Substitution*, John Wiley and Sons, Chichester, 1990.
16. E. J. Bourne, M. Stacey, J. C. Tatlow and J. M. Tedder, *J. Chem. Soc.*, 1952, 1695.
17. K. Smith, T. Gibbins, R. W. Millar and R. P. Claridge, *J. Chem. Soc., Perkin Trans. 1*, 2000, 2753.
18. H. Suzuki and T. Mori, *J. Chem. Soc., Perkin Trans. 2*, 1994, 479.
19. H. Suzuki, S. Yonezawa, N. Nonoyama and T. Mori, *J. Chem. Soc., Perkin Trans. 1*, 1996, 2385.
20. I. Schumacher, *Eur. Pat. Appl. No. 0053031*, 1981; A. Germain, T. Akouz and F. Figueras, *J. Catal.*, 1994, **147**, 163: ; A. Germain, T. Akouz and F. Figueras, *Appl. Catal. A: Gen.*, 1996, **136**, 37.
21. K. Smith, S. Almeer and S. J. Black, *Chem. Commun.*, 2000, 1571.
22. K. Tanabe and W. F. Hölderich, *Appl. Catal. A: General*, 1999, **181**, 399.
23. A. Katayama, M. Toba, G. Takeuchi, F. Mizukami, S.-i. Niwa and S. Mitamura, *J. Chem. Soc., Chem. Commun.*, 1991, 39.
24. P. P. B Notte, G. M. J. L. Poncelet, M. J. H. Remy, P. F. M. G. Lardinois and M. J. M. Van Hoecke, *EP Patent*, 0528096, 1993.
25. Z. Liu, P. Moreau and F. Fajula, *Appl. Catal. A: General*, 1997, **159**, 305.
26. K. Smith and S. D. Roberts, *Catalysis Today*, 2000, **60**, 227.

NOVEL LEWIS-ACIDIC CATALYSTS BY IMMOBILISATION OF IONIC LIQUIDS

M. H. Valkenberg*, C. deCastro, W. F. Hölderich

Department of Chemical Technology and Heterogeneous Catalysis, Worringerweg 1,
52074 Aachen, Germany; e-mail: hoelderich@rwth-aachen.de
Tel.: ++49-241-806560/61, Fax.: ++49-241-8888291

1 INTRODUCTION

Highly Lewis-acidic chloroaluminate ionic liquids (ILs) are well known to be both versatile solvents and effective catalysts for Friedel-Crafts reactions [1,2]. Tailoring the physical and chemical properties of the ILs to the needs of a specific reaction allows for a high diversity of applications [3,4]. We could show that immobilising these ILs on inorganic supports yields very active catalysts for alkylation reactions. The immobilisation of ionic liquids leads to novel Lewis-acidic catalysts (NLACs). The methods presented include the method of incipient wetness (method 1, further on called NLAC I), which has been introduced in detail by Hoelderich et al. [5], but focus of this presentation lies on the methods 2 (NLAC II) and 3 (NLAC III).

2 EXPERIMENTAL

To ensure reaction conditions free of water, all experimental steps have to be carried out under an inert atmosphere. Before impregnation the supports were calcined for three hours at 550 °C and stored under argon.

2.1 Immobilisation

The grafting of organic cations on MCM 41 and the preparation of organically modified MCM 41 were done according to the methods known from literature [6,7,8]. $AlCl_3$ was added to the supports in dried toluene, afterwards the excess of $AlCl_3$ was extracted with boiling dichloromethane in a Soxhlet apparatus.

2.2 Alkylation reactions

As a test reaction, the alkylation of benzene with dodecene was chosen. The batch reactor system consisted of a round-bottomed flask (slurry reactor) provided with a cooler, gas-inlet valve, and sampling exit. A magnetic stirrer equipped with a thermostat and a silicon oil bath were used to maintain the reaction temperature and ensure the homogeneity of reactants. The catalyst was weighed in the reactor and then benzene was added according to the proportion desired, followed by the addition of dodecene.

Undecane was added as an internal standard to follow the reaction kinetics and to calculate the mass balance. Samples were taken periodically and analysed by gas chromatography in order to quantify conversion and selectivity of the reaction. In all reactions presented here, a ratio of 1 mol dodecene to 10 mol of benzene was used.

2.3 Analytics

Solid state MAS-NMR measurements were carried out on a Bruker Avanza 500. X-ray diffractometrie (XRD) was done on a D 5000 from Siemens, using a copper tupe FL Cu 4KE. FT-IR was performed on a Nicolet Protégé 460. For BET-measurements an ASAP 2000 machine from Micromeritics was used. The concentrations of Si and Al in the samples were determined by ICP-AES, using a Spectro-Flame D machine from Spectro. CHN analysis was performed on a Elementar Vario EL. GC analysis was done on a Siemens RGC 202, using a 30 m SE 54 column from Hewlett Packard.

3 RESULTS AND DISCUSSION

Different methods for the immobilisation of ionic liquids have been tested. These can be divided into two different categories: a) via the inorganic anion (method of incipient wetness and b) via the organic cation, here a methyl-alkyl-imidazolium cation. The method of incipient wetness was presented by Hoelderich [5], we concentrate on the immobilisation of ionic liquids by the methods of grafting or by synthesising organically modified MCM 41 materials (Sol gel method) (seeFigure 1).

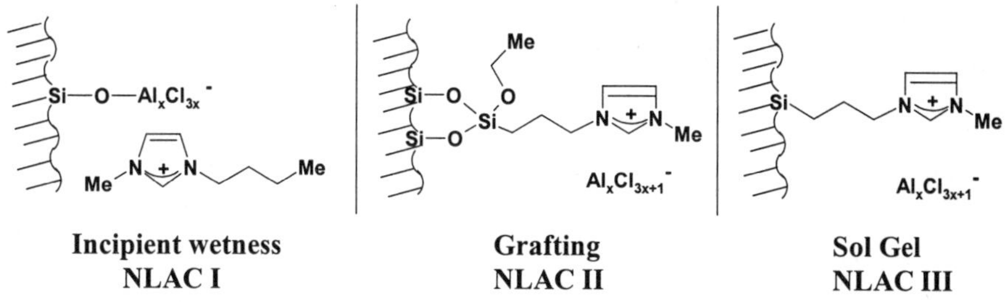

Incipient wetness	**Grafting**	**Sol Gel**
NLAC I	**NLAC II**	**NLAC III**

Figure 1 *Three different methods for the preparation of immobilised ionic liquids*

Method 1 is known as the method of incipient wetness, because the ionic liquid is added to the support until the mixture starts to lose the appearance of an dry powder. This is the most simple of the presented methods, allowing the immobilisation of high amounts of chloroaluminate liquids on any given silica support. Unfortunately, during the immobilisation step HCl is created which leads to a decomposition of zeolites and MCM 41 type supports. This problem could be overcome by a modification of the immobilisation method. The supported ILs synthesised this way show a high catalytic activity in Friedel-Crafts reactions.

Table 1   *BET data of different supports*

	BET-surface area [m²/g]	Max. pore diameter [Å]	Average pore diameter [Å]
MCM 41	1539	20 – 25	22.51
MCM 41, Grafted	1058	18 – 23	22.4
MCM 41, Sol gel	448	20 – 25	52.3

The catalysts synthesised according to method 2 and 3 are more suitable when working with MCM 41 type materials. In both cases the ionic liquids are supported via the organic cation. While the cations are grafted onto the support in method 2 [4], in method 3 the MCM 41 is synthesised as an organically modified material [5]. Both methods allow the immobilisation of high amounts of ILs without attacking the support. Furthermore, they facilitate the immobilisation of less acidic ILs which cannot be supported via the anion. Here the anion is added to the support in a second step. The supports used here were in all cases all silica MCM 41 type materials. BET data showed a decrease in surface area and average pore size for the grafting product. This is to be expected since the surface is covered with imidazolium chloride (see Table 1). The high average pore diameter found for the NLAC III (MCM 41, Sol gel method) can be explained by a slightly increased amount of mesopores in the range of 200 – 400 Å.

The chloroaluminate catalysts prepared according to method 2 show even higher activity in Friedel-Crafts reactions. This can be explained by the fact that here an ionic liquid is simulated on the surface of the support. The hydroxyl groups on the surface of the support, which would otherwise react with AlCl₃ are now used for the grafting of the organic cation. As shown in Figure 2, this is supported by NMR data [9,10].

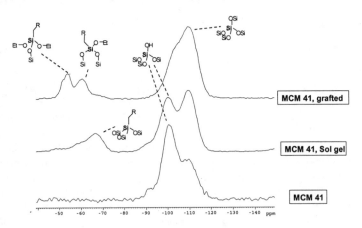

Figure 2   *Crosspolarised 29Si NMR spectra of different supports*

Table 2 *Results for the alkylation of benzene with dodecene, catalysed NLACs*
Reaction time: 1h, molar ratio benzene/dodecene: 10 : 1

Catalyst	T [°C]	Catalyst [wt%]	$C_{12}^=$ conversion	$C_{12}^=$ isomers	Sel. monoalkyl.	Sel. Heavier prod.
Al-IL, pure	80	6	4.5	86.7	13.3	0
MCM 41, pure	80	6	0.1	100	0.0	0
SiO₂/AlCl₃	80	6	15.7	80.2	19.8	0
SiO₂/ NLAC I [1]	80	6	99.4	0	99.7	0.3
SiO₂/NLAC II [2]	80	6	92.3	22.3	76.2	1.5
MCM 41/ NLAC II [2]	80	6	100	12.6	85.6	1.8
MCM 41/ NLAC III [3]	80	6	74	17.8	82.2	0
SiO₂/ NLAC I [1]	20	1	10.9	7.7	89.9	2.3
SiO₂/ NLAC II [2]	20	1	46.8	7.6	90.7	1.8
MCM 41/ NLAC II[2]	20	1				

[1] Incipient wetness method
[2] Grafted
[3] Sol gel

The catalysts prepared according to the different methods were used in Friedel-Crafts alkylation reactions. Table 2 shows the results for some of the batch reactions. The high amount of isomers of the dodecene found for the reactions catalysed by 6 wt% of the grafted materials indicates a possible contamination with water. Since a comparably high amount of heavier products was found for these reactions, the reaction conditions were changed to a lower temperature and a lower catalyst ratio. Here the grafted materials proved to be far superior to the materials prepared by the method of incipient wetness. All other catalysts gave either no conversion, or conversion only to the isomerisation products of dodecene under these conditions. The support itself is an all-silica MCM 41 which has no catalytic activity in the alkylation reaction, but leads to a certain amount of isomerisation of the dodecene. The catalytic activity of the pure ionic liquid is very low. This is due to the low solubility of ionic liquid in benzene. By optimising the reaction conditions better results might be obtained.

The lower catalytic activity of the catalyst based on organically modified MCM 41 can be explained by the high amount of hydroxyl groups found on the surface of the catalyst. A part of the aluminum(III)chloride will react with these. Due to the organic groups on the surface this AlCl₃ will not be accessible for the reactants, leading to a lower acidity of the surface and the observed low conversion and selectivity.

Further work will include the preparation of immobilised non acidic ILs, using e.g. tetrafluoroborate anions. While these ionic liquids have no catalytic activity, they are known to be rate enhancing solvents for some reactions. This is probably due to their coordinating properties.

4 CONCLUSIONS

The supported ionic liquids presented are promising catalysts for organic reactions. Their acidity can be varied over a wide range. By modifying the organic cation, the surface of the catalyst can be adjusted to the needs of the reaction. This gives access to the versatility of ionic liquids in a heterogeneous system.

5 ACKNOWLEDGEMENTS

This work was carried out as a part of the BRITE-Euram project BE 906-3745. The authors are grateful to the European Commission for the funding of this work. Our thanks also go to our consortium partners in the project for stimulating and helpful discussions.

Furthermore, we express our sincere thanks to Dr. Marko Bertmer, RWTH-Aachen, for his help in carrying out, discussing and interpreting the MAS-NMR measurements.

6 REFERENCES

[1]   K. R. Seddon, *Kinet. Catal.*, 1996, **37**, 693-697.
[2]   J. S. Wilkes, J. A. Levisky, R. A. Wilson, C. L. Hussey, *Inorg. Chem.*, 1982, **21**, 1263-1264.
[3]   T. Welton, *Chem. Rev.*, 1999, **99**, 2071-2084.
[4]   K. R. Seddon, *J. Chem. Technol. Biot.*, 1997, **68**, 351-356.
[5]   W. F. Hoelderich, H. H. Wagner, M. H. Valkenberg, *4th Int. Symp. on Supported Reagents* (St Andrews), Immobilised catalysts and their use in the synthesis of fine and intermediate chemicals, 2000.
[6]   D. Brunel, A. Cauvel, F. Fajula, F. DiRenzo, *Stud. Surf. Sci. Catal.*, L. Bonneviot, S. Kaliaguine, Vol. 92:Zeolites: A Refined Tool for Designing Catalytic Sites, 1995, Elsevier, 173-180.
[7]   D. J. Macquarrie, *Chem. Comm.*, 1996, **16**, 1961-1962.
[8]   D. J. Macquarrie, *Green Chem.*, 1999, **4**, 195-198.
[9]   H. Landmesser, H. Kosslick, W. Storek, R. Fricke, *Solid State Ionics*, 1997, **101-103**, 271-277.
[10]   C. W. Jones, T. Katsuyuki, M. E. Davis, *Nature*, 1998, **393**, 52-54.

HETEROGENEOUS ENANTIOSELECTIVE HYDROGENATION OF TRIFLUORO-METHYL KETONES

M. von Arx, T. Mallat and A. Baiker

Laboratory of Technical Chemistry
Swiss Federal Institute of Technology, ETH-Zentrum
CH-8092 Zurich, Switzerland

1 INTRODUCTION

The synthesis of optically pure chiral compounds is of great importance in various fields of chemistry such as pharmaceuticals, agrochemicals, flavours and fragrances[1]. The synthesis of highly enantioenriched compounds through heterogeneous catalysis is however still an intriguing task. Beside some immobilized homogeneous catalyst systems, the Ni - tartaric acid and the Pt - cinchona alkaloid systems are the two most succesful solid enantioselective hydrogenation catalysts[2-6]. Both catalysts combine the advantages of achieving enantioselectivities over 90 %, easy separation and product isolation, as well as reusability of the catalyst.

Systematic investigation of the Pt - cinchona alkaloid system, discovered by Orito et al. in 1979[7], has led to a better understanding of the α-ketoester-modifier-Pt interaction. It has been therefore possible to increase the enantioselectivities of the original reaction and, more importantly, to broaden the application range of cinchona-modified platinum.

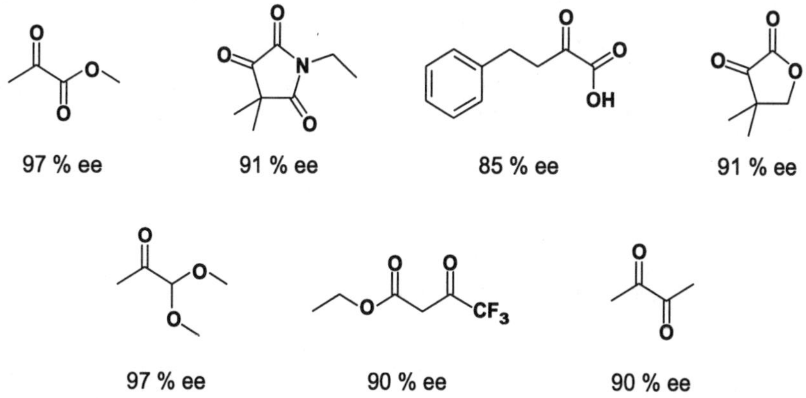

97 % ee	91 % ee	85 % ee	91 % ee

97 % ee	90 % ee	90 % ee

Scheme 1 *The best enantiomeric excesses in the hydrogenation of activated ketones with the Pt - cinchona alkaloid system*

Today, α-ketoesters[8,9], α-ketoamides[10], α-ketoacids[11], α-keto acetals[12,13], α-diketones[14], trifluoromethyl ketones[15] and ketopantolactone[16] can be reduced with up to 97 % ee (Scheme 1).

Here we focus on the hydrogenation of trifluoromethyl ketones over a Pt/alumina catalyst modified by cinchonidine (CD) and some of its simple derivatives. Our recent research has revealed interesting similarities and dissimilarities in the enantioselective hydrogenation of trifluoromethyl ketones and other activated ketones, which observations may be useful for a future mechanistic study.

2 EXPERIMENTAL

A 5 wt% Pt/Al$_2$O$_3$ catalyst (Engelhard 4759) was prereduced in flowing hydrogen for 90 min at 400 °C. The catalyst was then cooled to room temperature under hydrogen and transferred into the autoclave under exclusion of oxygen. The mean metal dispersion after heat treatment was 0.27 as determined from TEM images. All substrates and solvents, cinchonidine and cinchonidine hydrochloride (CD·HCl) were used as received (Aldrich, Fluka). O-methylcinchonidine (OMeCD) was synthesized as described before[17]. N-methylcinchonidinium chloride (NMeCD) was synthesized by the following procedure: To a solution of CD in methanol, 1.5 equivalents of methyliodide were slowly added. Crystalline N-methylcinchonidinium iodide was obtained after removal of the solvent and recrystallisation from a methanol:ethyl acetate (3:2) mixture. The salt was dissolved in aqueous methanol (33 vol%) and stirred for 24 h together with 2 equivalents of AgCl. The pure N-methylcinchonidinium chloride was obtained after filtration and recrystallisation from an ethyl acetate:methanol (4:1) mixture. The product was characterized by NMR spectroscopy and elemental analysis.

The hydrogenation reactions were carried out in a 100-ml stainless-steel autoclave equipped with a 50-ml glass liner and PTFE cover to provide clean conditions. The reactor was magnetically stirred (n = 1000 min^{-1}). The pressure was held at a constant value by a computerized constant volume - constant pressure equipment (Büchi BPC 9901). By monitoring the pressure inside the vessel and the injected pulses, the hydrogen uptake could be followed. Under standard conditions, 42 ± 2 mg prereduced catalyst, 1.84 mmol substrate, 6.8 μmol modifier and 5 ml solvent were used at 10 bar and room temperature.

Conversion and ee were determined using an HP 6890 gas chromatograph and a Chirasil-DEX CB (Chrompack) capillary column. The enantioselectivity is expressed as ee (%) = 100 x |(R-S)| / (R+S). The ee values were determined at full conversion if not otherwise stated.

3 RESULTS AND DISCUSSION

3.1 The Role of (Surface) Hydrogen Concentration

In the hydrogenation of 2,2,2-trifluoroacetophenone **1**, a significant effect of pressure and substrate concentration was observed (Figure 1). Using CD as modifier, the ee could be enhanced by 10 % by decreasing the pressure and doubling the substrate concentration. It seems that low hydrogen availability at the Pt surface is favourable for enantioselection. The surface hydrogen concentration can be further decreased by increasing the catalyst concentration in the slurry reactor and working in the hydrogen transport limited regime (Table 1). Triple amount of catalyst, in the presence of CD·HCl as modifier, afforded 74 % ee. This is the highest value achieved so far in the enantioselective hydrogenation of **1** over cinchona modified Pt[18-20].

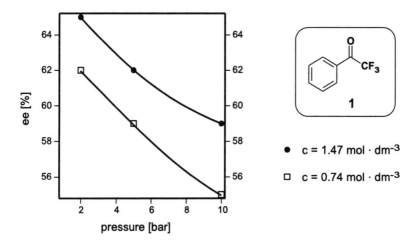

Figure 1 *Effect of pressure in the hydrogenation of trifluoroacetophenone **1**. Standard conditions, in 1,2-dichlorobenzene with CD*

Similar but smaller effects of hydrogen concentration were observed with other aromatic and alkyl-aromatic trifluoromethyl ketones such as **2** and **3** (Table 2). On the contrary, in the hydrogenation of the aliphatic ketone **4** medium to high hydrogen pressure favoured the enantiodifferentiation. Note that this correlation is typical for the hydrogenation of α-ketoesters and other activated ketones over the Pt-CD system[4].

Table 1   *Influence of catalyst amount and modifier on the enantioselective hydrogenation of **1**. Standard conditions, 1.47 mol dm^{-3} **1** in 1,2-dichlorobenzene*

modifier	catalyst [mg]	ee [%]
CD	42	65
CD	89	68
CD	130	68
CD·HCl	130	74

Table 2 *Effect of pressure in the hydrogenation of trifluoromethyl ketones. Standard conditions, in toluene with CD*

pressure [bar]	2 ee [%]	3 ee [%]	4 ee [%]
2	63	19	53
10	60	18	70
50	52	16	67

A feasible explanation for the considerable effect of hydrogen concentration is that hydrogen competes for the surface active sites on Pt and can change the adsorption strength and geometry of reactant and product. It has been shown recently that the adsorption geometry of methyl pyruvate on Pt changes from perpendicular to a tilted position due to coadsorption of hydrogen[21].

3.2 Structural Requirements to the Reactants

It has been proposed earlier[22] that only those compounds can be hydrogenated with good ee over the Pt-CD system which contain two carbonyl groups in α-position. The successful hydrogenation of **1**[20], and recently some α-ketoacetals[12,13], proved that the crucial requirement to the reactant ketone is an electron-withdrawing group in α-position.

In this context it is surprising, that the hydrogenation of the double activated ethyl-trifluoropyruvate **5** affords only 5 % ee at moderate conversion under standard conditions. The reaction is unusually slow at ambient temperature and 10 bar pressure (35 % conversion after 3 h). A closer inspection of this reaction revealed the reason for this failure. An NMR study of the reaction of **5** with 2-propanol (1:1 molar ratio), mimicking the secondary alcohol function of CD, showed a complete and fast transformation of **5** to the corresponding hemiketal **6** (^{19}F peak at -82 ppm) (Figure 2). A similar almost complete consumption of the starting material by hemiketal formation (new peak at -81.5 ppm) was observed, when CD was used instead of 2-propanol. This analogy implies that CD is consumed by the reactant during the hydrogenation reaction and the resulting hemiketal **7** is a poor modifier for the hydrogenation of **5**.

There is another reason for the poor ee in the hydrogenation of **8**. The steric difference between the two groups on both sides of the keto carbonyl group is minor. Hence, efficient enantioselection by the Pt-CD system requires considerable steric differences between the two alkyl (and aryl) groups on both sides of the carbonyl group. This is illustrated by some examples in Scheme 2.

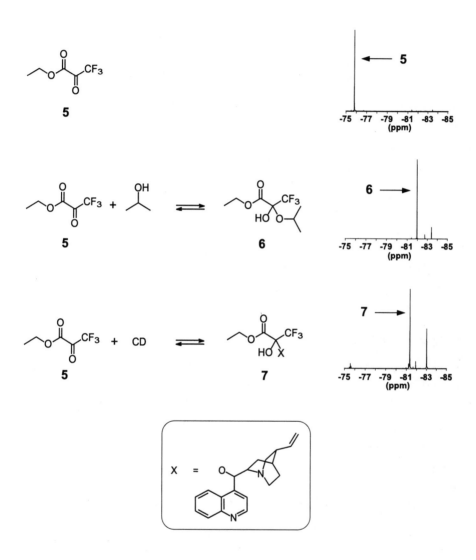

Figure 2   *NMR study of the hemiketal fomation from* **5**. *The spectra correspond to the pure reactant (***5***), or the product(s) formed by the interaction of* **5** *with 2-propanol or CD*

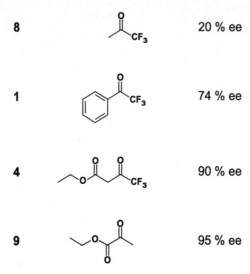

8		20 % ee
1		74 % ee
4		90 % ee
9		95 % ee

Scheme 2   *Efficiency of the Pt-CD system as a function of the bulkiness of the side chains of activated ketones*

3.3 Hydrogenation with Cinchonidine Derivatives

In order to reveal the importance of the basic quinuclidine N and the OH group of CD in the enantioselection, the efficiency of CD as chiral modifier of Pt was compared to those of some CD derivatives (Table 3). Two typical trifluoromethyl ketones, 1 and 4, were hydrogenated in an apolar solvent and in acetic acid. Data on ethyl pyruvate 9 hydrogenation are also schown in Table 3 for comparison.

Protonation of the quinuclidine N by hydrochloric acid had only minor effect on the ee but quaternization led to a complete loss of enantioselection in all three reactions. This is a strong indication that the quinuclidine N of CD is involved in the crucial interaction between reactant and modifier. A H-bond (N-H-O type interaction) has been proposed for the CD - ethyl pyruvate complex[3], which interaction is not possible after methylation of the basic N atom of CD.

Methylation of the hydroxyl group of CD in OMeCD does not affect all reactions in the same way. In the hydrogenation of ethyl pyruvate 9 the influence is only marginal, independent of the solvent used. A similar small effect was observed in the hydrogenation of 4 in apolar solvent, while the ee increased by 20 % in acetic acid, using OMeCD instead of CD. On the contrary, O-methylation of CD led to a remarkable drop in ee in the hydrogenation of 1 in apolar solvents such as 1,2-dichlorobenzene. In the latter reaction, acetic acid is a poor solvent, affording low ee's with both CD and OMeCD.

The data in Table 3 suggest that the nature of enantiodifferentiation in the hydrogenation of α,α,α-trifluoromethyl ketones is only partly similar to that of α-ketoester hydrogenation. Apparently, the OH group of CD has some role in enantioselection during trifluoromethyl ketone hydrogenation, and in this respect there is a considerable difference also among the various trifluoromethyl ketones.

Table 3 *Enantioselective hydrogenation of some activated ketones in different solvents, in the presence of CD or some of its simple derivative. Standard conditions*

reactant	solvent	CD ee [%]	CD·HCl ee [%]	OMeCD ee [%]	NMeCD ee [%]
1	1,2-dichlorobenzene	55	60	17	<2
1	acetic acid	6	-	11	-
4	toluene	46	42	43	<2
4	acetic acid	70	-	90	-
9	toluene	86	89	80	<2
9	acetic acid	94	-	95	<2

4 CONCLUSIONS

The Pt-cinchonidine system is a promising catalyst for the enantioselective hydrogenation of α,α,α-trifluoromethyl ketones to the corresponding alcohols. The ee varies in a broad range between 5 and 90 %, depending on the chemical nature of the R group in the general structure F_3C-CO-R. The remarkable differences may be well explained by steric and electronic effects.

The study of the role of surface hydrogen concentration and that of the OH group of cinchonidine indicates, that (i) the nature of enantiodifferentiation in the hydrogenation of trifluoromethyl ketones and α-ketoesters is partly different, and (ii) concerning the reactant-modifier interaction there are important differences also among the various trifluoromethyl ketones.

References

1. *Chirality in Industry: The Commercial Manufacture and Applications of Optically Active Compounds*, eds. A. N. Collins, G. N. Sheldrake and J. Crosby, John Wiley, Chichester, 1997.
2. T. Osawa, T. Harada and A. Tai, *Catal. Today*, 1997, **37**, 465.
3. A. Baiker, *J. Mol. Catal. A*, 1997, **115**, 473.
4. A. Baiker and H. U. Blaser, In *Handbook of Heterogenous Catalysis*, eds. G. Ertl, H. Knözinger and J. Weitkamp, VCH, Weinheim, 1997, Vol. 5, p. 2422
5. P. B. Wells and A. G. Wilkinson, *Top. Catal.*, 1998, **5**, 39.
6. A. Pfaltz, and T. Heinz, *Top. Catal.*, 1997, **4**, 229.
7. Y. Orito, S. Imai and S. Niwa, *J. Chem. Soc. Japan*, 1979, 1118.

8   B. Török, K. Balazsik, G. Szöllösi, K. Felföldi and M. Bartok, *Chirality*, 1999, **11**, 470.

9   X. Zuo, H. Liu, D. Guo and X. Yang, *Tetrahedron*, 1999, 7787.

10 N. Künzle, A. Szabo, M. Schürch, G. Wang, T. Mallat and A. Baiker, *Chem. Commun.*, 1998, 1377.

11 H. U. Blaser and H. P. Jalett, in *Heterogeneous Catalysis and Fine Chemicals III*, eds. M. Guisnet et al., Elsevier, Amsterdam, 1993, p. 139.

12 M. Studer, S. Burkhardt and H. U. Blaser, *Chem. Commun.*, 1999, 1727.

13 B. Török, K. Felföldi, K. Balazsik and M. Bartok, *Chem Commun.*, 1999, 1725.

14 M. Studer, V. Okafor, and H. U. Blaser, *Chem. Commun.*, 1998, 1053.

15 M. von Arx, T. Mallat and A. Baiker, *J. Catal.*, 2000, **193**, 307.

16 M. Schürch, N. Künzle, T. Mallat and A. Baiker, J. Catal., 1998, **176**, 569.

17 K. Borszeky, T. Bürgi, Z. Zhaouhui, T. Mallat and A. Baiker, *J. Catal.*, 1999, **187**, 160.

18 M. Bodmer, T. Mallat and A. Baiker, in *Catalysis of Organic Reactions*, ed. F. E. Herkes, Marcel Dekker, New York, 1998, p. 75.

19 K. Balazsik, B. Török, K. Felföldi, and M. Bartok, *Ultrasonics Sonochem.*, 1999, **5**, 149.

20 T. Mallat, M. Bodmer and A. Baiker, *Catal. Lett.*, 1997, **44**, 95.

21 T. Bürgi, F. Atamny, A. Knop-Gericke, M. Hävecker, T. Schedel-Niedrig, R. Schlögl and A. Baiker, *Catal. Lett.*, 2000, **66**, 109.

22 K. E. Simons, P. A. Meheux, S. P. Griffiths, I. M. Sutherland, P. Johnston, P. B. Wells, A. F. Carley, M. K. Rajumon, M. W. Roberts and A. Ibbotson, *Rec. Trav. Chim. Pays-Bas*, 1995, **113**, 465.

STRUCTURAL AND REACTIVE PROPERTIES OF SUPPORTED TRANSITION METAL TRIFLATES

Karen Wilson and James H. Clark

Green Chemistry Group,
Department of Chemistry,
University of York,
Heslington,
York, YO10 5DD

1. INTRODUCTION

Conventional homogeneously catalysed processes utilised by the fine and speciality chemicals industries produce vast amounts of waste on removal of the catalyst from the reaction. Tightening environmental legislation has led to a drive to develop new heterogeneous systems that offer ease of catalyst separation combined with high activity and selectivity, thus reducing the production of waste[1,2,3]. One major area of concern is the wide range of organic synthetic routes, that currently rely on the use of inorganic or mineral acids (e.g $AlCl_3$, $ZnCl_2$, H_2SO_4), which would benefit hugely from the use of cleaner, reusable solid acid catalysts.

Solid acid catalysts are generally categorised by their Brønsted and/or Lewis acidity, the strength and number of these sites, and the morphology of the support (e.g. suface area, pore size). The synthesis of pure Brønsted and pure Lewis acid catalysts attracts a great degree of academic interest, although the latter is more of a challenge due to Brønsted acidity arising from Lewis acid-base complexation, as illustrated below in **scheme 1**.

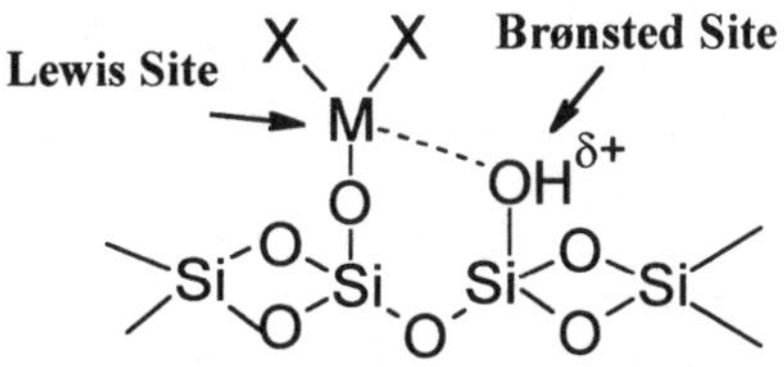

Scheme 1: *Brønsted acidity arising from inductive effect of Lewis acid centre coordinated to a silica support*

Traditionally Lewis acid catalysed reactions often utilise metal halides, however in recent years, metal trifluoromethanesulfonates or triflates having the general formula $M^{n+}(SO_2CF_3)_n$ have been reported as a new and interesting type of Lewis acid[4]. These possess stronger Lewis acidity and higher water tolerances than their halide counterparts, which readily form the hydroxide or oxide[5]. A summary of some of the main metal triflates employed to homogeneously catalyse organic reactions is shown in **table 1.**

REACTION	METAL TRIFLATE
Aldol Reaction	$Cu(OTf)_2$[6]; $Sc(OTf)_3$[7]; $La(OTf)_3$[5]; $Pr(OTf)_3$[5]; $Nd(OTf)_3$[5]; $Sm(OTf)_3$[5]; $Eu(OTf)_3$[5]; $Gd(OTf)_3$[5]; $Yb(OTf)_3$[5]
FC Alkylation	$B(OTf)_3$[8]; $Al(OTf)_3$[8]; $Ga(OTf)_3$[8]; $Cu(OTf)_2$[6]; $Hf(OTf)_4$[9]
FC Acylation	$B(OTf)_3$[8]; $Al(OTf)_3$[8]; $Ga(OTf)_3$[8]; $Cu(OTf)_2$[6]; $Bi(OTf)_3$[10]; $Sc(OTf)_3$[5]; $Yb(OTf)_3$[5]; $Hf(OTf)_4$[9]
Allylation Reaction	$Sc(OTf)_3$[5]
Barbier Reaction	$La(OTf)_3$[11]
Diels Alder Reaction	$Bi(OTf)_3$[12]; $Sc(OTf)_3$[5]; $Pr(OTf)_3$[11]; $Nd(OTf)_3$[11]; $Yb(OTf)_3$[11,14]
Esterification Reaction	$Sc(OTf)_3$[13]; $La(OTf)_3$[13]; $Pr(OTf)_3$[13]; $Eu(OTf)_3$[13]; $Yb(OTf)_3$[13]
Michael Reaction	$Cu(OTf)_2$[6]; $Sc(OTf)_3$[5]; $Yb(OTf)_3$[11,14]
Mannich Reaction	$Yb(OTf)_3$[11]

Table 1: *Summary of some of the reactions catalysed by metal triflates*

Recently we reported on the structural and reactive properties of a new silica supported $Zn(OTf)_2$ catalyst in the rearrangement of α-pinene oxide to campholenic aldehyde[15], an intermediate used by the fragrance industry for the manufacture of santalol. The current homogenous systems used in this reaction are $ZnCl_2$ and $ZnBr_2$, however rapid irreversible catalyst deactivation is observed with turnover numbers as low as 20 being reported[16]. Development of a solid acid for this reaction is quite a challenging problem develop a solid acid for, with over 100 products being observed depending on the reaction conditions[17].

In this subsequent report, we now wish to discuss how the performance of $Zn(OTf)_2/SiO_2$ catalysts compares with supported and unsupported $ZnCl_2$.

2. EXPERIMENTAL

Silica supported $Zn(OTf)_2$ catalysts with loadings of 2mmolg^{-1} (7.5 wt % Zn) and 0.01mmolg^{-1} (0.06 wt % Zn) were prepared by wet impregnation of amorphous SiO_2 (K100 (Merck) – S.A = 300m^{2}g^{-1}), or Hexagonal Mesoporous Silica (HMS - S.A=

1100 m²g⁻¹) with a solution of Zn(OTf)$_2$ (98% Aldrich) in 100 ml methanol. The slurry was stirred for 2 hours at 25°C, prior to removal of the excess methanol at 50°C on a rotary evaporator. HMS$_{24}$ was prepared according to the sol gel route of Pinnavaia *et al* [18] by condensing 62.5g of tetraorthoethoxysilicate in a mixture of 123g ethanol and 160g H$_2$O with 15.3g n-dodecylamine as the surfactant template. The resulting gel was stirred for 24 hours, filtered, dried and then calcined at 600°C for 6 hours to remove the template. Silica supported ZnCl$_2$ catalysts were prepared in an analogous method.

Catalyst characterisation was performed using thermal analysis and DRIFTS. Thermal analysis was performed using a NETZSCH 409 STA with a temperature ramp of 10°C/min. Acid site determination was performed using pyridine titration in conjunction with DRIFTS (Bruker Equinox 55 FTIR). Pyridine vapour was adsorbed over a period of 24 hours prior to recording the DRIFTS spectrum.

The rearrangement of α-pinene oxide was performed in a glass batch reactor at both 25°C and 85°C, using 100 ml of 1,2 dichloroethane (Aldrich 98%) as solvent and 1g α-pinene oxide (Aldrich 99%) with 0.5g decane (Aldrich 99%) as internal standard. The reaction products were analysed using a HP5890 Gas Chromatograph fitted with a 25m HP1 capilliary column. Identification of major reaction products was performed using authentic samples, and a VG Autospec Mass Spectrometer.

3. RESULTS AND DISCUSSION

3.1 Catalyst Characterisation

Following catalyst preparation the physical properties of the silica supported Zn(OTf)$_2$ samples were examined using thermal analysis and acid site characterisation by pyridine titration [19]. Differential thermal analysis was used to determine the thermal stability of the Zn(OTf)$_2$, Zn(OTf)$_2$/SiO$_2$ and triflic acid treated SiO$_2$ samples. A comparison of these spectra is shown in **figure 1** which indicates that while Zn(OTf)$_2$ does not decompose until 540°C, Zn(OTf)$_2$/SiO$_2$ decomposes at a lower temperature, with a major weight loss at 475°C and a shoulder at 390°C. The spectrum obtained from physisorbed triflic acid on SiO$_2$ shows that triflic acid desorbs at 370°C, indicating that the weight loss observed at 475°C from Zn(OTf)$_2$/SiO$_2$ cannot be assigned to physisorbed triflic acid formed during catalyst preparation. Therefore, the weight

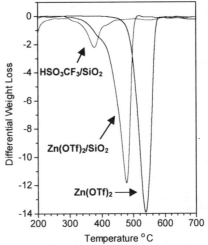

Figure 1: *DTA comparison of Zn(OTf)$_2$, Zn(OTf)$_2$/SiO$_2$ and HSO$_3$CF$_3$/SiO$_2$*

loss observed from both supported and unsupported Zn(OTf)$_2$ can be assigned to thermal decomposition of triflate groups. The reduction in thermal stability of the silica supported material may be ascribed to the reaction of highly dispersed Zn(OTf)$_2$ with surface hydroxyl groups on silica, which will hydrolyse the Zn-(OTf) bond. Similar effects have been observed with other supported acid catalysts, where the -OH groups on silica can promote decomposition of the acid site[20].

Figure 2 shows the DRIFT spectra recorded following adsorption of pyridine on a 2mmolg^{-1} loading Zn(OTf)$_2$/SiO$_2$ sample. The four intense peaks observed at 1450, 1490, 1580 and 1605 cm^{-1} are assigned to the presence of Lewis acid sites, indicating that zinc triflate maintains its Lewis acidity following adsorption on SiO$_2$. The weaker band at 1540 cm^{-1} also indicates the presence of a low level of Brønsted acidity, which most likely results from the interaction of the triflate groups with the hydroxyl groups on SiO$_2$. The nature of these acid sites is unchanged for lower loading samples.

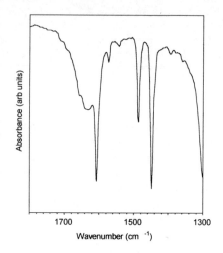

Figure 2: *Pyridine Titration of 2mmolg^{-1} Zn(OTf)$_2$/SiO$_2$*

The physical properties of silica supported ZnCl$_2$ will not be discussed here, but have been widely studied elsewhere, shown to be a pure Lewis acid[21].

3.2 Rearrangement of α-pinene oxide

The reactivity of a 2mmolg^{-1} ZnCl$_2$/SiO$_2$ catalyst in the rearrangement of α-pinene oxide is shown in **figure 3**. When using 0.5g of the as prepared catalyst, it can be seen that after 30 minutes reaction time the conversion of α-pinene oxide is only 78% compared to 100% obtained using an equivalent amount of homogeneous ZnCl$_2$. However when the ZnCl$_2$/SiO$_2$ catalyst is pre-treated at 200°C under a flow of N$_2$ conversion and selectivity comparable to the homogeneous system can be obtained.

Unfortunately ZnCl$_2$/SiO$_2$ cannot be recycled after reaction, due to irreversible catalyst deactivation which is believed to occur via hydrolysis of the Zn-Cl bond. Indeed chlorinated campholenic aldehyde was observed to form during the reaction, which would occur through reaction with HCl from hydrolysed ZnCl$_2$. It was anticipated that the enhanced water stability of the metal triflate would thus be advantageous for this reaction.

Initial experiments were performed using 2mmolg^{-1} Zn(OTf)$_2$/SiO$_2$ which was observed to react rapidly with α-pinene oxide. When using only 58mg of catalyst 100% conversion of the oxide was observed in under 30 seconds at both 85 and 25°C, and even on catalyst reuse. Zn(OTf)$_2$/SiO$_2$ is also considerably more active than a comparative amount of unsupported Zn(OTf)$_2$, which only reaches 100% conversion of α-pinene oxide after 30 minutes. These results in addition to those obtained using 0.2g of 0.01mmol^{-1} Zn(OTf)$_2$/SiO$_2$ are summarised in **figure 4.** One of the problems of having such an active catalyst for α-pinene oxide rearrangement is that the products formed are themselves highly reactive, and consecutive reactions will decrease the observed selectivity towards campholenic aldehyde.

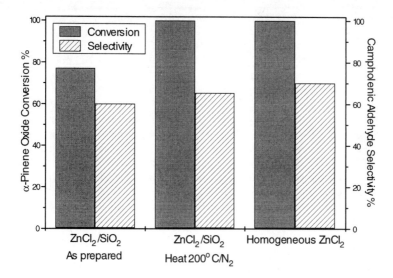

Figure 3: *Comparison of 2mmolg⁻¹ ZnCl₂/SiO₂ and ZnCl₂ in the rearrangement of α-pinene oxide after 30 minutes reaction*

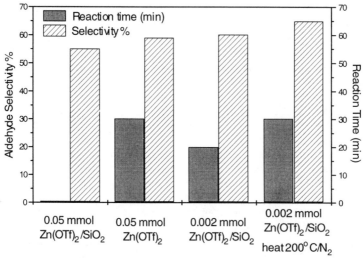

Figure 4: *Comparison of reaction time required to reach 100% conversion of α-pinene oxide at 85°C using Zn(OTf)₂/SiO₂ and Zn(OTf)₂*

By using 200mg of 0.01mmolg-1 Zn(OTf)₂/SiO₂ the reaction rate was slowed so that 100% conversion was reached after 20 minutes, and a selectivity to campholenic aldehyde of 60% was obtained. On reuse only a slight decrease in catalyst activity was observed in DRIFT spectra and by thermal analysis, even after prolonged washing of the catalyst.

The selectivity towards campholenic aldehyde can be boosted further to 65% if the 0.01mmolg^{-1} Zn(OTf)$_2$/SiO$_2$ catalyst is pretreated under N$_2$ at 200°C prior to use. This increase in selectivity is attributed to loss of Brønsted acid sites by dehydration of the catalyst surface, which in turn reduces the amount of side reactions.

During the reaction, the selectivity to the other main products formed, p-cymene, trans-sobrerol, trans-carveol does not change. It should be noted that the formation of trans-sobrerol is not observed in ZnCl$_2$ and ZnCl$_2$/SiO$_2$ catalysed rearrangement of α-pinene oxide, and is formed as a consequence of the slight Brønsted acidity of Zn(OTf)$_2$/SiO$_2$. The selectivity towards campholenic aldehyde appears to be essentially dictated by subsequent reactions of campholenic aldehyde. Indeed a control reaction at 85°C of pure campholenic aldehyde with 0.01mmolg^{-1} Zn(OTf)$_2$/SiO$_2$ shows that complete conversion of the aldehyde to a range of other products is achieved over 24 hours.

It should also be noted that while a 65% selectivity to campholenic aldehyde is lower than obtained using homogeneous ZnCl$_2$, the superior turnover numbers of >3000 obtained with Zn(OTf)$_2$ gives this catalyst a greater production of campholenic aldehyde per active site. With careful tuning of reaction conditions and catalyst composition it should be possible to boost this selectivity further. Indeed a further improvement in the selectivity to campholenic aldehyde can be achieved if HMS$_{24}$ is used as a support. It was anticipated that the smaller more regular pore structure of the HMS$_{24}$ material (~ 40Å compared to 100 Å) and different surface polarity would alter the diffusion rates of the reactants and products through the support. An improvement in both conversion rate and selectivity can be achieved using 200 mg of 0.01mmolg^{-1} Zn(OTf)$_2$/HMS$_{24}$, with the selectivity to the aldehyde increasing to 64%. Further improvements in selectivity to 69% can be achieved if the reaction is performed at 25°C, but under these conditions the reaction rate is slower with 100% conversion being reached after 1 hour. Surprisingly no improvement in selectivity was observed if Zn(OTf)$_2$/SiO$_2$ is used at 25°C.

The improved selectivity obtained with the HMS$_{24}$ supported catalyst can be ascribed to its lower surface polarity compared to K100-SiO$_2$. This arises in part due to the different thermal pre-treatments used during catalyst preparation, with HMS$_{24}$ receiving a high temperature (600°C) calcination step to remove the template. Infra red analysis on these materials indicates that the surface hydroxyl group density is reduced compared to conventional silica. This will lower the surface polarity[22], weaken the interaction of polar molecules with the catalyst surface and in turn increase the selectivity towards campholenic aldehyde by reducing the number of consecutive reactions.

To summarise, we have demonstrated that supported Zn(OTf)$_2$ catalysts are highly active for the rearrangement of α-pinene oxide, offering enhanced turnover frequencies compared to ZnCl$_2$ which is thought to be a consequence of their greater stability towards hydrolysis. These supported metal triflates are new and exciting solid acid catalysts that are currently being assessed in other traditional Lewis acid catalysed reactions, with preliminary results suggesting that they will be promising catalysts for esterification, alkylation and acylation reactions.

References

1. J.H.Clark, D.J.Macquarrie, *Chem.Comm.*, 1998, 853
2. R.Sheldon, *Chem. Ind.*, **1**, (1997), 12
3. W.F.Hölderich, *Stud.Surf.Sci.& Catal.*, 1993, **75**, 127
4. S.Kobayashi, *Chem.Lett*, 1991, 2187
5. S.Kobayashi, *Synlett*, 1994, 689
6. C.Hertweck, *J.Prackt. Chem.*, 2000, **342**, 316
7. T.Tsuchimoto, K.Tobita, T.Hiyama, S.Fukuzawa, *J.Org.Chem.*, 1997, **62**, 6997
8. G.A.Olah, O.Farooq, S.M.F.Farina, J.A.Olah, *J.Am.Chem.Soc.*, 1998, **110**, 2560
9. I.Hachiya, M.Moriwaki, S.Kobayashi, *Bull.Chem.Soc.Jpn.*, 1995, **68**, 2053
10. S.Repichet, C.LeRoux, J.Dubac, J.Desmurs, *Eur. J.Chem*, 1998, 2743
11. J.B.F.N. Engberts, B.L.Feringa, E.Keler, S.Otto, *Recl.Trav.Chim.Pays.Bas,* 1996, **115**, 457
12. H.Laurent Robert, C.LeRoux, J.Dubac, *Synlett*, 1998, 1138
13. A.G.M.Barrett, D.C.Braddock, *Chem.Comm.,* 1997, 351
14. S.Kobayashi, I.Hachiya, T.Takahori, M.Araki, H.Ishitani, *Tet.Lett.*, 1992, **33**, 6815
15. K.Wilson, A.Renson, J.H.Clark, *Catalysis Letters*, 1999, **61**, 51
16. J.Kaminska, M.A.Schwegler, A.J.Hoefnagel, H.van Bekkum, *Recl.Trav.Chim, Pays- Bas*, **111**, 1992, 432
17. W.F.Hölderich, J.Röseler, G.Heitmann, A.T.Liebens, *Catal.Today*, 1997, **37**, 353
18. P.T.Tanev, T.J.Pinnavaia, *Science*, 1995, **267**, 865
19. K.Tanabe, M.Misono, Y.Ono, H.Hattori, *Stud.Surf.Sci.& Catal.*, **51**, chap. 2
20. K.Wilson, J.H.Clark, *Chem.Comm.*, 1998, 2135
21. Y.Okamoto, T.Imanaka, S.Teranishi, *Bull.Chem.Soc.Jpn*, 1973, **47**, 464
22. E.F.Vansant, P.van der Voort, K.C.Vrancken, *Stud.Surf.Sci.& Catal*, **93**, chap.3

SOLUBLE FLUOROPOLYMER CATALYSTS FOR HYDROFORMYLATION OF OLEFINS IN FLUOROUS PHASES AND SUPERCRITICAL CO$_2$

W. Chen, A. M. Banet-Osuna, A. Gourdier, L. Xu and J. Xiao*

Leverhulme Centre for Innovative Catalysis
Department of Chemistry
University of Liverpool
Liverpool L69 7ZD

1 INTRODUCTION

Soluble polymers have attracted recent attention in catalysis and combinatorial chemistry.[1-3] When used in catalysis by organometallic compounds, soluble polymer ligands offer the following advantages: The reaction is homogeneous in nature and separation of catalysts can be easily achieved by filtration or precipitation. We have now developed a new class of polymer ligands based on fluoroacrylate-arylphosphine copolymers for catalysis in perfluorocarbon solvents and supercritical CO$_2$ (scCO$_2$).

Catalysis in perfluorocarbon solvents, that is, fluorous biphase catalysis (FBC), and catalysis in scCO$_2$ represent one of the current frontier research areas in homogeneous catalysis.[4,5] For catalysis in such non-conventional solvents, modifications to commonly used ligands need to be made in order to make their metal complexes soluble. A well-established strategy for catalysis by organometallic complexes in scCO$_2$ and fluorous phases is to modify phosphine ligands with fluorinated groups such as perfluoroalkyls, which are known to exhibit exceptionally high solubility in these solvents. A number of phosphines bearing fluorinated ponytails have been reported and shown to be effective for FBC and for catalysis in scCO$_2$.[4-8] However, application of these ligands in catalysis is not without a problem, as they in general display finite solubility in common organic solvents, thus making catalyst recycle more difficult than one would expect if they were negligibly soluble in the organo phases. This is particularly true for fluoroponytail-modified arylphosphines such as P(4-C$_6$H$_4$CH$_2$CH$_2$C$_6$F$_{13}$)$_3$. These are more soluble in organic solvents than are analogous alkylphosphines such as P(CH$_2$CH$_2$C$_6$F$_{13}$)$_3$ and thus less useful in FBC.[4,9] Ironically, it is the former class of phosphines that are far more extensively used in homogeneous catalysis. Herein we describe soluble fluoropolymer-supported arylphosphine ligands as an alternative to soluble molecular ligands for FBC and the application of the new ligands in fluorous biphase hydroformylation of higher and functionalised olefins. We anticipated that by incorporating an arylphosphine into a fluoropolymer the solubility of the former in normal solvents would be minimised. We also describe the application of the polymer ligands in homogeneous hydroformylation in scCO$_2$. A fluoroacrylate copolymer containing alkylarylphosphine ligands has recently been reported and shown to be effective in rhodium catalysed hydrogenation of olefins, while this work was in progress.[1,10]

Hydroformylation of higher olefins is an important commercial process for producing aldehydes and alcohols and has been addressed in a number of publications.[11,12] The focal point is to search for more active and selective rhodium catalysts in conjunction with easy catalyst separation and reuse. Non-conventional solvents such as perflorocarbons and scCO$_2$ have also been examined as a means for easier catalyst separation and recycle.[13-17] Amidst many forms of catalysts, polymer-supported rhodium catalysts have been investigated for more than two decades.[18,19] A recent entry into this area is the use of water-soluble polymers as ligands for rhodium for hydroformylation in aqueous phases.[20-22] These catalysts, while potentially having the advantages of both homogeneous and heterogeneous catalysts, usually display low catalytic activities.

2 RESULTS AND DISCUSSION

2.1 Preparation of Soluble Fluoropolymer Ligands

The poly(fluoroacrylat-*co*-styryldiphenylphosphine) ligands **1** and **2** were prepared by free radical copolymerisation of 1H,1H,2H,2H-perfluorodecylacrylate with styryldiphenylphosphine at 65 °C in the presence of AIBN in α,α,α-trifluorotoluene (Scheme 1). For the fluoropolymer **1**, an acrylate to styryldiphenylphosphine molar ratio of 5:1 was used. For **2**, the ratio was 9:1. After removing the solvent, the resultant solid was washed with hot toluene, affording the polymers as white powders in greater than 90% yields. The IR spectrum of both polymers showed disappearance of the absorption due to C=C stretching, in line with the lack of resonance due to olefinic protons in the ^{1}H NMR spectrum. The C=O absorption appeared at 1738 cm^{-1} for **1** and 1740 cm^{-1} for **2** in the IR spectrum. The ^{31}P NMR spectrum of each polymer in α,α,α-trifluorotoluene (with 5% CDCl$_3$) displayed a relatively sharp singlet at around δ -6.2. Surprisingly, the singlet became much broader in perfluoro-1,3-dimethylcyclohxane, with half-height line width Δ = 200 Hz as opposed to Δ = 58 Hz in α,α,α-trifluorotoluene. The polymers are soluble in both solvents. One possible explanation for the line broadening is that in perfluorinated solvents the polymers, which contain both fluorous-soluble and insoluble segments, may aggregate, whereas in partly fluorinated solvents such aggregation may not be favoured due to favourable solvent-solute interactions. The phosphorus content of the polymers was estimated to be 1.2% for **1** and 0.8% for **2** by ^{31}P NMR using bis(diphenylphosphino)methane

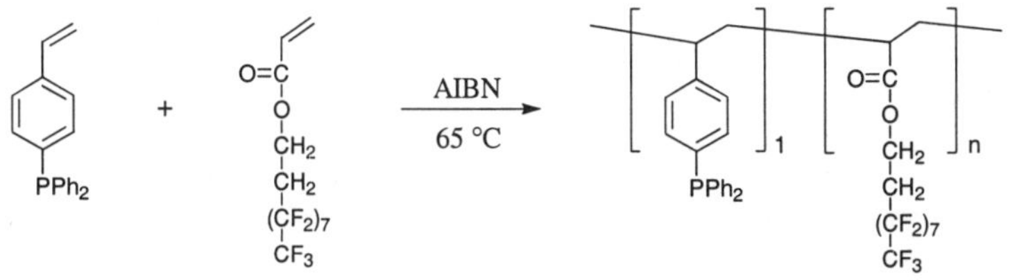

n = 5, **1**; 9, **2**

Scheme 1

Table 1   *Partition Coefficients of Phosphines in Organo-Fluorous Phases* [a]

Phosphine	Solvent [b]	f [c]
1	Toluene/FDMC	26.9
	THF/FDMC	27.7
$P(4\text{-}C_6H_4C_6F_{13})_3$	Toluene/FDMC	4.4
	THF/FDMC	2.2
$P(4\text{-}C_6H_4CH_2CH_2C_6F_{13})_3$	Toluene/FDMC	0.9
	THF/FDMC	0.2

[a] Measured at 24 °C by first saturating an organo/perfluorocarbon solvent mixture with the ligand under question followed by syringing equal volume solution from the two phases and by weighing the dried and solvent-free ligand. [b] FDMC = perfluoro-1,3-dimethylcyclohexane. [c] f = weight of ligand in fluorous phase/weight of ligand in organo phase.

as an internal standard. These values are close to the values of 1.1% and 0.6% calculated on the basis of the monomer ratios, and are consistent with the high yields of polymer synthesis.

As is expected, the fluorinated polymer ligands are much less soluble in common organic solvents than are molecular arylphosphines modified with fluoro ponytails. This is evident from the approximate partition coefficient f shown in Table 1 above. The partition coefficient is defined as the ratio of the weight of a polymer in a fluorous phase *vs.* that of the same polymer in an organic solvent under equilibrium conditions. The partition coefficient for **1** is *ca.* 27 and does not vary significantly with the organic solvents. In contrast, the values of f for the molecular ligand $P(4\text{-}C_6H_4C_6F_{13})_3$ are much smaller, and are still smaller for the ethylene-spaced $P(4\text{-}C_6H_4CH_2CH_2C_6F_{13})_3$.[23-26] The latter is in fact more soluble in the organic solvents than in the perfluorinated solvent. As can be expected, the two molecular ligands are more soluble in THF than in toluene, due probably to the higher polarity of the former. Clearly, for FBC involving arylphosphine ligands, **1** and **2** would offer a better choice for retaining a meal catalyst in the fluorous phases.

2.2 Fluorous Biphase Hydroformylation with Soluble Fluoropolymer Catalysts

The catalytic performance of the fluoropolymer ligands **1** and **2** was first tested in the fluorous biphase hydroformylation of 1-alkenes, styrene and *n*-butyl acrylate. The reaction was conducted in a batch reactor in a 40/20/40 vol% hexane/toluene/perfluoromethylcyclohexane solvent mixture (10 mL). The catalyst was formed *in situ* by adding $[Rh(CO)_2(acac)]$ (5 μmol, P/Rh = 6) to the polymer-containing solvent mixture followed by introduction of syngas (30 bar, CO/H_2 = 1/1). Table 2 summarises the results obtained. The salient features of the results are: Firstly, the activity of the fluorous soluble polymer catalysts are significantly higher than that reported for solid polymer- and aqueous soluble polymer-supported rhodium catalysts.[18-22] For example, the average turnover frequency (TOF) for the fluorous biphase hydroformylation of 1-decene is 136 mole aldehyde h^{-1} per mol of rhodium catalyst with an aldehyde selectivity of 99%. In comparison, a rhodium catalyst supported on the

Table 2 *Fluorous Biphase Hydroformylation of Olefins by Soluble Polymer Catalysts* [a]

Olefin	Polymer	Olefin/Rh	Conversion (%)	Selectivity [b] (%)	L/B [c]
1-Decene	1	2120	97	99	4.8/1
	2	2120	90	99	5.9/1
1-Hexadecene	1	2100	78	98	4.8/1
	2	2100	59	99	5.0/1
Styrene	1	3500	85	>99	1/6.2
	2	3500	80	>99	1/5.4
n-Butyl acrylate	1	2800	100	>99	B [d]
	2	2800	100	>99	B [d]

[a] Reaction conditions: 5 μmol [Rh(CO)$_2$(acac)], P/Rh = 6, 30 bar CO/H$_2$ (1:1), 100 °C for 1-decene and 1-hexadecane, 80 °C for styrene and *n*-butyl acrylate, hexane/toluene/perfluoromethylcyclohexane = 4/2/4 (mL), 15 h reaction time. The products were analysed by ^{1}H NMR and the conversion and selectivity confirmed by GC. [b] To aldehyde, olefin isomerisation accounts for the product balance. [c] Linear to branched aldehyde ratio, determined by ^{1}H NMR. [d] The branched product was a 1:1 mixture of enol and aldehyde, the linear aldehyde was < 1%.

water soluble polymer poly(enolate-*co*-vinylalcohol-*co*-vinylacetate) gave a TOF of 56 (100 °C, 41 bar) with an aldehyde selectivity < 70% in the aqueous hydroformylation of 1-octene.[22] Secondly, as with solid polymer-supported catalysts,[18,19] the linear/branched (L/B) aldehyde ratio is markedly higher than achievable with similar P/Rh ratios when using homogeneous rhodium-phosphine catalysts, e.g., [RhH(CO)(PPh$_3$)$_3$], which yielded a L/B ratio of 2.9 only in the presence of excess PPh$_3$ (P/Rh = 19) in the hydroformylation of 1-pentene in benzene (100 °C, 27 bar).[27] Thirdly, smaller olefins appear to give higher turnovers, probably owing to better miscibility between the olefins and the fluorous phase. In fact, when 1-hexene was hydroformylated under conditions identical to those for 1-decene, a conversion of 70% with an aldehyde selectivity of 98% and a L/B ratio of 4.4 was obtained in 1 h reaction time, corresponding to a remarkable TOF of 1454. The olefin isomerisation was again low, only at 1.7%.

The activity and stability of the soluble fluoropolymer catalysts may also be judged by the hydroformylation of 1-hexene when the olefin/Rh ratio was increased to 200000. At 100 °C and 50 bar syngas with polymer **1** as the supporting ligand, the catalyst afforded a turnover number (TON, mole of aldehyde per mol of rhodium) *ca.* 140000 with a 98% selectivity to aldehyde (L/B = 4.4; 2% isomerisation) for 58 h reaction time. We also examined the recyclability of the fluoropolymer catalysts taking the reaction of 1-hexene as a model example. The other substrates and related products should be easier to separate under the fluorous biphase conditions as they are less miscible with the perfluoro solvent. At 100 °C and 50 bar with olefin/Rh = 48000, three consecutive hydroformylation reactions were run, giving an excellent combined TON of 70000 and an average aldehyde selectivity of 99%. A 1 ppm loss of rhodium accompanied with a 6% decrease in conversion in each recycle experiment was measured. This loss in rhodium and in catalyst activity appears to be largely due to the finite miscibility of the substrate/product with the perfluoro solvent. At the end of the third run, all the perfluoromethylcyclohexane had leached to the product phase, thus making the polymer catalyst partially soluble in the product. By optimising the operating conditions, *e.g.* by varying the organic solvent, the problem of rhodium leach can be minimised.

2.3 Hydroformylation with Soluble Fluoropolymer Catalysts in scCO$_2$

We have also extended the above reactions to scCO$_2$, in which the polymer ligands and their rhodium complexes are soluble. A potential advantage for using such ligands in scCO$_2$ is that catalyst separation from product could be easier to perform than using molecular, scCO$_2$-soluble ligands, as the fluoropolymer catalysts would spontaneously separate from the product upon releasing the CO$_2$ or lowering its pressure in the end of a reaction. Much to our surprise, when the hydroformytion of 1-alkenes was carried out in scCO$_2$ (180 bar), much lower rates were observed. Thus, the hydroformylation of 1-decene at 100 °C and 30 bar syngas by [Rh(CO)$_2$(acac)] in the presence of **1** led to a conversion of only 39% to the aldehydes with L/B = 3.6 for 15 h reaction time (P/Rh = 6, olefin/Rh = 2150), which contrasts sharply with the value of 97% obtained under FBC (Table 2). The contrast in the case of 1-hexene is even more significant; the average TOF observed is 94 in scCO$_2$ *vs.* 1454 under the FBC conditions aforementioned. The higher rate under the FBC conditions could be due to the higher concentration of 1-alkenes in perfluoromethylcyclohexane than in scCO$_2$ (1.1 *vs.* 0.25 M in the case of 1-decene). However, research carried out in liquid solvents has shown that while the rate of hydroformylation with Rh-PPh$_3$ catalysts is first order in the concentration of 1-alkenes at low concentrations, beyond a certain limit the rate shows a negative order or independence with regard to the olefin concentration.[28,29] For example, kinetic analysis by Chaudhari shows that the rate of the hydroformylation of 1-hexene catalysed by Rh-PPh$_3$ in ethanol decreases when the olefin concentration is higher than 0.4 M.[29] Furthermore, when **1** was replaced with P(4-C$_6$H$_4$C$_6$F$_{13}$)$_3$, the TOF rose to 1961 for 1-hexene and 2202 for 1-decene in scCO$_2$ under otherwise identical reaction conditions, indicating that the low rate with **1** is due probably mainly to the polymeric ligand. Similar observations were made with 1-hexadecene, styrene and vinyl acetate, that is, all the reactions were slower in scCO$_2$ than under FBC conditions when using **1**, and the reactions were much slower with **1** than with P(4-C$_6$H$_4$C$_6$F$_{13}$)$_3$ in scCO$_2$.

An exception was observed when alkyl acrylates were hydroformylated in scCO$_2$. Thus, *n*-butyl acrylate reacted with syngas at 30 bar and 80 °C to give aldehydes with an average TOF of 742 in the case of **1** and 1494 in the case of P(4-C$_6$H$_4$C$_6$F$_{13}$)$_3$ (P/Rh = 6, olefin/Rh = 2700 - 2800). Similar observation was made with methyl acrylate, that is, the hydroformylation of acrylates with **1** was only *ca.* 2 fold slower, a sharp contrast to the over 20 fold difference when 1-hexene was hydroformylated using **1** and the same molecular phosphine.

The slower rates with **1** could stem from probable formation of aggregates by the polymer ligand in scCO$_2$, with the core of the aggregates containing the "CO$_2$-phobic" arylphosphines and the shell containing the "CO$_2$-philic" fluoro tails. If aggregates do form, the slower rates for 1-alkenes and styrenes could be a result of partitioning of the olefins between the bulk scCO$_2$ and core of the aggregate leading to a lower concentration inside the aggregates where catalysis occurs, and/or a result of coordination by the carbonyl groups of **1** to rhodium blocking the access of olefins to the rhodium atom. With the polar acrylates, higher concentration within the aggregates and easier coordination to rhodium are more probable compared with the unfunctionalised 1-

alkenes and styrenes, thus resulting in faster hydroformylation rates. The possible formation of aggregates by **1** in perfluorinated solvents is already mentioned above. And the self-assemble into micellar structures of block copolymers consisting of $scCO_2$-soluble and insoluble segments has been demonstrated in $scCO_2$.[30] In the case of FBC, the increasing miscibility between the organic and fluorous phases with temperature would make the formation of such aggregates less likely.

3 CONCLUSIONS

We have introduced a fluorous soluble polymer ligand for FBC and shown the arylphosphine-containing ligand, when combined with rhodium, to be highly active and selective in the fluorous biphase hydroformylation of various olefins. The same soluble polymer ligand has also been investigated in the hydroformylation in $scCO_2$. However, the reaction is in general slower in $scCO_2$ than in perfluoro solvents and still slower in comparison with that where fluoroponytail-modified molecular phosphines are used as ligands in $scCO_2$. These observations may be accounted for by assuming that the fluoropolymers aggregate in $scCO_2$. Given the easy availability of various vinyl monomers that can be used for fluoropolymer synthesis, better-performing soluble polymer catalysts coupled with efficient phase separation could be envisioned for catalysis in fluorous and $scCO_2$ phases and for fluorous combinatorial chemistry.

Acknowledgments

We thank the EPSRC (WC), the Leverhulme Centre for Innovative Catalysis (ABO), the Socrates programme (AG) and the University of Liverpool Graduates Association (Hong Kong) (LX) for financial support.

References

1. D. E. Bergbreiter, *Catal. Today*, 1998, **42**, 389.
2. D. C. Sherrington, *Chem. Commun.*, 1998, 2275
3. P. Wentworth Jr. and K. D. Janda, *Chem. Commun.*, 1999, 1917.
4. I. T. Horváth, *Acc. Chem. Res.*, 1998, **31**, 641.
5. P. G. Jessop, T. Ikariya and R. Noyori, *Chem. Rev.*, 1999, **99**, 475.
6. L. P. Barthel-Rosa and J. A. Gladysz, *Coord. Chem. Rev.*, 1999, **192**, 587.
7. E. de Wolf, G. van Koten and B. J. Deelman, *Chem. Soc. Rev.*, 1999, **28**, 37.
8. R. H. Fish, *Chem. Eur. J.*, 1999, **5**, 1677.
9. D. Sinou, G. Pozzi, E. G. Hope and A. M. Stuart, *Tetrahedron Lett.*, 1999, **40**, 849.
10. D. E. Bergbreiter, J. G. Franchina and B. L. Case, *Org. Lett.*, 2000, **2**, 293.
11. M. Beller, B. Cornils, C. D. Frohning and C. W. Kohlpaintner, *J. Mol. Catal.*, 1995, **104**, 17 and references therein.
12. B. Cornils and W. A. Herrmann, eds., *Applied Homogeneous Catalysis with Organometallic Compounds*, VCH, Weinheim, Germany, 1996.
13. I. T. Horváth, G. Kiss, R. A. Cook, J. E. Bond, P. A. Stevens, J. Rábai and E. J. Mozeleski, *J. Am. Chem. Soc.*, 1998, **120**, 3133.
14. D. Koch and W. Leitner, *J. Am. Chem. Soc.*, 1998, **120**, 13398.

15. I. Bach and D. J. Cole-Hamilton, *Chem. Commun.*, 1998, 1463.

16. D. R. Palo and C. Erkey, *Ind. Eng. Chem. Res.*, 1999, **38**, 2168.

17. N. J. Meehan, A. J. Sandee, J. N. H. Reek, P. C. J. Kamer, P. W. N. M. van Leeuwen and M. Poliakoff, *Chem. Commun.*, in press.

18. C. U. Pittman, Jr., in *Comprehensive Organometallic Chemistry*, eds. G. Wilkinson, F. G. A. Stone and E. W. Abel, Pergamon Press, Oxford, 1982,Volume 8, p. 553.

19. F. R. Hartley, *Supported Metal Complexes. A New Generation of Catalysts*, Reidel, Dordrecht, 1985.

20. T. Malmström, C. Andersson and J. Hjortkjaer, *J. Mol. Catal.*, 1999, **139**, 139.

21. A. N. Ajjou and H. Alper, *J. Am. Chem. Soc.*, 1998, **120**, 1466.

22. J. Chen and H. Alper, *J. Am. Chem. Soc.*, 1997, **119**, 893.

23. B. Betzemeier and P. Knochel, *Angew. Chem., Int. Ed. Engl.*, 1997, **36**, 2623.

24. S. Kainz, D. Koch, W. Baumann and W. Leitner, *Angew. Chem., Int. Ed. Engl.*, 1997, **36**, 1628.

25. P. Bhattacharyya, D. Gudmunsen, E. G. Hope, R. D. W. Kemmitt, D. R. Paige and A. M. Stuart, *J. Chem. Soc., Perkin Trans. 1,* 1997, 3609.

26. W. Chen, L. Xu and J. Xiao, *Org. Lett.*, 2000, **2**, 2675.

27. C. U. Pittman, Jr. and R. M. Hanes, *J. Am. Chem. Soc.*, 1976, **98**, 5402.

28. B. M. Bhanage, S. S. Divekar, R. M. Deshpande and R. V. Chaudhari, *J. Mol. Catal.*, 1997, **115**, 247.

29. R. M. Deshpande and R. V. Chaudhari, *Ind. Eng. Chem. Res.*, 1988, **27**, 1996.

30. J. B. McClain, D. E. Betts, D. A. Canelas, E. T. Samulski, J. M. DeSimone, J. D. Londono, H. D. Cochran, G. D. Wignall, D. Chillura-Martino and R. Triolo, *Science,* 1996, **274**, 2049.

Subject Index

RETURN TO: CHEMISTRY LIBRARY

100 Hildebrand Hall • 642-3753

LOAN PERIOD 1	2	3
4	**1-MONTH USE** 6	

ALL BOOKS MAY BE RECALLED AFTER 7 DAYS.

Renewable by telephone.

DUE AS STAMPED BELOW.

NON-CIRCULATING UNTIL: _1/28/02 2PM_		
DEC 19 2002		

FORM NO. DD 10   UNIVERSITY OF CALIFORNIA, BERKELEY
3M   3-00   Berkeley, California 94720–6000